Materials and Dematerialization

Materials and Dematerialization

Making the Modern World

Vaclav Smil

WILEY

Registered Offices
John Wiley & Sons, Inc., 111 River Street, Hoboken, NJ 07030, USA
John Wiley & Sons Ltd, The Atrium, Southern Gate, Chichester, West Sussex, PO19 8SQ, UK

For details of our global editorial offices, customer services, and more information about Wiley products visit us at www.wiley.com.

Wiley also publishes its books in a variety of electronic formats and by print-on-demand. Some content that appears in standard print versions of this book may not be available in other formats.

Library of Congress Cataloging-in-Publication Data applied for
Paperback ISBN: 9781394181209

Cover Design: Wiley
Cover Images: World Map: © cienpies/Getty Images
Steel Pipe: © Rost-9D/Getty Images
Wood Texture: © alfexe/Getty Images
CPUs: © AlterYourReality/Getty Images
Wheat Berries: © Viorika/Getty Images

Set in 10.5/13.5pt Times by Straive, Pondicherry, India

SKY10047790_051523

Also by Vaclav Smil

China's Energy

Energy in the Developing World
  (edited with W. Knowland)

Energy Analysis in Agriculture
  (with P. Nachman and T. V. Long II)

Biomass Energies

The Bad Earth

Carbon Nitrogen Sulfur

Energy Food Environment

Energy in China's Modernization

General Energetics

China's Environmental Crisis

Global Ecology

Energy in World History

Cycles of Life

Energies

Feeding the World

Enriching the Earth

The Earth's Biosphere

Energy at the Crossroads

China's Past, China's Future

Creating the 20th Century

Transforming the 20th Century

Energy: A Beginner's Guide

Oil: A Beginner's Guide

Energy in Nature and Society

Global Catastrophes and Trends

Why America Is Not a New Rome

Energy Transitions

Energy Myths and Realities

Prime Movers of Globalization

Japan's Dietary Transition and Its
  Impacts (with K. Kobayashi)

Should We Eat Meat?

Harvesting the Biosphere

Made in the USA

Making the Modern World

Power Density

Natural Gas

Still the Iron Age

Energy Transitions (new edition)

Energy: A Beginner's Guide
  (new edition)

Energy and Civilization: A History

Oil: A Beginner's Guide (new edition)

Growth

Numbers Don't Lie

Grand Transitions

How the World Really Works

Invention and Innovation

Size

Contents

Preface: Why and How

The story of humanity – evolution of our species, the prehistoric shift from foraging to permanent agriculture, rise and fall of antique, medieval and early modern civilizations, economic advances of the past two centuries, mechanization of agriculture, diversification and automation of industrial protection, enormous increases in energy consumption, diffusion of new communication and information networks and impressive gains in quality of life – would not have been possible without an expanding and increasingly intricate and complex use of materials. Human ingenuity has turned materials first into simple clothes, tools, weapons and shelters, later into more elaborate dwellings, religious and funerary structures, pure and alloyed metals, and in recent generations into a still increasing variety of designs, machines and extensive industrial and transportation infrastructures, megacities, even as silicon, doped with small amounts of other elements, has been turned into substrate for solid-state devices that have enabled the new electronic world.

This material progress has not been a linear advance but it consisted of two unequal periods. First was very slow rise that extended from prehistory to the beginnings of rapid economic modernization, that is until the 18th century in most of Europe, until the 19th century in the US, Canada and Japan, and until the latter half of the 20th century in most of Asia. An overwhelming majority of people lived in those premodern societies with only limited quantities of simple possession that they made themselves or that were produced by artisanal labor as unique pieces or in small batches – while the products made in larger quantities, be they metal objects, fired bricks and tiles or drinking glasses, were too expensive to be widely owned.

The principal reason for this limited mastery of materials was the energy constraint: for millennia our abilities to extract, process and transport biomaterials and minerals were limited by the capacities of animate prime movers (human and animal muscles) aided by simple mechanical devices and by only slowly improving capabilities of the three ancient mechanical prime movers, sails, water wheels and wind mills. Only the conversion of chemical energy in fossil fuels to inexpensive and universally deployable kinetic energy of mechanical prime movers (first by external combustion of coal to power steam engines, later by internal combustion of liquids and gases to energize gasoline and Diesel engines and, later still, gas turbines) brought a fundamental change and ushered in the second, rapidly ascending, phase of material consumption, an era further accelerated by generation of electricity and by the rise of commercial chemical syntheses producing an enormous variety of compounds ranging from fertilizers to plastics and drugs.

As a result, the world has become divided between the affluent minority that commands massive material flows and embodies them in long-lasting structures as well as in durable and ephemeral consumer products – and the low-income majority whose material possessions amount to a small fraction of material stocks and flows in the rich world. Now the list of products that most of the Americans claim they cannot live without includes cars, home air conditioning, microwave ovens, dish-washers, garburators, clothes dryers, home computers and mobile phones (Taylor et al. 2006; Langlois 2020) – and they have forgotten how recent many of these possessions are because 60 years ago many of them were rare or nonexistent. In 1960 fewer than 20% of all US households had dishwasher, clothes dryer or air conditioning, first color TVs had just appeared, and (before the first microprocessors were made in 1971) there were no personal computers, mobile phones and other portable electronic devices – and also no SUVs (they began their rise to market dominance only during the late 1980s).

In contrast, those have-nots in low-income countries who are lucky to have their own home often live in a poorly-built small earthen brick or wooden structure with as little inside as a bed, a few benches and cooking pots and some worn clothes. Those readers who have no concrete image of this great material divide should look at Peter Menzel's *Material World: A Global Family Portrait* where families from 30 nations were photographed in front of their dwellings amidst all of their household possessions (Menzel 1995). The book was published nearly three decades ago, and during the intervening time hundreds of millions of people (mostly in Asia) have been lifted from the deepest poverty to a more dignified existence, but its message still resonates. The latest World Banka data show that by the early 2020s large shares of national populations in Asia (about 20% in India, Bangladesh and Pakistan) and Africa (40% in Nigeria, 60% in Congo) still live below poverty line, beyond the reach of adequate material consumption (World Bank 2022a).

And this private material contrast has its public counterpart in the gap between extensive and expensive infrastructures of the rich world (transportation networks, functioning cities, agricultures producing large food surpluses, largely automated manufacturing) and their inadequate and failing counterparts in poor countries. These contrasts make it obvious that a further substantial material mobilization and transformation will be needed just to narrow the gap between these two worlds. And an even larger demand for old and new materials will arise from the unfolding energy transition.

The world's primary energy supply remains dominated by fossil fuels (they provided 86% of all primary energy in 2000 and still 83% in 2020) and a new (as yet uncertain) pattern will emerge during the coming decades, consisting of a mixture of electricity generated without carbon combustion (mostly by wind turbines, photovoltaic cells and nuclear reactors), biofuels and (much more importantly) fuels produced by using non-carbon electricity (for electrolysis to make hydrogen used directly for combustion or in fuel cells and in ammonia synthesis) or (less likely) by syntheses relying on carbon from captured CO_2.

And new energy converters necessarily accompanying this transition – ranging from electric vehicles and other means of transportation relying on batteries to heat pumps and new ways of energy uses by industries (with electricity displacing direct fuel combustion – will create further substantial material needs, including much higher demand for cobalt, copper, lithium and nickel as well as new substantial demand for steel, aluminum and cement needed for requisite infrastructures (ranging from new high voltage lines to water electrolysis, and from massive wind turbine foundations to new hydrogen pipelines).

This new demand surge will only intensify a truly global extent of environmental pollution and degradation resulting from extraction, processing and use of materials and it will involve some unprecedented challenges. As for the extraction, even the last intact domain, deep ocean floor, will see considerable amount of activity before 2050 and at the opposite end of the chain we will have to come up with new, effective ways of recycling and disposal of hundreds of thousands of massive plastic blades (some are now longer than 100 m), millions of PV panels and hundreds of millions of discarded vehicular batteries. In the absence of such measures our use of indispensable materials would pose even more worrisome threats on scales ranging from local degradation and contamination to concerns about the integrity of the biosphere.

These impacts also raise the questions of analytical boundaries: their reasoned choice is inevitable because including every conceivable material flow would be impractical and because there is no universally accepted definition of what should be included in any fairly comprehensive appraisal of modern material use. This lack of standardization is further complicated by the fact that some analyses have taken the maximalist (total resource flow) approach and have included every conceivable input and waste stream, including waste flows (sometimes called hidden flows) associated with the extraction of minerals and with crop production as well as oxygen required for combustion and the resulting gaseous emissions and wastes released into waters or materials dissipated on land.

In contrast, other studies have restricted their accounts to much more reliably quantifiable direct uses of organic and inorganic material inputs that are required by national economies. I will follow the latter approach, as I will focus in some detail on key materials consumed by modern economies, an approach easily justified by their magnitude or their irreplaceable properties. Their huge material claims lead us to ask a number of fundamental questions. How much further will the affluent world push its already often excessive material consumption? To what extent is it possible to divorce economic growth and improvements in average standard of living from increased material consumption – in other words, how far we can push relative dematerialization?

This reduction in the use of materials is most often expressed per unit of product (standard soft drink aluminum can gets lighter) or per unit of economic output (less copper or steel is needed per unit of GDP), and it has been a common phenomenon that has been well documented in sectors ranging from construction to transportation and with products ranging from small consumer items to large high-bypass jet engines. Ultimately, relative dematerialization runs into fundamental physical limits: a standard soft drink container cannot be made to weigh just one gram, and the law of conservation of mass requires that in every chemical synthesis the total mass of the reactants must equal the total mass of products. For example, ammonia synthesis requires a molecule of nitrogen and three molecules of hydrogen to produce two molecules of NH_3 ($N_2 + 3H_2 \rightarrow 2NH_3$) and this means that to synthesize one ton of ammonia we will always need 176.47 kilograms of hydrogen, no more but no less.

Synthesis of ammonia cannot be decoupled from hydrogen, but the use of nitrogenous fertilizers per unit of harvest can be reduced by improving the rate of their uptake by crops (a difficult task to do since the compounds are subject to leaching, volatilization and denitrification). The only way to uncouple it completely would be to endow cereal and oil and sugar crops with the ability to supply their own nitrogen as legumes do in symbioses with nitrogen-fixing bacteria – but this is not coming anytime soon to wheat or sunflower field near where you live. The more realistic dematerialization questions in food production are thus to ask to what extent we can limit the use of fertilizers by growing less feed for animals and eating less meaty (and less dairy-rich) diets, and how much we can reduce the current, unacceptably high but persisting, level of food losses?

But before I get to answer such questions in convincing manner, I must review first the evolution of human material uses, describe all the principal materials, their extraction and production and their dominant applications, and take a closer look at the evolving productivities of material extraction, processing, synthesis, finishing and distribution and at the energy costs and environmental impact of rising material consumption. And as always in my books, I will not offer any time-specific forecasts regarding the future global and national use of materials. Instead, I will look at possible actions that could reduce our dependence on materials while maintaining good quality of life and narrowing the gap between affluent and low-income economies.

We must realize that in the long run even the most efficient production processes, the least wasteful ways of design and manufacturing and (for those materials that can be recycled) the highest practical rates of reuse may not be enough to result in absolute dematerialization rates that would be high enough to negate the rising demand for specific materials generated by continuing population growth, improving standards of living and universal human preferences for amassing possession. And any dreams of circular economy are just that: we should strive for maximum practicable rates of recycling and reuse but it is impossible to have a closed global material economy akin to the reuse of carbon, nitrogen and sulfur by grand global biogeochemical cycles. This makes it highly likely that in order to reconcile our wants with the preservation of biosphere's integrity we will have to make deliberate choices that will help us to reduce absolute levels of material consumption and thereby redefine the very notion of modern societies whose very existence is predicated on incessant and massive material flows.

1
What Gets Included

Marcos Mello/Adobe Stock

Any study aiming to elucidate the complexity of material flows of modern societies, their prerequisites, and their consequences should be as comprehensive as possible and its coverage should be truly all-encompassing. But this easily stated aspiration runs immediately into the key categorical problem: what constitutes the complete set of modern material uses? There is no self-evident choice, no generally accepted list, only more or less liberally (and also more or less defensively) defined boundaries of a chosen inclusion, a reality best illustrated by reviewing the selections made by the past comprehensive studies and adopted by leading international and national databases of material flows.

The first comparative study of national resource flows (Adriaanse et al. 1997), subtitled *The Material Basis of Industrial Economies*, excluded water and air but included all agricultural harvests (not just raw materials but all food and feed as well), all forestry products, aquatic catches, extraction of minerals, and fossil fuels but also hidden (waste) flows accounting for extraction, movement, or losses of materials that create environmental impacts but have no acknowledged economic values. These hidden flows are dominated by overburden materials (soil and rocks that have to be removed before reaching mineral deposits, obviously most massive with open-cast coal and ore mining), processing wastes (particularly tailings, massive flows associated with separation of relatively rare metals from rocks), soil, sand, and rocks that have to be removed and shifted during large construction projects, and soil erosion originating from fields and permanent plantations.

Waste flows are not monitored, their quantification is, at best, a matter of approximate estimates, more often of just informed guesses – but their volume and mass have been increasing, both because we have been exploiting minerals in deeper overcast mines (more massive overburden) and because we require more metals (from Co to Zn) whose ores are not as rich as the ores that we extract to produce the world's two dominant metallic materials, iron and aluminum. While hematite, the most commonly exploited iron ore, contains 50–60% of the metal (when pure it is about 70% Fe) and bauxite (the only commercially exploited aluminum ore) contains 15–25% aluminum, copper ores that dominate the metal's extraction in the early 2020s have only 0.3–1.7% Cu, and in Chile, the world's largest producer, they average 0.6% Cu (Schlesinger et al. 2022). Mass of materials wasted during the extraction phase is thus roughly equal to iron's output, it is as much as nearly seven times larger than the production of aluminum, and it is about 170 times larger than the output of Chilean copper.

Thanks to the coming mass-scale electrification of transport and of many industries, it will be copper whose production will grow faster than any other of the five metals now produced in largest quantities (Fe, Al, Cu, Zn, Pb). Uncertainties about mass flows are even greater with the annual totals for hidden flows associated with imported raw materials: obviously, this reality will make the greatest difference in the case of large affluent economies that import a wide range of raw materials, including precious metals, from scores of countries. For example, in 2020 US imports of gold, silver, platinum, and diamonds equaled 3.4% of all purchases abroad, a share three times as high as the imports of integrated circuits (OEC 2022). But that gold came mostly from Switzerland, an intermediate source whose gold imports come mostly from other intermediaries (Hong Kong, UAE, Thailand, UK), making it exceedingly difficult to trace the flow to its origin in order to determine the total mass of waste flows behind these transactions.

Not surprisingly, Adriaanse's study resorted to using worldwide averages for these calculations: for example for overburden it applied the rate of 0.48 t for a ton of bauxite and 2 t per ton of iron ore, global generalizations that must result in considerable errors when used as national averages. Soil erosion rates are even more variable, their detailed national studies are rare, annual soil losses (depending on precipitation, extent of drought periods, wind speed, cultivation methods, deforestation) can differ by up to an order of magnitude even within relatively small regions, and yet the study used only the rates derived from the US inventory. Another highly uncertain inclusion was quantifying the mass of grass grazed by cattle (other animal feed was included in crop harvests): obviously an average Maasai cow in Kenya will consume only a fraction of grasses digested every year by beef cattle in Alberta or Colorado.

Three years after this first comparative study came another project led by the World Resources Institute (WRI), *The Weight of Nations* (Matthews et al. 2000). That study presented material flows for the four nations included in the original work (US, Japan, Germany, and the Netherlands) as well as for Austria and that extended the accounting period from 1975 to 1996 (the original ended in 1993). Its subtitle, *Material Outflows from Industrial Economies*, indicated the report's concern with outputs produced by the metabolism of modern societies. As its predecessor, this study included all fossil fuels, estimates of hidden material flows (dominated by the removal of overburden in surface coal mining), as well as the totals of all processing wastes.

The report had also quantified earth moved during all construction activities (highway, public, and private and also for dredging), soil erosion losses in agriculture and waste from synthetic organic chemicals and from pharmaceutical industry. But unlike the original study, the 2000 report also includes data on additional inputs (oxygen in combustion and in respiration) and outputs, including the total output of CO_2 from respiration and water vapor from all combustion and it separates waste streams into three gateways, air, land and water. The air gateway quantified gaseous emissions (CO_2, CO, SO_x, and NO_x, volatile organic carbohydrates) including oxygen from all combustion, the outputs to land include municipal solid waste, industrial wastes, and dissipative flows to land (manure, fertilizers, salt spread on roads, worn tire rubber, evaporated solvents), and water outputs trace organic load and total nitrogen and phosphate burdens.

Eurostat has been publishing annual summaries of domestic material consumption for all EU countries since the year 2000, disaggregating the total flows into fossil fuels, biomass (crops and forest products), metal ores, and nonmetallic minerals (Eurostat 2022a). Eurostat's methodological guides for economy-wide material flow accounts offer detailed procedures for the inclusion of biomass (food, feed, fodder crops, grazed phytomass, wood, fish, hunting, and gathering activities), metal ores, and nonmetallic minerals and for all forms of fossil fuels as well as for all dissipative uses of products, including organic and mineral fertilizers, sewage sludge, compost, pesticides, seeds, road salt, and solvents (Eurostat 2018).

Eurostat aggregates also include unused materials (mining overburden, losses accompanying phytomass production, soil excavation, dredging, and marine by-catch) and quantify emissions (CO_2, water disposal, and landfilled wastes) but leave out oxygen and water. The latest compilations at the time of writing, for the year 2021 (Eurostat 2022a), show the expected recovery from the Covid-induced lows of 2020 and equally expected long-term decline in the EU's fossil fuel extrac-tion (down to about 1.1 Gt from just over 1.5 Gt in 2012) but continued growth in the mobilization of nonmetallic minerals (about 3.3 Gt in 2021 compared to 2.9 Gt in 2012). OECD publishes annual estimates for its 34 member states and for 170 other countries and city states, with some data going back to 1970. These totals include domestic consumption of all materials originating from natural resources and forming the bases of economies: all metals, nonmetallic minerals, biomass (wood and food), and fossil fuels (OECD 2022).

In 1882, the US Congress mandated annual collection of statistics for mineral commodities produced and used in the country. The US Geological Survey became the first agency responsible for this work, then the US Bureau of Mines and since 1995 the task reverted to the USGS. This statistics were the basis for preparing the first summary of America's material flows aggregated by major categories and covering the period between 1900 and 1995 (Matos and Wagner 1998). The series was subsequently extended and by 2022 updates for most commodities are available until 2018–2019 (Matos 2009; Kelly and Matos 2016 with updates). The latest data on individual elements, compounds, and materials are updated annually in *Mineral Commodity Summaries* (USGS 2022a).

The USGS choice of items included in its national material accounts is based on concentrating only on the third class of the material triad by leaving out food and fuel and aggregating only the materials that are used in all branches of the economy. The series offers annual totals for domestic production, exports, imports, and domestic consumption; it excludes water, oxygen, hidden material flows, and all fossil fuels; and it includes all raw materials produced by agricultural activities (cotton, seeds yielding industrial oil, wool, fur, leather hides, silk, and tobacco), materials originating in forestry (all kinds of wood, plywood, paper, and paperboard), metals (from aluminum to zinc), an exhaustive array of nonmetallic minerals (be they extracted in their natural form, such as gypsum, graphite, or peat, processed before further use, such as crushed stone or cement or synthesized, such as ammonia), and nonrenewable organics derived from fossil fuels (asphalt, road oil, waxes, oils, and lubricants and any variety of solid, liquid, or gaseous fossil fuel used as feedstocks in chemical syntheses).

Very few of these inputs are used in raw, natural form as virtually all of them undergo processing (cotton spinning, wood pulping, ore smelting, stone crushing or cutting, and polishing) and, in turn, most of these processed materials become inputs into manufacturing of semifinished and finished products (cotton turned into apparel, pulp into paper, smelted metals into machine parts, crushed stone mixed with sand and cement to make concrete). This compilation of agriculture- and forestry-derived products, metals, industrial minerals, and nonrenewable organics gives a fairly accurate account of annual levels and long-term changes in the country's material flows. While all imports and exports of raw materials are accounted, the series does not include materials that were contained in traded finished goods: given their mass and variety their tracking would be very difficult.

Where does this leave us? Those material flow studies that conceive their subject truly *sensu lato*, as virtually any substance used by humans, include everything with a notable exception of water, that is not only biomaterials used in production of goods, all metals, nonmetallic minerals, and organic feedstocks but also all agricultural phytomass (harvested food and feed crops, their residues, forages, and grazed plants), all (biomass and fossil) fuels and oxygen needed for combustion. Slightly more restrictive studies exclude oxygen and all food and feed crops, and they consider only those agricultural raw materials that undergo further processing into goods but include all phytomass and fossil fuels. In contrast, the USGS series exemplifies a *sensu stricto* approach as it includes only raw biomaterials used for further processing and as it excludes oxygen, water, all fuels (phytomass and fossil), and all hidden (and always tricky to estimate) material flows. My preferences for setting the analytical boundaries are almost perfectly reflected by the USGS selection but instead of simply relying on that authority I will briefly explain the reasons behind my exclusions.

Leaving out oxygen required for combustion of fuels is a choice easily defensible on the basis of free supply of a virtually inexhaustible atmospheric constituent. Claims about danger of serious O_2 depletion through combustion were refuted long time ago (Broecker 1970; Liu et al. 2019). Complete combustion of 1 kg of carbon consumes 2.67 kg of oxygen and burning of 1 kg of methane (CH_4), the simplest hydrocarbon, requires 4 kg of O_2. This means that in 2021 the global combustion of more than 11 Gt of fossil carbon (as coal, refined oil products, and natural gas) claimed about 40 Gt of O_2 (Liu et al. 2019) – or about 0.0027% of the atmosphere content of 1.5 Pt of the gas. Even a complete combustion of generously estimated global resources of fossil fuels (a clear impossibility, just a theoretical consideration) would lower the atmospheric O_2 content by no more than 2%.

There is thus no danger of any worrisome diminution of supply (to say nothing of exhaustion) of the element, and yet once the choice is made to include it in material flow accounts, it will dominate the national and global aggregates. For example, as calculated by the comparative WRI study, oxygen was 61% of the direct US processed material output in 1996, and in Japan in the same year the element's share was 65% (Matthews et al. 2000). Consequently, magnitudes of national material flows that would incorporate oxygen needs would be nothing but rough proxies for the extent of fossil fuel combustion in particular economies.

Reasons for excluding waste flows from the accounts of national material flows are no less compelling: after excluding oxygen they would dominate total domestic material output in all countries that have either large mineral extractive industries (especially surface coal and ore mining) or large areas of cropland subject to soil erosion. They are dominated by unusable excavated earth and rocks, mine spoils, processing wastes, and eroded soil, while earth and rocks moved around as a part of construction activities will make up a comparatively small share. Not surprisingly (after excluding oxygen), in the WRI analysis these hidden flows accounted for 86% of the total domestic material output in both the United States and Germany, but with much less mining and with limited crop cultivation, the rate was lower (71%) in Japan (Matthews et al. 2000).

Daily flow of materials a large copper mine illustrates the cumulative immensity of these waste flows (GRID 2017). Two-thirds of the 270,000 t of solid rock dug out daily (180,000 t) are dumped directly, while the processing of 90,000 t of ore requires 114,000 t of water and it yields 1,750 t of concentrate ready for smelting. Just over 200,000 t (88,250 + 114,000) of material are tailings retained behind dams that must be large enough to accommodate this waste flow for some two decades of operation: when the mine is closed, it leaves behind some 1.3 Gt of waste rock and more than 600 Mt of solid tailings, nearly 2 Gt of material that can be never recycled and that is most unlikely to be reused in any other way.

But the principal problem with the inclusion of hidden flows is not their unsurprising dominance of domestic output of materials in all large, diversified economies, but the indiscriminate addition of several qualitatively incomparable flows. Unusable mass of stone left in a quarry after it ceased its operation may be no environmental burden, not even an eyesore. Moreover, once the site is flooded to create an artificial lake those waste flows may become truly hidden as a part of a new and pleasing landscape. On the other hand, bauxite processing to extract alumina (to give one of many possible common examples) leaves behind toxic waste (containing heavy metals) that is also often slightly radioactive and acidic and its worst recent accidental release (in 2010 when about 1 Mm^3 spread over an area of some 40 km^2 in northern Hungary, killing 10 people and injuring 120) can cause serious long-term environmental damage (Gelencsér et al. 2011).

And no less fundamental is the difference between in situ hidden flows generated by mineral extraction (abandoned stone, gravel, and sand quarries, and coal and ore mines with heaps, piles, layers or deep holes or gashes full of unusable minerals or processing waste) and by rain- and wind-driven land erosion that transports valuable topsoil or desert sand not just tens or hundreds but as much as thousands of kilometers downstream or downwind. The first kind of hidden flows may be unsightly but not necessarily toxic and its overall environmental impacts beyond its immediate vicinity may be negligible or nonexistent.

In contrast, surface erosion is globally important, often regionally highly worrisome and locally devastating process that reduces (or destroys) the productivity of crop fields, silts streams, contributes to eutrophication of fresh and coastal waters, creates lasting ecosystemic degradation and substantial economic losses, or drives large masses of fine dust right across the Atlantic Ocean carrying persistent organic pollutants, metals, and microbes to the Caribbean (Garrison et al. 2006) or deposits Saharan dust over the Alpine snow (Di Mauro et al. 2018). In any case, magnitudes of these associated flows and their often undesirable environmental impacts dictate that they should not be ignored when analyzing particular extractive or cropping activities: as long as we remember that the flows cannot be quantified with high accuracy, we should try to include them in specific analyses of future material demand (I will return to this point when assessing material needs of the unfolding energy transition).

My reasons for excluding water are based on several considerations that make this indispensable input better suited for separate treatment rather than for inclusion into total material requirements of modern economies. The most obvious reason is, once again, quantitative: with the exception of desert countries, water's inclusion would dominate virtually all national material flow accounts and it would misleadingly diminish the importance of many inputs whose annual flows are a small fraction of water withdrawals but whose qualitative contribution is indispensable. For example, in 2015 (the date of the latest detailed nationwide USGS estimate), the total water withdrawals in the United States were about 445 Gt, while all materials directly used by the country's economy (the total dominated by sand, gravel, and stone used in construction) added up to less than 1% of the withdrawn water mass (USGS 2018).

At the same time, there are fundamental qualitative differences between these two measures that make any direct comparisons highly misleading. The most voluminous water withdrawal in the United States (accounting for 41% of the total in 2015), that of cooling water for thermal electricity-generating stations, is not a consumptive use: a small part of that flow is evaporated to become available later (downwind, after condensation and precipitation) and most of that water becomes available almost instantly after it is discharged (slightly warmed) for further downstream uses. In contrast, materials that become embedded in long-lasting structures and products are either never reused or are partially recycled only after long period of being out of circulation.

And most of the second most voluminous water use in the United States (37% used for irrigation), is also nonconsumptive: all but a tiny fraction of the irrigation water is evaporated and transpired by growing plants, and (as with the cooling water) after re-entering the atmosphere it is eventually condensed again and it is precipitated, often after a long-distance transport downwind. And if the inclusions of water were driven by resource scarcity concerns, then a critical distinction should be made between water supplied by abundant precipitation and water withdrawn at a high cost from deep and diminishing aquifers that cannot be replenished on a civilizational time scale.

At this point it might be useful to note yet another (comparatively minor) problem with aggregate measures of material flows that is usually neglected by the assemblers of national and global accounts, namely that of water content of sand and of harvested biomass. Even when looking just at those biomaterials that are used as industrial inputs, their water content is from less than 15% for raw wool to more than 50% for freshly cut tree logs (the range is wider for food crops, ranging from only about 5% for some seeds and less than 15% for harvested cereal grains to more than 90% for fresh vegetables).

Freshly excavated sand can contain more than 30% of water, purified sands have 15–25% of moisture, storage in drainage bins reduces that level to about 6% and drying in rotary bins or in fluidized bed dryers expels all but about 0.5% of moisture for sands used in such processes as steel castings or hydraulic fracturing under pressure. Moreover, sand used in hydraulic fracturing is also coated with resins reinforced with nanomaterials in order to alter its surface wetting properties, crush strength, and chemical resistance. The best solution would be to report the masses of any moisture-containing materials in terms of absolutely dry weight in order to make their flows comparable to those of materials that contain no moisture. This is not the case in practice, and hence all national material aggregates contain far from negligible shares of water.

Foodstuffs and fuels are obviously indispensable for the survival of any civilization, and their flows have been particularly copious in modern high-energy societies enjoying rich and varied diets, while traditional biofuels remain important in many low-income countries. Moreover, unlike with water or oxygen, their inclusions would not dwarf all other material flows combined: for example, even in the fuel-rich United States, the mass of annually consumed coal, crude oil, and natural gas is equal to about 50% of all non-energy minerals. So why to leave them out? Exclusion of food and fuel is justified not only because these two large consumption categories have been traditionally studied in separation (resulting in rich literature on achievements and prospects of energy and food production) but because they simply are not *sensu stricto* materials, substances repeatedly used in their raw state or transformed into more- or less-durable finished products used in all sectors of the economy.

Unlike raw biomaterials (wood, wool, cotton, leather, silk), metals, nonmetallic minerals and nonrenewable organics (asphalt, lubricants, waxes, hydrocarbon feedstocks), foodstuffs, and fuels are not used to build long-lasting structures and are not converted or incorporated into a still-increasing array of ephemeral, as well as durable industrial, transportation, and consumer items. Foods are rapidly metabolized to yield energy and nutrients for human growth and activity; fuels are rapidly oxidized (burned) to yield, directly and indirectly, various forms of useful energy (heat, motion, light): in neither case they increase the material stock of modern societies. And, a critical difference to which I will return later when noting the impossibility of circular economy, energy flows of any kind (fuels, electricity, food) cannot be recycled.

Finally, I must defend a conceptual change that concerns the handling of materials put by the EU's material balances into the category of dissipative flows. According to the EU definition, the eight categories of dissipative losses are a collection of disparate residuals: some of them add up to small total flows (think about solvents escaping from dry cleaning or about rubber tires wearing-away on roads), others are more substantial (leaching and volatilization of manures, sewage sludge, and composts applied to cropland) but dissipative losses contributed by both of these material categories are not monitored and are very difficult to quantify. The USGS approach accounts for the largest flows in this category (salt and other thawing materials, including sand and grit, spread on winter roads, nitrogenous and phosphatic fertilizers and potash applied to crops and lawns) by including them in the industrial minerals group.

While salt and sand are abundant materials whose production is not energy-intensive, inorganic fertilizers are critical material inputs in all modern societies that cannot be ignored and that will receive a closer look when I examine advances in the production of synthetic materials. But I would argue that most of the remaining dissipative flows add up to relatively small amounts whose inherently inaccurate quantification appears to outweigh any benefits of including them in any grand total of consumed materials. And while manures and sludges represent relatively large volumes to be disposed of, they do not recycle biomass but rather the products of its decomposition: water, carbon, and small amounts of nutrients (above all nitrogen); sludge contains at least 80% water, fresh manures 70–85%, but only a few percent of nitrogen. Moreover, in many instances sewage sludge should not be recycled as it contains heavy metals, pathogens, pesticide, and drug residues.

This leaves me with an argument for a single addition to the USGS list for the inclusion of industrial gases. Although air (21% oxygen) is needed for combustion of fossil fuels, the dominant energizer of modern civilization, adding air to the total material input would have (as I have already explained) a skewing and confusing effect similar to that of counting all uses of water - but assessing the use of gases separated from the air in order to enable many industrial processes is another matter. In simple mass terms, the global use of oxygen, hydrogen, nitrogen, and rare gases such as argon or xenon constitutes only a minor item, but in qualitative terms their use is indispensable in industries ranging from steelmaking (basic oxygen furnaces now dominate the production of the metal) to synthesis of ammonia (using nitrogen separated from air and hydrogen liberated from methane) and efficient lighting.

And although there is no way to anticipate accurately the global trajectory of hydrogen - an energy carrier whose ascendance has been promised for generations but whose production without carbon ("green hydrogen" liberated by electrolysis of water using only electricity from renewable conversions) began receiving both widespread and intensive consideration during the early 2020s (Green Hydrogen Systems 2022) - there is no doubt that without the introduction of substantial volumes of hydrogen into the global energy supply we cannot think about mass-scale decarbonization of future industrial and transportation energy uses. And in addition to green hydrogen, there has been also rising interest in green ammonia both as an industrial feedstock and as a possible transportation fuel: I will have more to say on both of these materials when I look at the unfolding energy transition.

2

How We Got Here

AlexAnton/Adobe Stock

Materials and Dematerialization: Making the Modern World, Second Edition. Vaclav Smil.
© 2023 John Wiley & Sons Ltd. Published 2023 by John Wiley & Sons Ltd.

sunsinger/Adobe Stock

The Earth's biosphere teems with organisms that use materials for more than just their metabolism. Moreover, in aggregate mass terms the material flows commanded by the humanity do not appear to be exceptionally high when compared with the work of marine biomineralizers. But it is the combination of the overall extent, specific qualities, and increasing complexity of material uses (extraction, processing, and transformation into a variety of inputs destined for structures, infrastructures, and for myriads of finished products) that is a uniquely human attribute. To set it into a wider evolutionary perspective, I will first note some of the most remarkable ways of material uses by organisms ranging from marine phytoplankton to primates, particularly those distinguished either by the magnitude of their overall fluxes or by their unique qualities.

Afterward I will proceed with concise chronological surveys of human use of materials, focusing first on the milestones in our prehistory, above all on those still poorly explained feats of megalithic construction that required quarrying, transportation, and often remarkably accurate placement of massive stones. Then I will review and quantify some notable deployments of traditional materials (dominated by stone and wood) during the antiquity, the Middle Ages and the early modern era (1500–1800), concentrating above all on the advances in building roads, aqueducts, ceremonial, and religious structures and ships, as well as on the origins and developments in metallurgy and on materials used by households.

I will end this chapter by two closely related sections that will describe the creation of modern material civilization during the nineteenth century and its post-1900 spatial expansion and growth in complexity. I will focus on key quantitative and qualitative advances in the use of materials that laid the foundations to the 20th societies as they supported fossil fuel extraction, industrialization, urbanization, and evolution of modern transportation modes on land, water, and in the air. These developments were based on materials whose production required high energy inputs and whose introduction and use have been dynamically linked with enormous advances in scientific and technical capabilities. In turn, new materials have been the principal drivers of increased food production and improvements in sanitation that led to unprecedented gains in quality of life. They also expanded capabilities for mechanized and automated production and for long-distance travel, information sharing, and telecommunication.

2.1   Materials Used by Organisms

Inevitably, all organisms use materials: that is the essence of metabolism. Global photosynthesis, the foundation of life in the biosphere, creates new biomass by incorporating annually more than 60 Gt of carbon absorbed as CO_2 from the atmosphere (Smil 2013; Ryu et al. 2019) and millions of tons of the three key macronutrients (nitrogen, phosphorus, and potassium, absorbed by roots) that are incorporated into complex compounds forming plant tissues and organs. But these metabolic necessities – mirrored by the nutritional requirements of heterotrophs, be they herbivorous, carnivorous, or omnivorous organisms – are not usually included in the category of material uses that is reserved for active, extrasomatic processes.

In terms of the initial acquisition, these material uses fall into five major categories. The rarest, and in aggregate material terms quite inconsequential, category is the use of collected natural materials as tools. The second category with limited aggregate impact is the use of secreted materials to build protective or prey-catching structures; the latter use has been mastered, often spectacularly, by web-making spiders. The next one is the removal of specific biomass tissues (branches, twigs, leaves, flowers), and now also discarded man-made materials (bits of plastics, paper, glass, and metals) and their purposeful emplacement to create remarkably designed structures ranging from beaver dams to intricate nests or, in the case of birds of paradise, often elaborate decorated mating bowers. Then comes the removal and repositioning of soils and clays, invisible as intricate rodent burrows and prominent in termite mounds. And, finally, the most massive endeavor is the extraction of minerals from water, mostly to build exoskeletons, the process dominated by marine biomineralizers including phytoplankton, corals, and mollusks.

2.1.1   Tools and Construction

Tool-using activities have been well documented with species as diverse as otters, seagulls, elephants, and finches (Bentley-Condit and Smith 2009; Shumaker et al. 2011; Sanz 2013), but they have reached the greatest complexity, and have gone as far as resulting in specific cultures, among chimpanzees who use blades of grass or twigs to collect termites or as honey-dipping sticks, leaves, or moss sponges to extract mineral-rich liquids at natural clay-pits, and small stones and stone anvils to crack open nuts, with studied populations displaying some "cultural" differences in prevailing practices (Wrangham et al. 1996; Boesch and Tomasello 1998; Whiten et al. 1999; Gruber et al. 2011; Lamon et al. 2018; Bessa et al. 2021).

Spider silk (made almost entirely of large protein molecules) is certainly the most remarkable secreted material: some strands have tensile strengths comparable to steel and some silks are nearly as elastic as rubber, resulting in toughness two to three times that of such synthetic fibers as Nylon or Kevlar (Römer and Scheibel 2008; Brunetta and Craig 2010). On the other end of secretion spectrum are frothy nests excreted by spittle bugs. Use of collected materials is quite widespread among heterotrophs. Even some single-cell amoebas can build portable, intricate, ornate sand grain houses whose diameter is mere 150 µm (Hansell 2007, 2011). And perhaps the most remarkable collecting activity among insects is that of leafcutter ants (genus *Atta*) as they harvest leaves, drag them underground into elaborately excavated nests in whose chambers they cultivate fungus (Hölldobler and Wilson 1990). Garrett et al. (2016) estimated that 2.9 (±0.3) km of leaf-cutting with mandibles was needed to reduce a square meter of leaf to fungal substrate, with nearly 90% of the cutting taking place inside nests.

Beavers are active harvesters of wood used to build their dams, and when wood is not sufficient, they use stones (up to 30 cm in diameter) combined with branches stacked in layers. Most of the dams are less than 10 m wide, with head differences below 1.5 m, but the record sizes are equivalents of engineered structures up to 850 m long with heads up to 5 m (Müller and Watling 2016). But birds, rather than mammals, provide the most varied and sometime spectacular examples of construction using collected materials; they range from simple and rather haphazard assemblies of twigs or stems to intricate constructs produced by *Ploceidae*, family of tropical weaver birds, and they may use a single kind of a collected material or are made from an assortment of tissues (Gould and Gould 2007; Burke 2012).

Birds use not only a wide range of collected plant tissues (slender blades of grass to heavy twigs used by storks and eagles) but also feathers of other species and spider silk (most passerine birds), and some nests may contain thousands of individual pieces. Use of mud (by swallows) is not that common but many ground-nesting birds (including penguins) collect small stones, while elaborate structures prepared by some bower birds of Australia and New Guinea to attract females may include not only such colorful natural objects as shells, berries, leaves, and flowers but also discarded bits of plastic, metal, or glass, and some species even make courts creating forced visual perspective for the courted females (Endler et al. 2010). Some insect species also use collected material to build their nests: paper wasps cut tiny pieces of wood and mix them with their salivary secretions, and mud wasps shape mud into cylindrical nests. In contrast, primates, our closest animal predecessors, use branches and leaves to build only simple, temporary structures on the ground or in the trees.

2.1.2   Soil Movements

Soil-displacing species engage mostly in digging tunnels, burrows, and nest but also in using soils and clay to build above-ground structure range from insects to mammals. The earliest burrow constructs date to the pre-Cambrian (650–700 million years ago) oceans, coinciding with the emergence of macropredation (Turner 2000). As demonstrated by Darwin in his last published book, earthworms are capable of such prodigious effort of earth displacement (passing the particles through their guts and excreting the worm casts on the surface) that they can bury monuments of human activity in remarkably brief periods of time (Darwin 1881). Rodents are diligent builders of often extensive subterranean networks of tunnels and nests that may also help with temperature control and ventilation and that facilitate escape.

Termites are the greatest aggregate excavators and movers of soils in subtropical and tropical environments. They construct their often impressively tall and voluminous mounds by removing and piling-up soil to build their underground nests sheltering their massive colonies. Internal structure of mounds makes it clear that they provide induced ventilation driven by pressure differences (Turner 2000). Biomass densities of these abundant warm-climate insects range from 2 g/m² in the Amazonian rainforest (Barros et al. 2002) to around 5 g/m² in Australia's Queensland (Holt and Easy 1993) and 10 g/m² in arid Northeast Brazil, in Sao Paulo state as well as in dry evergreen forest of northeast Thailand (Vasconcellos 2010), while in African savannahs their total fresh-weight biomass can be more than twice the biomass of elephants (Inoue et al. 2001).

Species belonging to genus *Macrotermes* move clay particles to build conical mounds that are usually 2–3 m tall but can reach 9 m, with typical basal diameter of 2–3 m, but much wider mounds are not uncommon. In Northeast Brazil, mounds created by the excavation of vast tunnel networks by *Syntermes dirus* have persisted for nearly four millennia and they consist of some 200 million soil cones typically 2.5 m tall and 9 m in diameter covering about 230,000 km² and adding up to about 10 km³ of volume (Martin et al. 2018). Typical mass of termite mounds (wall and nest body) is between 4 and 7 t but spatial density of mounds varies widely, with as few as 1–2 and as many as 10/ha (Fleming and Loveridge 2003; Tilahun et al. 2012).

As a result, the total mass of termite mounds varies widely, from just 4–8 t/ha to as much as 15–60 t/ha. A very conservative estimate of the clay mass used to build termite mounds (assuming average of 5 t/ha and area of about 10 million km² of tropical and subtropical grasslands inhabited by mound-building insects) would be 5 Gt, but the actual total may be several times larger. In any case, this means that the mass of soil displaced annually by these tiny heterotrophs would be of the same order of magnitude as our civilization's global extraction of metallic ores and other non-fuel minerals at the beginning of the twenty-first century.

In aggregate terms both the mass of materials collected by vertebrate animals to build structures and the mass of soils displaced by burrowing heterotrophs, earthworms, and termites are negligible compared to the mass of compounds excreted by species capable of biomineralization, above all by phytoplankton, protists, and invertebrates. Biomineralization evolved independently across phyla, transcending obvious biological differences: more than 30 biogenic minerals (two-thirds of them being carbonates) are produced by a small number of vascular plants (belonging to Bryophyta and Trachaeophyta), animal species ranging from Porifera to Chordata, some fungi, many protists, and some Monera (Lowenstam 1981; Boskey 2003; Gilbert et al. 2022). Some biomineralizers deposit the minerals on organic matrices but most of them produce extracellular crystals similar to those precipitated from inorganic solutions.

2.1.3 Biomineralizers

In mass terms by far the largest users of natural materials are marine biomineralizers able to secrete inorganic compounds they produce from chemicals absorbed from water. Marine biomineralizers use dissolved $CaCO_3$ to form calcite or aragonite shells, two identical minerals that differ only in their crystal structure. Reef-building corals (Anthozoa belonging to the phylum Cnidaria) are the most spectacular communal biomineralizers, while coccolithophores (calcareous marine nanoplankton belonging to the phylum *Prymnesiophyceae*) encase themselves with elaborate calcitic microstructures (smaller than 20 µm), and foraminifera (amoeboid protists of the eponymous phylum) create pore-studded micro shells (tests). Unicellular coccolithophores are abundant throughout the photic zone in nearly all marine environments of the Northern hemisphere and up to about 50°S in the Southern Ocean where their blooms account for a major share of global marine $CaCO_3$ production and export to the deep sea (O'Brien et al. 2012; Hernández et al. 2020).

They also form massive ocean blooms that last for weeks, cover commonly 10^5 km^2 of the ocean surface, and are easily identifiable on satellite images. Coccolithopores cover themselves by coccoliths formed inside the cell and extruded to form a protective armor; many coccoliths detached from cells are also floating freely in water. Coastal blooms have coccolith to coccolithophore cell ratios of 200–400, but the ratios for open waters are much lower, between 20 and 40. *Syracosphaera, Umbellosphaera,* and *Gephyrocapsa* are common genera, but *Emiliania huxleyi* is the biosphere's leading calcite producer (Stanley et al. 2005; Boeckel and Baumann 2008; Monteiro et al. 2016). The species is also unusual because it produces fairly large coccoliths at a high rate, sheds about half of them, and, unlike many other planktonic species, is a relative newcomer that originated only about 270,000 years ago.

Photic zone can extend from just a few meters to about 200 m; densities of coccolithophores can be as low as a few 1,000 cells/l, in blooms they surpass 100,000/l; daily calcification rates range from less than 10 to 80 pg of calcite per cell a day; and the largest blooms can cover 10^5 km^2 for periods of weeks and their annual extent adds up to about 1.5 million km^2 (Lampert et al. 2002; Boeckel and Baumann 2008; Hopkins et al. 2015). The most comprehensive compilation of carbonate production (based on nearly three thousand measurements) ranged across six orders of magnitude (Daniels et al. 2018). Given these enormous natural variabilities, it is impossible to offer any reliable estimate of coccolithophore – mediated annual global calcification in the ocean – but a conservative set of assumptions illustrates the magnitude of the process at its likely lower bound.

Assuming continuous coccolithophore production in 60% of the world's ocean in depths only up to 50 m, with average concentration of just 25,000 cells/l and calcification rate of just 10 pg of calcite per cell a day, would result in annual global sequestration of about 900 Mt of calcite. More liberal assumptions (50,000 cells/l, 20 pg/cell a day) would yield an annual withdrawal of roughly 3.7 Gt of calcite. The best conservative estimate is thus annual coccolithophoride-mediated calcification on the order of a few Gt/year, a low rate in geological perspective because high Mg/Ca ratio and low absolute concentration of calcium in the modern ocean limit the production by most extant species (Stanley et al. 2005). The most obvious testimony to high productivities of coccolithophores in the past are immense sequestrations in Cretaceous and Tertiary chalk deposits (including the white cliffs of Dover).

In comparison to this ocean-wide process the periodic blooms, no matter of how spectacular, make a minor contribution. A bloom covering 250,000 km^2 in the northeastern Atlantic in June 1991 had calcification rates up to 1.5 mg C/m^3/h and in less than a month it had sequestered about 1 Mt of carbon in calcite (Fernández et al. 1993). That would be just over 8 Mt of calcite and if a similar rate were to apply to about 1.5 million km^2 coccolithophore blooms that cover the ocean every year, the total sequestration would be just on the order of 50 Mt of calcite. The same order of magnitude is obtained by assuming 50 m of highly productive photic zone, 150,000 cells/l, daily calcification at 70 pg/cell, and average bloom duration of 30 days: that combination yields annual sequestration of about 25 Mt of calcite in coccolithophoride blooms. Recent studies indicate that coccolithophores will become more abundant but less calcified as CO_2 increases (Krumhardt et al. 2019).

Silicon is the other mineral that is massively assimilated by marine microorganisms, above all diatoms, silicoflagellates, and radiolarians: they use silicic acid $(Si[OH]_4)$ to create their elaborate opal (hydrated, amorphous biogenic silica, $SiO_2 \cdot 0.4H_2O$) structures. Tréguer et al. (1995) estimated the rate of that uptake at about 7 Gt Si a year. This means that the total mass of calcareous and siliceous materials sequestered by marine phytoplankton is on the order of 10 Gt/year larger than the total extraction of all metallic ores and larger than the annual production of coal. In comparison to marine processors of calcium and silica, aggregate use of these elements by other organisms is orders of magnitude smaller but its forms are often not only elaborate but also quite beautiful.

This is true about many mollusk shells: some of them are quite simple but others show remarkable geometric properties even as the mechanisms of biomineralization remain speculative (Abbott and Dance 2000; Furuhashi et al. 2009). Carbonates are also bioprecipitated by reptiles and birds to build their eggs, and by snails to form their shells. Curiously, a leading student of structures built by animals had deliberately left out such activities from his surveys: Hansell (2007) acknowledged that Great Barrier Reef may be (as its common description claims) the world's largest structure built by living animals, but in his writings, he focused on building that requires behavior, an attribute that is obviously absent in coral polyps or coccolithophores, organisms that just secrete their skeletons.

2.2 Materials in Prehistory

Evolution of hominins (the human clade that diverged from chimpanzees more than five million years ago and that had eventually produced our species) should be more accurately seen as a dynamic coevolution of several traits that have made us human: upright walking, endurance running, cooperative hunting, eating meat, symbolic language – and tool making, using natural materials to fashion objects that provided simple but practical extensions and multipliers of human physical capacities. That is why the archaeology of hominin evolution traces nearly as much the developments in making stone tools as it does the changes in bone structure, diets, and socialization.

2.2.1   Tools

Until 2010 the oldest stone tools of the lower Paleolithic, discovered in East Africa (Olduvai Gorge in Tanzania, Gona in Ethiopia), were dated to about 2.6 million years ago, but it was clear that earlier origins cannot be excluded (Davidson and McGrew 2005). In 2010 cut marks found on animal bones at Dikika in Ethiopia were dated to 3.4 million years ago, and a year later stone tools of similar age (3.3 million years ago) were discovered at Lomekwi west of Kenya's Lake Turkana (Harmand et al. 2015). The tools, somewhat larger than the Oldowan artifacts, were clearly made by knapping (intentional flaking) and did not arise as accidental rock fractures. They predate the appearance of genus Homo (now shifted to as far back as 2.8 million years ago) and were most likely made by australopithecine (species made famous by the discovery of "Lucy" skeleton).

For millions of years, the aggregate mass of stones used to make Paleolithic tools (the era that ended about 10,000 years ago) and then Neolithic artifacts remained minuscule as hominin population counted in mere tens or hundreds of thousands. Genetic studies and demographic models indicate 1.2 million years ago there were only about 20,000 hominins, fewer than the combined total of chimpanzees and gorillas (Huffa et al. 2010). A quarter million years ago the hominin numbers reached 50,000, and 10,000 years ago there could have been about 100,000 individuals of *Homo sapiens*. Subsequent population growth was reduced by cooling caused by the expansion of ice sheets during the last glacial maximum. European population declined from more than 300,000 people 30,000 years ago to about 130,000 individuals 23,000 years ago and then grew to about 400,000 by the end of the last ice age (Tallavaaraa et al. 2015).

Given these low population totals, it is obvious that using stones for tools was not a matter of quantity and mass supply but of specific quality and manufacturing skills to produce desired shapes and edges. Obsidian (volcanic glass formed by rapid cooling of magma) and flint (crystalline quartz) were the best materials to produce sharp cutting edges and piercing points by knapping; tools for pounding and pulverizing were made by grinding from basalt, rhyolite, and greenstone. Perhaps the most notable innovation in making stone tools was to modify stone properties by heating: there is clear evidence that 164,000 years ago our species began to use fire as an engineering tool to heat-treating stones in order to improve their flaking properties (Brown et al. 2009).

As is attested by numerous finds in both Old and New World, the combination of expert stone selection, appropriate manufacturing methods (heat treatment and skilled flaking, knapping, or grinding), produced adzes, axes, hammers, awls, arrows, tips, and knives that were ingeniously designed, esthetically pleasing, admirably ergonomic, and practical. In contrast, wooden tools have been preserved only in rare instances, particularly when buried in anoxic layers. There is no doubt that hominins used wooden sticks to dig up roots and wooden clubs to kill small animals but carefully crafted wooden weapons arrived relatively late. Although some massive herbivores could be killed without any tools – carefully planned stampeding of buffalo herds over cliffs, whose best known example is Alberta's Head-Smashed-In Buffalo Jump near Fort Mcleod in Alberta (Frison 1987) – killing of megafauna (from mammoths to elands) had generally required projectile weapons.

The earliest, and surprisingly well-preserved, throwing spears were found in Germany in 1996: six 2.25 m long spruce-wood spears were dated to between 380,000 and 400,000 years ago (Thieme 1997), that is nearly 200,000 years before the appearance of our species (Trinkhaus 2005). And a new evidence from a site of Kathy Pan 1 in South Africa (fracture types, modification near the base, distribution of edge damage) indicates that stone points could have been hafted to spear tips about 500,000 years ago, pushing the time of first hafted multicomponent tools about 200,000 years further than previously thought (Wilkins et al. 2012).

Other designs of food-procuring wooden tools are much more recent. Dating of the first bows (from the most suitable species of wood, above all yew, white ash, black locust, osage orange) and arrows (merely sharpened wood or stone-tipped) remains elusive but they were common in the late Paleolithic. Hunting sticks made of wood or mammoth tusks were used in Eurasia long before the invention of Australian boomerang (returning stick) some 10,000 years ago. The earliest use of fishing nets is impossible to estimate as they were undoubtedly woven from flexible plant stems or boughs that decayed rapidly; the oldest known net, made from slim willow branches and found in Finnish Karelia, is about 10,000 years old (Miettinen et al. 2008).

Foragers living without permanent abodes left no traces of their temporary shelters built from branches, grasses, reeds, or palm fronds. The oldest preserved components of shelters are, naturally, stones arranged for protection (walls, roofs), mammoth bones, and tusks made into walls, timbers, and hides. In contrast to the vanished living spaces of foragers (save for the caves, some adorned with remarkable paintings of animals made by the Neolithic hunters), archeologists have found thousands of foundations belonging to houses, sheds, and store rooms of many pre-agricultural or early agricultural societies. Perfect reconstructions of most of these Neolithic structures are usually impossible and we can only speculate about the actual use of wood (slender trunks of small trees, branches), reeds, straw, or clay in their walls and ceilings. Neolithic foragers also built the first wooden vessels: the oldest excavated examples of the simplest design, canoes dug out from a single tree, are nearly 10,000 years old.

2.2.2   Processed Materials

Although the prehistoric societies managed without any metals, some of them succeeded to change the properties of minerals by producing fired pottery. Excavations of some of the earliest permanently inhabited sites indicate that they used quicklime (burnt lime), the material whose production required processing almost as sophisticated as smelting of ores. In order to extract quicklime (CaO) from limestone ($CaCO_3$) by heating ($CaCO_3 \rightarrow CaO + CO_2$), the rocks had to be first crushed by hand to a fairly uniform small size (but tiny pieces and dust had to be avoided in order to keep the charge porous); the earliest firing took place in pits filled with layers of firewood and crushed limestone, and later small stone structures (kilns) were built to confine the process and to concentrate the heat needed for calcination.

After prolonged firing came the most hazardous operation, removal of the highly caustic quicklime whose reaction with water (that is with moist skin, eyes, or lungs) produces severe irritation and eventually burn, while ingestion of the dust brings abdominal pain and vomiting. Controlled addition of water to quicklime produced hydrated (slaked) lime ($Ca(OH)_2$, calcium hydroxide), the key ingredient of whitewash, mortar, and plaster. Producing lime is clearly a complex process that requires considerable amount of planning (collecting sufficient firewood, fashioning a suitable pit or a simple kiln) and management (temperature inside a kiln must reach at least 825 °C and it must be maintained for a prolonged period) – but lime products dating to 9600 BCE were excavated at Göbekli Tepe (Courland 2011). Kilning limestone was thus the first successful industrial process dependent on a chemical reaction – and one whose fundamentals had remained nearly identical until the nineteenth century.

Many of these wood-and-stone gatherer-hunter societies had also mastered the firing of shaped clays to produce pottery, the first process used by human to convert commonly available minerals into utilitarian or ornamental objects (Cooper 2000; Tsetlin 2018). Firing removes water from shaped clays, sets their shape, and strengthens their final form. Firing clay to produce a small object requires relatively low temperatures of 500–600 °C, and the oldest such piece of ceramics is the famous Venus of Dolní Věstonice in Moravia, 11.1-cm tall statue of a nude female that was made by Paleolithic foragers between 25,000 and 29,500 years ago (Vandiver et al. 1989). To make larger, utilitarian pieces of simple earthenware temperatures must reach about 1,000 °C and that could be achieved (for brief periods of time) in covered fire pits or mounds but is much better done in stone kilns. The earliest pottery pieces were undoubtedly pit-fired, but the dates of the oldest fired objects have been receding with new finds, particularly in Asia.

Findings from Xianrendong Cave (in northern Jiangxi province) show that the Upper Paleolithic hunters may have been firing simple small vessels in China as far as 19,000–20,000 years ago (Wu et al. 2012). Japan's Jōmon (cord-marked) pottery (first various round-bottom bowls up to 50 cm tall, then larger flat-bottom vessels for food storage and cooking) is up to 12,000 years old (Habu 2004). After 6000 BCE, ceramic objects – vessels, cups, vases, chalices, plates, figurines, many of them finely ornamented, others fired at higher temperature to produce more durable stoneware – became common throughout Europe as attested by numerous finds of post 5500-BCE *Bandkeramik* in Germany, Czech lands, and Austria (Petrasch 2020). In the Middle East Egyptian, ceramic was particularly diverse, both in terms of raw materials used and shapes produced. That was also the time when potter's wheel (allowing, in skilled hands, a superior execution of an intended shape) came into widespread use.

2.2.3 Megaliths

Pre-agricultural and proto-agricultural societies have also left behind almost inde-structible record of megalithic structures, beginning with remarkable stone circles with carved animal reliefs at Göbekli Tepe (Caletti 2020). Western Europe – from Sweden and Denmark through North Germany, the Low Countries, British Isles, northwest France, northern Spain, and Portugal – and parts of the Mediterranean region (southeastern Spain, southern France, Corsica, Sardinia, Sicily, Malta, Balearics, Apulia, northern Italy) and Switzerland – have tens of thousands of meg-aliths (Daniel 1980). These include tombs, menhirs (single standing stones), stone circles, and alignments (such as famous parallel arrays at Carnac in southern Brittany and Stonehenge nearby Avebury), and megalithic buildings or temples with graves oriented toward the east or southeast, toward the rising Sun. Radiocarbon dating suggest that these megaliths graves emerged during the second half of the fifth millennium BCE in northwest France, the Mediterranean, and the Atlantic coast of Iberia and that they spread along the sea route (Paulsson 2019).

Despite centuries of speculation and decades of interdisciplinary scientific stud-ies, we still do cannot offer any definitive explanation concerning the specific meth-ods of quarrying these massive stones, transporting them across often rough terrain, and erecting them in predetermined fashion, sometimes in remarkably accurate alignments with periodical configurations of celestial bodies. Transportation arrangements remain particularly unclear as both the unit masses and the straight-line distances were often considerable. The inner circle at Stonehenge was made of bluestones (43 remain, the original number is unknown) weighing 2–4 t. They had to brought from Preseli Hills in southwestern Wales about 230 km from the Salisbury plain (and hence their move almost certainly involved shipments in coastal vessels), and although the outer circle came from Marlborough Downs only 32 km north of the site moving 50-t stones was an even greater logistic challenge (Thorpe and Williams-Thorpe 1991; Ixer and Bevins 2017).

2.2.4 Textiles

Finally, there is no doubt that the first textiles are of prehistoric origin but the poor preservation of natural materials used for clothing makes it impossible to offer any reliable chronology (Ginsburg 1991). Bone sewing needles used to make warm clothes needed to survive the Ice Age temperatures (and also fishing nets and bags) became common during the Solutréan phase of the Upper Paleolithic Europe (21,000–17,000 years ago). The first deliberately cut, processed, and shaped body covers made of plant material (leaves, bark fibers) and animal skins had long decayed, precluding any dating. Skin tanning, needed to keep the hides supple in low temperatures, is definitely of prehistoric origin but identification of specific processes (using plant materials or minerals) becomes possible only during the Mesopotamian and Egyptian antiquity. Quality and functionality of prehistoric clothes made by Neolithic foragers is best demonstrated by *annuraangit*, Inuit skin-and-fur garments that provided excellent protection in the Arctic environment (Oakes 2005).

Flax weaves were produced even before the plant was domesticated in the Middle East more than 10,000 years ago and seed size indicates that by the 3rd millennium BCE different forms of flax (for oil and for fiber) were cultivated in Europe north of the Alps (Karg 2011). Cotton cultivation began in Asia about 7,000 years ago, much later in Mesoamerica, and phylogenetic analysis shows that the African-Asian species and the New World species stem independently from wild progenitors and converged morphologically under domestication (Wendel et al. 1999). Primitive looms used to make rough linen fabric came around 6,000 years ago, rough wool cloth was spun and woven more than 10,000 years ago, woolly sheep were domesticated about 8,000 years ago but it took at least another 2,000 years before the appearance of first woven wool fabric (Broudy 2000).

2.3 Ancient and Medieval Materials

Material world of the oldest settled societies – whose crop productivity and population growth sufficed to establish cities and support very slow economic expansion – was determined by their immediate environment. As a result, living quarters included caves excavated in silt (in China's Shaanxi) or in limestone (in central Anatolia), light wooden frames with walls of bamboo and clay (in Japan), unbaked brick structures (in sub-Saharan Africa), painstakingly assembled stone houses (in the Mediterranean and high-lying parts of Asia), massive log houses (in Scandinavia and Russia), and sturdily built structures using fired bricks or mortared stones (throughout the continental Atlantic Europe).

Environment made yet another, strategic, difference. As Adshead (1997) pointed out, a key reason for China's (and more generally East Asia's) choice of ephemeral wooden housing in contrast to Europe's preference for sturdy construction using stone and bricks was the high frequency of destructive earthquakes and I would also add of annual typhoons and major floods. The other reasons were the Chinese preference for low initial capital outlay and high cost of maintenance, the reverse of European approach. Heritage also mattered, with medieval and Renaissance Europe admiring and emulating many examples of monumental Roman stone structures.

Some of the earliest sedentary societies soon mastered the construction of monumental structures whose erection required not only ingenious quarrying and transport of stones to building sites but, unlike with pre-historic megaliths, also careful (and often detailed and extremely accurate) cutting and intricate assembly of massive structures or production and emplacement of prodigious quantities of bricks. Some ancient societies had also developed remarkably ingenious infrastructures designed to serve their growing cities, with the Roman aqueducts (commonly tens of kilometers long, with extended sections carried on stone arches) being perhaps the best illustration of these capabilities. The first sedentary societies of the Middle East, Mediterranean, and East Asia were also the first complex civilizations to smelt metals, initially only copper and tin on limited scales (for ornamental or limited military uses), later mastering also the production of zinc, lead, iron, mercury, silver, and gold.

In the Middle East, India, and Europe iron became the dominant metal after roughly 1200 BCE and its artisanal production was eventually on a scale sufficient to supply relatively inexpensive nails for house and ship construction, shoes for horses, variety of hand tools and machine components, and a wide range of weapons. These metallurgical advances led Christian Thomsen (1836) to divide material evolution into Stone, Bronze, and Iron Ages but that was not a universal sequence as some societies (perhaps the best example is Egypt before 2000 BCE) had a pure copper era, while others skipped the bronze age and moved directly from stone to iron artifacts. In any case, in terms of total material flows all pre-industrial societies remained in wooden age: in all forested environments no materials were as ubiquitous as timber for buildings and wood for tools, utensils, implements, and machines.

2.3.1 Wood

First, woody phytomass used in construction consisted of small trunks, tree branches, and bushes, and only the availability of stone adzes (and later good metal axes) and acquisition of requisite carpentry skills made it possible to build first massive log houses made of shaped and joined timbers; later still came metal saws that could be used to produce long boards and other precisely cut components. Most preindustrial wooden houses were poorly built and had short life spans, but properly constructed wooden structures were surprisingly durable, even in monsoon climate where they have been exposed to annually heavy rains: wood for Japan's Hōryūji pagoda in Nara, the world's oldest extant wood building, was cut nearly 1,400 years ago (Yamato 2006).

Remarkably, East Asian (Japanese, Korean, and Chinese) buildings did not use any fasteners, but their frames, beams, and roofs relied on a variety of precise mortise-and-tenon joints that not only allowed the wood to expand and contract in reaction to changing atmospheric humidity but also to withstand frequent earthquakes. Moreover, tall pagodas were essentially earthquake-proof because of *shin-bashira*, a central pole that does not support the structure but acts as a massive stationary pendulum (Atsushi 1995). Its action is equivalent to tuned mass damping in modern skyscrapers with their large concrete or steel blocks suspended at the tops of tall buildings.

Wood was also the only material used during the antiquity and the Middle Ages for the hulls and masts of ocean-going vessels. Small river boats were made of a single-tree trunk but Homer credited Odysseus with cutting down 20 trees to build his ship (*Odyssey* 5:23): that would translate to mass of 10–12 t. Egyptian boats used to transport stone obelisks were much heavier but the ships that made the first Atlantic crossings were remarkably light: a Viking ship (based on a well-preserved Gokstad vessel built around 890 CE) required wood of 74 oaks, including 16 pairs of oars (Bill 2011). Typical Mediterranean vessels also remained small and had hardly grown during the millennium between the demise of the Roman Empire and the first great intercontinental voyages of the late fifteenth century.

2.3.2 Stone

Stone's durability made it the preferred material for utilitarian structures built to last (aqueducts, roads) for funerary (pyramids, tombs) and religious or ceremonial (temples, gates, palaces, towers, obelisks, statues) monuments. Many stone structures impress due to their mass but among them Egyptian pyramids at Giza are unique: Khufu's pyramid not only remains the largest stone structure ever built (195 m high), it required 2.5 million stones whose average weight was 2.5 t and yet this mass of more than 6 Mt of stone occupying 2.5 Mm^3 was laid down with an admirable degree of accuracy (Lehner and Hawass 2017). In comparison, great Mesoamerican pyramids (Teotihuacan and Cholula) are not that imposing, not only because Teotihuacan's flat-topped Pyramid of the Sun rises only to just over 70 m (including the temple) but also because its core is a mass of soil, rubble, and adobe bricks, and only its exterior was clad with cut stone and plastered with lime mortar.

Other stone structures are distinguished by their ubiquity, other yet by complexity and lightness that belies the heaviness of the material (medieval cathedrals), and some have shown an almost incredibly tight fit and accurate emplacement. During the Roman Empire, the largest mass of stones went into well-engineered aqueducts (including elevated bridges, inverted siphons, and distribution networks within cities) as well as in their extensive system of roads (*cursus publicus*), but most of their monumental structures (triumphal arches being a major exception) were marble-clad bricks (Adam 1994). Intricately designed stone monuments dominated in South and Southeast Asia (including Angkor Wat, the world's largest Hindu temple in Cambodia, and Borobudur, the world's largest Buddhist monument in Java) as well as the Mesoamerican (Maya, Aztec) and South American (Inca) ceremonial sites. And Africa's most massive stone structure, 240-m long oval enclosure of Great Zimbabwe whose construction began after 1100 CE, required an estimated one million of stone blocks (Ndoro 1997).

Medieval Europe brought stone architecture to unprecedented heights (literally and figuratively) with its bold and intricate designs of large cathedrals with arched vaults, flying buttresses, and tall spires (Fitchen 1961; Schutz 2002). Prak (2011) showed that this was achieved by relying on surprisingly simple underlying principles of modular design (transferred on a personal basis by designers/builders) augmented with on-site experimentation. The most obvious commonalities of stone construction are elaborate quarrying (the task all the more remarkable during the earliest period when, as in ancient Egypt, the masons had only copper chisels and dolerite mallets), often highly accurate dressing, frequent needs for long-distance land- or water-borne transport to often distant building sites and ingenious ways of emplacing massive components. Even today we are impressed by the inventiveness and skills used in quarrying, transporting, shaping these massive stones, and in designing and organizing their final emplacement, often with admirably tight tolerances.

That is particularly true in case of monoliths (used for statues or as steles or obelisks: Rome has eight of them imported from ancient Egypt) by traditional civilizations; we still can only speculate about many specific artisanal techniques and construction management required to build the pyramids on the Giza plateau or to fit the polygonal stones of massive Inca walls at Sacsayhuamán and Ollantaytambo with incredible tightness. Nor do we know exactly how the ancient Egyptians orchestrated the transport of a 1,000-t stone statue 270 km by a Nile ship from Aswan to Thebes. Not surprisingly, specific modalities of building many of these structures remain elusive: the only certainty is that their construction required careful planning and extraordinary supply and worksite management, particularly for those structures that were completed in what appear to be incredibly short periods of time.

Egypt's largest pyramid took only 20 years to complete (2703–2683 BCE). The Parthenon, a simple post-and-lintel building but one possessing perhaps the best-ever proportions, required only 15 years (447–432 BCE). And the astonishing Hagia Sophia in Constantinople, a relatively light structure deriving its impact from enormous vaults, did not take even five years to complete (527–532). In contrast, centuries might have elapsed before laying the foundation stone and the consecration of many European cathedrals: construction of Notre Dame took nearly two centuries, from 1163 to 1345; St. Vitus, dominating the Prague Castle, was begun in 1344 and it was completed only in 1929.

Nor can we, obviously, quantify the mass or volume of construction stone used annually in antique or medieval societies, but revealing calculations can be made for the requirements of some large projects. Correct order of magnitude of materials deployed by the Roman to build their roads can be estimated by assuming average width of 4 m (surviving stretches of Italian *viae* have widths between 2.4 and 7.5 m) and the depth of at least 1 m (with paving stones topping layers of gravel, pebbles, clays, and sand): leaving aside subsequent repairs, the initial construction of 85,000 km of major roads would have required 425 Mm^3 of aggregate (sand and gravel) and quarried stone (Smil 2017).

2.3.3 Clays

In alluvial plains, clays turned to bricks became the dominant material of construction: Mesopotamian kingdoms built their ziggurats, palaces, and walls from bricks (both adobe and better quality made from fired clays). Bricks were also indispensable for the building boom in the imperial Rome that needed relatively inexpensive and rapidly (and locally) produced material in order to build massive buildings whose exteriors were then clad in marble. The first bricks were made with puddled clay (just mixing clay and water with additions of sand, chopped straw, or dung), these mixtures were shaped in wooden molds and bricks were left to dry in the sun (Fernandes et al. 2010).

Early Mesopotamian cultures (Sumer, Babylon, Assyria) produced prodigious amounts of these bricks (Babylonian standard was a square-shaped prism, 40×40×10 cm), which they used to build their houses, palaces, and towers. Poor quality of such bricks turned even the most massive structures into mounds of clay: a few exceptions aside – above all Chogha Zanbil, a partially preserved Elamite structure in Iran's Khuzestan with three of the five stories still standing (Porada 1965) – Mesopotamia's stepped flat-topped temple towers (ziggurats) built after 2200 BCE are now nothing but hillocks on an arid plain. Bricks burnt in kilns fueled with wood or charcoal were used first in ancient Mesopotamia, they were common in both European and Asian antiquity and remained the most important quotidian construction material in many Old World cultures until the arrival of concrete and structural steel.

Slim, oblong Roman bricks ($45 \times 30 \times 3.75$ cm) were clad by marble in all monumental buildings, but even after that protective layer was stripped away they have survived fairly well in dry Mediterranean climate where their poor heat conductivity helped to keep the structures cool in summer (Brodribb 1987). Romans also used bricks to construct remarkably expansive vaulted ceilings. There were many different methods of laying down the bricks and combining them with stones and wood (common Roman choices were *opus reticulatum, vittatum, mixtum,* and *testaceum*), all of them seen in the ruins stripped of its marble cladding.

Structures built with fired bricks proved more durable (naturally, with some repairs) even in Asia's tropical climates: many stūpas (meaning, literally, heaps in Sanskrit), huge mound-like constructs built to house Buddhist relics have survived for more than a millennium (Longhurst 1979). Mahabodhi stūpa (Bodhgaya temple), built at the site of Buddha enlightenment 2,500 years ago, great sandstone-clad Sāñcī stūpa of Bhopāl (36 m base diameter, 15 m tall), Phra Pathom Chedi stūpa in Thailand, and 122-m tall Jetavanaramaya in Anurādhapura (Sri Lanka), built some 1,700 years ago, are among the best examples of these massive fired-bricks structures (Bandaranayake 1974; Pant 1976). Number of bricks (made of 60% fine sand and 35% clay) required to build Jetavanaramaya was put at between 93 and 200 million (Leach 1959).

Fired earthenware and stoneware (the latter material having much lower water absorption after firing) was also used to make roof, floor, and decorative tiles. Ceramic roof tiles were commonly used around the Mediterranean since the antiquity, while the first example of glazed bricks and decorative tiles are found in even older Mesopotamian structures. Floor and wall tiles were eventually produced in a great variety of finishes (unglazed, glazed, mosaic, hand-painted), and particularly intricate tiling designs became one of the signature marks of medieval Muslim builders who excelled in matching colors and shapes; a less exuberant ancient tiling tradition had also evolved in a number of European countries, especially in Italy, Spain, and Portugal.

Clay amphoras were the most common containers of sea-born Mediterranean commerce (Twede 2002). These ceramic jars were used to transport wine, olive oil, and also processed foodstuffs such as *garum*, Roman fish sauce. Romans also used wooden barrels made by skilled craftsmen (coopers) from hardwood staves kept tight by iron hoops, with a typical barrel consuming 50–60 kg of wood (Twede 2005). During the Middle Ages, these containers dominated European trade in liquids as well as in pickled and salted foods. Porcelain, the highest-quality material made of fired clay (specifically kaolin, soft white kaolinite clay indispensable for making fine ceramics), was invented around the sixteenth century BCE (during the Shang Dynasty) in China, soon made in Korea and Japan, and this East Asian monopoly was broken only during the early eighteenth century with the production of true porcelain in Europe, first in Germany and France, and soon afterward in England (Atterbury 1982).

Durable construction using bricks (and tiles) required good bonding material, but the first clay and mud mortars were weak; only the calcination of limestone ($CaCO_3$) resulting in quicklime (CaO) whose hydration produced $Ca(OH)_2$, slaked lime, made a superior material. As already noted, the kilning of limestone predates the era of the oldest centralized empires and slaked lime was used routinely in ancient Egypt. Davidovits (2002) caused a controversy by claiming that the blocks of Giza pyramids were not quarried stones but mixtures of granular limestone aggregate and alkali alumino-silicate-based binder that were *cast* in situ. Although this hypothesis was rejected by most Egyptologists, it continues to have adherents even among material scientists.

Romans are credited with the invention of concrete but that is an inaccurate attribution. Concrete is a mixture of cement, aggregates (sand, pebbles), and water, and cement is a finely ground mixture of lime, clay, and metallic oxides fired in kilns at high temperature. There was no cement in Roman *opus cementitium* and hence the sturdy mixture, strong enough to build large vaults and domed structures, was not the material known as concrete. *Opus cementitium* contained aggregates (sand, gravel, stones, broken bricks, or tiles) and water, but its bonding agent was lime mortar (Adam 1994). Only the combination of slaked lime and volcanic sand from the vicinity of Puteoli near the Mount Vesuvius (*pulvere puteolano*, later known as *pozzolana*) produced a superior mixture that could harden even under water and that could be used to build not only massive and durable walls but also spectacular vaults. The coffered ceiling of Pantheon (118–126 CE) with dome span of 43.2 m remained unsurpassed for nearly two millennia (although San Pietro's dome, designed my Michelangelo and completed in 1590, came close with 41.75 m).

2.3.4 Glass

The last ancient material invention that required heat processing was glass (Macfarlane and Martin 2002). Its earliest small Mesopotamian pieces are about 5,000 years old, the art of making art objects was advanced during the New Kingdom era of the ancient Egypt, glass finds are not uncommon at Roman sites – but it became an important material only during the Middle Ages, both in construction (as attested both by intricately designed multicolored cathedral windows) and in everyday use (as shown by elaborate glass goblets of Venetian and Bohemian glass).

Crown glass production was the only practical way to produce pans of limited sizes until the mid-nineteenth century. Molten material was mouth-blown to create a large bubble and then spin it into a disc (with a circle in the middle, up to 1 m in diameter) that was, after cooling, cut into panes. This glass was too expensive for glazing of ordinary homes, and glass windows became common only during the early modern era.

But the most consequential material development in the antiquity was not a routine use of a wide variety of biomaterials (wood, bones, hides, plant- and animal-textiles) and common construction (stones, clays, sand, concrete) and ornamental (tiles, glass) materials, but an ability to smelt and to shape a growing array of metals. Ore mining and metal smelting resulted in an epochal advance that began with the use of copper and its alloys and was followed by smelting of iron ores making iron the dominant metal of the ancient Greece and Rome, the two great Mediterranean civilization whose accomplishments influenced so much of the subsequent European, and global, history.

2.3.5   Non-ferrous Metals

Copper, soft and malleable when pure and easy to alloy, was the first metal used by Stone Age societies as early as almost 10,000 years ago, initially without any smelting as the bits of outcropping pure metal were shaped cold or subjected to annealing, that is repeated heating and hammering (Forbes 1972). Copper smelting and casting took off more than 6,000 years ago, first in Mesopotamia (Gilgamesh, the great Sumerian epos [about 2500 BCE] refers to a copper box and a bronze bolt and excavations found copper knives, axes, and spears in the region's ancient empires), a millennium later in Egypt, and about 3,500 years ago it was common in China. Relative abundance of copper sulfide ores made the element the dominant metal between 3000 and 1000 BCE.

Ore reduction (Cu has a fairly high melting point, 1,083 °C) was done with wood or charcoal, first in clay-lined pits, later in small clay furnaces and the metal was purified by further heating. Producing copper from abundant sulfide ores (chalcopyrites) was more complicated: crushed ore (by hand, later by harnessed horses or waterwheel-driven hammers) had to be the first roasted to remove sulfur and various associated metals (As, Fe, Pb, Sn, Zn). Roasted ore was first smelted in shaft furnaces and then smelted once again to yield 95–97% pure metal. All of this devastated local and regional wood resources and copper smelting was a leading cause of Mediterranean deforestation, particularly in Spain and Cyprus (Smil 2017).

But annealed copper was a soft metal and it had low tensile strength. That is why a much stronger and much harder bronze became the first practical alloy in history and why Christian Thomsen chose it for his now classic periodization of Stone, Bronze, and Iron Ages (Thomsen 1836). Bronze, an alloy of copper with 5–30% of tin (10% being typical), has tensile strength nearly four times that of annealed copper and it is nearly six times harder, good enough for durable knives, swords, axes, and medals (as well as for bells and musical instruments). Brass is an alloy of copper (ranging from less than 50% to about 85%) and zinc; its smelting dates to the first century BCE but it became common only during the high Middle Ages. Zinc increases the alloy's tensile strength and hardness to about 1.7 times that of cold-drawn copper without reducing malleability and resistance to corrosion. Pewter is an alloy dominated by tin (about 90%) with added antimony and copper.

Besides copper, tin, bronze, and brass, the other color (nonferrous) metals whose production was mastered by ancient metallurgists included zinc, lead, mercury, silver, and gold. Analysis of lead in an ice core from Greenland made it possible to reconstruct the world's longest series of metal production. Hong et al. (1994) estimated that large lead–silver smelting by Greeks and Romans increased total lead output from about 250 t/year in 750 BCE to nearly 80,000 t by 50 CE. Lead's low melting point and easy malleability made it an early candidate for making water pipes in Roman cities but the greatest demand for the metal was for inverted siphons (U-shaped pipes connecting a header tank on one side with a lower-lying receiving tank on the opposite bank), a preferred way for the Roman engineers to cross valleys where stone bridges would have to be taller than 50–60 m. Post-Roman decline and stagnation kept lead output at only about 12,500 t thousand years later, and the global output surpassed the highest antique production only by the middle of the eighteenth century.

Gold and silver were used in antiquity for ornaments (Tutankhamun's burial mask, from about 1320 BCE, is perhaps the most famous golden object of the era) and jewelry in both the Old and the New World, and they were also used to mint coins, often debased by addition of cheaper metals: gradual debasement of the Roman *denarius* is perhaps the most notorious example of the process (Salmon 1999). Patterson (1972) made the following rough estimates of annual rates of Greco-Roman silver production: 25 t between 350 and 250 BCE, 200 t between 50 BCE and 100 CE and, again, 25 t during the fourth century CE. Mercury was incorporated into ointments and cosmetics (just about the worst possible use imaginable) and later became a key element in experimental alchemy.

2.3.6 Iron

Iron ores are distributed worldwide and many deposits exploited before the twentieth century were rich in metal: pure magnetite contains 72.4% of iron and pure hematite has 69.9% of the metal. Smelting iron ore requires higher temperatures

(1,204 °C) than producing copper or lead (1,084 °C and only 328 °C). The first iron artifacts date to around 5000 BCE, but the metal came into common use only less than 3,000 years ago, both in the ancient Egypt and in Asia (Smil 2016). Cast iron, an alloy of the metal with carbon, was the easiest ferrous metal to produce but its tensile strength was no higher than that of cold-drawn copper and its hardness was similar, or only slightly superior, to that of bronze. But the greater abundance of iron ores and slow advances in smelting the metal had eventually made iron the most important metal of antiquity and, in aggregate mass terms, it has kept its dominance ever since: as I will show in detail in the third chapter, ours is still the Iron Age, with the total consumption of other metals adding to a small fraction of iron use.

Due to its relatively high carbon content (between 2% and 4.3%), cast iron (traditionally called pig iron because it used to be cast into forms whose shape resembled small pig) has low tensile strength (at 200 MPa lower than bronze or brass), low impact resistance, and also very low ductility. But the alloy is strong in compression and this makes it suitable for a large variety of tools, utensils, and objects (ranging from nails to horseshoes and from pots to fireplace grates) and weapons (including heavy guns and cannon balls), but precluding its use wherever the metal was subject to greater tension: iron columns are fine, and iron beams could never support tall buildings.

Iron smelting began in simple, partially enclosed hearths, producing small (around 50 kg) slag-contaminated masses of the metal. Bloomery hearths were gradually transformed into small shaft furnaces whose development led to the first blast furnaces (during the fourteenth century in the Rhine-Meuse region) in whose columnar bodies smelting takes place as upward movement of hot CO-rich gases reduces ferrous oxides into iron at temperatures reaching more than 1,600 °C. For the next five centuries, blast furnace sizes were limited because charcoal could not support heavier charges of iron ore and limestone (used to remove impurities). Fundamental innovation that allowed mass production of inexpensive iron came only during the late eighteenth century with the replacement of charcoal by metal-lurgical coke (Smil 2016).

The oldest way to make steel was by carburizing (cementation), that is by adding carbon to practically carbon-free sponge iron (small porous masses of iron and slag from bloomery furnaces) by prolonged heating in charcoal; that produced an out-side layer of hardened steel and uniform carbon distribution (needed for steel des-tined for weapons) was achieved only by repeated forging (Birch 1967). Decarburizing (removal of carbon by oxygenation) from cast iron was first done in Han China, producing an alloy good enough to make chains for suspension bridges to span deep gorges. But such uses were exceptions as steel remained expensive and hence in limited use until the nineteenth century.

2.4   Materials in the Early Modern Era

I prefer a simple delimitation of the early modern era as the period of the three centuries between 1500 and 1800; an obvious alternative anchored by specific events would be 1492 and 1789. These were fascinating centuries, an amalgam of many persistent, old, medieval realities and new (and is clear in retrospect) fundamental departures that, undoubtedly, laid the foundation of the modern world. In material terms, the era was characterized by some new qualities but even more so by much increased quantities. No new materials were introduced either in construction or in (still overwhelmingly) artisanal manufacturing but the growing sizes of cities and ocean-going ships, more reliance on water wheels and windmills, larger fortresses, ingenious construction of new canals, ports, and roads, and (first in England and Wales) expanding mining of coal led to higher demand for the two basic high-quality construction materials, wood and iron, and required greater removal and emplacement of larger volumes of soil, sand, gravel, and stone.

The early modern era was the time of higher rates of population growth, incipient urbanization, and protoindustrialization, and these processes stimulated changes in material consumption. The best available reconstructions indicate that the global population increased by less than 60% during the 500 years between 1000 and 1500, but then it had more than doubled (from about 460 million to nearly a billion) by 1800 – but had remained overwhelmingly rural, with cities accounting for less than 5% of all humanity (Klein Goldewijk et al. 2010). Protoindustrialization relied on cheap rural cottage and urban workshop labor and its products reached wider national and even international markets (Mendels 1972).

2.4.1   Possessions

In Europe, the regions affected by this process included parts of England (Cotswolds, Ulster), France (Picardy), and Germany (Westphalia, Saxony and Silesia); in Asia large-scale artisanal production was concentrated in coastal areas of Qing China and Mughal India and in the cities of Tokugawa Japan; and an increasing share of manufactures from all of these regions found its way to other Asian and European markets. Large-scale artisanal production implies incipient mass consumption, and as Mukerji (1983) showed, materialism was not a consequence of industrialization because in parts of the Atlantic Europe it was quite evident during the sixteenth century.

Paintings of the Dutch artists active during the Golden Age (1581–1701, or essentially the entire seventeenth century) show many neat spacious houses in Amsterdam, Haarlem, or Delft, with small but neat backyards, clean tiled floors, large glass

windows, walls adorned with pictures or maps, musical instruments, and no shortage of well-built furniture and bedding. Interiors painted by Jan Molenaar, Pieter de
Hooch, or Jan Vermeer leave us with unmistakable impressions of material comforts and incipient affluence enjoyed by many Dutch burghers long before the
European industrialization ushered in the successive waves of modern mass consumption (Franits 2004; Shawe-Taylor and Buvelot 2015). Moreover, owners of
these houses were eager buyers of an expanding range of consumer goods ranging
from better cookware to finer clothes and from engravings to porcelain imported
from China and Japan (the Dutch having the trading monopoly with the Tokugawa
shogunate).

And the greatest source of information about the accomplishments of the early
modern era – the world's first encyclopedia published between 1751 and 1777 under
the editorship of Denis Diderot and Jean le Rond D'Alembert – is filled with
descriptions and engravings of a large variety of machines, including many complex
and intricate designs whose adoption presaged even greater advances of the
nineteenth-century industrialization (Diderot and D'Alembert 1751–1777). At
the same time, housing remained commonly primitive and inadequate: even in
France during the sixteenth century most village dwelling were just mud huts covered with straw or rushes, and in cities living quarters were often combined with
(or were adjacent to) workshops or stores (such as in Kyōtō's dark, long *machiya*
houses). Buildings were also poorly heated and badly lit: dominant fireplaces were
inconvenient and inefficient, and more efficient stoves (most notably German
Kachelofen, a Dutch and Scandinavian tiled designs) had diffused only slowly from
their areas of origin.

As a result, fuel-wasting fireplaces and braziers resulted in huge demand for fuelwood and charcoal to heat-expanding cities of pre-coal era. In Paris, the demand
rose from more than 400,000 loads of wood in 1735 to more than 750,000 loads in
1789 (about $1.6\,Mm^3$) and the same amount of charcoal, prorating to more than a
ton of fuel per capita (Roche 2000). Cramped rooms in rural houses were also lit-
tered with tools and even in urban dwellings furnishing was often minimal. Chair,
absent in most houses during the Middle Ages (people sat on the ground, on benches,
ledges, or cushions), became a common piece of furniture, but a good bed remained
very expensive: in France just before 1700 its value was at least 25% of the total for
all furniture in low-income families, and nearly 40% for servants (Roche 2000).

And even during the seventeenth century, it was still common to cook food and
eat it from the same pots. Louis XIV ate with his fingers early in his reign (during
the 1660s) but a century later even middle-class urbanites had on their tables an
increasing number of objects devoted to specific uses ranging from egg cups to
teapots. This advance was accompanied by a retreat of noble materials (fewer
objects made from silver or crystal glass) and rise of cheaper objects made from

common metals, ceramics and blown glass. This shift marked the onset of ephemeral consumption and rapidly changing fashions. Poverty was widespread but for increasing numbers of people daily living had gone beyond necessities, As Roche (2000, 77) put it, "a new pattern of cultural behavior, made up of aspiration to well-being and dignity, asserted itself."

2.4.2 Wood

More substantial rural housing began to appear in some countries, but its wood requirements continued to reflect the environment: light structure in earthquake-prone regions, and solid log houses in parts of the Atlantic Europe and eastern North America. Japan's *minka* (people's houses) used post and beam construction, clay or bamboo walls, sliding doors, paper partitions, and earthen floors. As a result, a 100-m^2 *minka* often required no more than 8 m^3 of pine, cedar, or cypress wood (Kawashima 1986) – while the equally small Scandinavian log house (*stock hus*) needed 100 m^2 of wood for walls, doors, ceilings, and roof and large farmhouses in Germany or Switzerland required commonly 1,000 m^3 of timber (Mitscherlich 1963). Frequent repairs and rebuilding are added to this wood demand.

Wood remained indispensable not only for building houses and transportation equipment (carts, wagons, coaches, boats, ships) but as iron smelting rose in parts of Europe, more of it was needed to produce charcoal for blast furnaces (substitution by coke began only during the latter half of the eighteenth century and it was limited to the United Kingdom). And as Europe's maritime powers (Spain, Portugal, England, France, and Holland) competed in building large ocean-going vessels, both commercial and naval, the increasing number of such ships and their larger sizes brought unprecedented demand for high-quality timber needed to build the hulls, decks, and masts. At the very beginning of the early modern era, Columbus's *Santa Maria* displaced about 110 t, Magellan's *Victoria*, the first ship to circumnavigate the world, displaced 85 t. With wooden hulls, masts, and spars being as much as 70% of the total mass (the remainder was divided among ballast, supplies, sails, armaments, and crew), those pioneering vessels contained 60–75 t of sawn timber (Fernández-González 2006).

By the end of the eighteenth century, large two-decked naval battleships (originally of French design) were about 54 m long, carried 74 guns, and crews of up to 750 men (Watts 1905). Such a ship needed about 3,700 loads (equal to 1.4 m^3) of oak timber or about (with density of 650 kg/m^3) 3,400 t of wood, roughly 50 times the mass used for the vessels used for first intercontinental sailing three centuries earlier. But because as much as 60% of all timber bought by the builders disappeared as smaller pieces of wood were commonly taken from shipyards by workers for fuel, for making simple furniture, or to be sold (Linebaugh 1993), the actual mass removed from a forest was typically more than 5,000 t per ship.

Naval ships of the early modern era also provide an excellent illustration of increasingly massive production of armaments. Ships of the early sixteenth century had typically fewer than 10 guns, in 1588 English vessels that defeat the Spanish Armada averaged only 12 guns, before the end of the seventeenth century large men-of-war carried up to 100 guns, and during the battle of La Hogue in 1692, the British and Dutch captains deployed a total of 6,756 guns (Anderson and Anderson 1926). And iron for naval armaments was not the only reason for the metal's rising consumption. Higher demand also came from the expansion of mining, growing production of nails, wire, horseshoes, and weapons for land armies. English statistics show pig iron output rising from only about 10,000 t in 1700 to 26,000 t by 1750 and to 156,000 t by 1800 (Bell 1884).

2.4.3   Metals

Iron production in small blast furnaces required enormous quantities of charcoal, and with inefficient wood-to-charcoal conversion this led to widespread deforestation of iron-smelting regions: by 1700 a typical English furnace consumed 12,000 t of wood a year (Hyde 1977). Reprieve came only with a slow adoption of metallurgical coke: it was first used in a blast furnace in 1709, but even in the United Kingdom it became dominant only a century later. Steel remained a commodity of restricted supply even by the mid-eighteenth century. During the late 1740s, Benjamin Huntsman (1704–1776) began producing his crucible cast steel by carburizing wrought iron, but that metal was used only for such specialized, limited-volume applications as expensive weapons (famous Damascene or Japanese swords), razors, cutlery, watch springs, and engineering (above all metal-cutting) tools whose quality justified higher prices (Bell 1884). Mining for all kinds of ores had expanded in Europe, with particularly notable innovations introduced in Germany, France, and Italy.

But the greatest change came from the exploitation of the New World's metals, thanks to an unprecedented expansion of gold and silver mining. The first questions put by Columbus to the natives after his landing in the New World were about gold, but the Spanish fortunes changed only with the discoveries of both gold and silver in Mexico (Zacatecas), Peru, and Bolivia (Potosí in 1545). During the next 250 years, this American silver not only enriched the monarchy through direct transfers but the trans-Pacific transfer of the metal to the Philippines and then to China created a true world system of exchange (Frank 1998).

Barrett's (1990) estimates show total annual silver flows to Europe rising from 40 t during the early 1500s to 600 t during the last four decades of the eighteenth century. Total annual silver flows to Asia (through the Dutch and English companies and through the Levant trade, and directly from Americas by trans-Pacific sailings) rose from 75 t during the first decade of the seventeenth century to nearly 170 t by the mid-eighteenth century. And Flynn and Giráldez (1995) argued that founding of Manila in 1571 (in order to link, for the first time in history, Asia and America with "substantial, direct and continuous trade") was the birthday of true world merchandize exchange.

2.4.4 Textiles

The eighteenth century had also seen a notable expansion of textile-making because the first category of consumer goods that saw an obvious and widespread increase of expenditures was clothing, starting in cities and spreading to the countryside, eventually creating what Roche (2000) called "the unification of sartorial habits." Vertical looms were used since the antiquity, and many traditional societies in Asia and America used portable body-tensioned looms. In either case, a weaver could produce cloth only as wide as her arm span and a wider cloth required cooperation of two weavers. This changed in 1733 with John Kay's invention of the flying shuttle that could be sent by flicks of the wrist from one end of the loom to another; the most important innovation allowing cheaper, mass-scale production of cloth came in 1785 with Edmund Cartwright's power loom initially powered by steam (O'Brien 2019).

A no less important innovation was the response to imports of Indian printed cotton fabrics (calico, produced in Calicut since the eleventh century) to Europe. Adoption of Indian techniques by European craftsmen began in France before 1650, and before 1700 both French and English workshops were able to produce lasting colors and the practice spread to the Netherlands, Germany, Switzerland, and Austria. A century later Lancashire calico industry was gaining a clear competitive advantage over the India product, especially after Thomas Bell's invention of printing by copper rollers was adopted in 1785 (Jenkins 2003). Soon the trade was radically reversed and the British statistics show that by 1835 total number of cloths pieces (plain and dyed) sent from Britain to India was 169 times of that imported to Britain (National Archives 2012).

2.4.5 Stone

Finally, a few paragraphs about stone and other construction materials in the early modern era. Cut stone remained the dominant material for monumental buildings, opulent private structures, and religious architecture. Splendors of the late Renaissance and baroque Rome are perhaps the best examples of this deployment in palaces, basilicas, churches, and colonnades designed by such masters as Michelangelo Buonarroti, Carlo Maderno, Gian Lorenzo Bernini, Francesco Borromini, and Girolamo Rainaldi. New markets for cut stone opened up with the expansion of urban housing. Perhaps the most notable example of this development is the expansion of Paris. For centuries, the Lutetian cut limestone (*pierre de taille*) was supplied by subterranean quarries on the city's outskirts but the city's expansion above the mined galleries led to collapsing tunnels and surface subsidence (Blanc et al. 1998).

A commission set up by Jean-Baptiste Colbert, France's minister of finance under Louis XIV (between 1665 and 1683), identified limestone from Saint-Maximin quarries in the Val d'Oise only about 40 km north of the city's center as a nearly perfect color match for the city's major stone monuments (MB Stone International 2022). The soft stone is easy to cut and yet it is fairly resistant to weathering, and since the late seventeenth century, massive blocks for weight-bearing foundation floors and slimmer cuts for the façades have been giving Paris its inimitable look in a few subtle shadings ranging from whites to faint yellows. The first great wave of this construction took place between 1715 and 1752 when 22,000 new substantial houses were added, nearly every fifth one with *porte cochère* (Brice 1752). That remarkable pace of construction was surpassed only a century later some 40,000 houses added during Haussmann's reshaping of Paris between 1853 and 1870 (Des Cars 1988).

But in parts of Europe, stone gave way to brickworks in new, and more material-intensive, type of fortresses that emerged in response to greater capability of long-distance artillery. While typical medieval fortifications relied on compact (and very often elevated) sites and were built with high and thick stone walls, the new designs were radically different: they were variations of star-shaped polygons (hexagons, octagons) adapted to local (and often entirely flat) terrain but always with relatively low masonry walls backed by massive earthen embankments that were to absorb artillery shelling, to conceal and protect, and, at the same time, to establish clear lines of defensive fire.

Sébastien Le Prestre Vauban, military engineer and eventually Marshal of France, was the most prominent architect of these fortifications requiring unprecedented masses of bulk construction materials. During 40 years between 1667 and 1707, he upgraded fortifications of some 300 cities and built 37 new fortresses along the western, northern, and eastern borders of France (they are now on UESCO's list of World Heritage sites), including such spectacular projects as citadels of Le Palais and Mount Louis, city of Besançon, and an island redoubt of Saint Martin-de-Ré (Duffy 1985; Hebbert 1990). Lille fortress required 60 million bricks, and his largest project, fortifications of Longwy in northeastern France, consists of some 640,000 m³ of rock and earth and 120,000 m³ of brick masonry (Anderson 1988).

2.5 Creating Modern Material Civilization

When judged in basic existential terms – that is looking at prevailing diets, longevities, typical living conditions, fuel consumption, prime movers, and everyday choices, uses, and abundance of materials – even advanced pre-industrial societies of the late eighteenth century (China during Kangxi's rule, late Tokugawa Japan, pre-revolutionary France, Russia under Catherine the Great) were not radically different from their late medieval precursors. Famines were infrequent, but prevailing diets were barely adequate, monotonous, and overwhelmingly cereal-based; high infant mortality reduced the average life expectancy at birth to less than 40 years; housing was crowded, unhygienic, and uncomfortable; wood (and charcoal made from it) and crop residues (mostly cereal straws) were the dominant fuels; human and animal muscles were the most important prime movers; and wood, stone, and clays (shaped and fired as bricks) furnished the basic building materials, while possession of objects made from metals and alloys (iron, copper, bronze, brass) was uncommon.

Only in England coal displaced wood as the dominant fuel already before the end of the seventeenth century, by 1780s James Watt's (still quite inefficient) steam engines began to offer the first competition to windmills and waterwheels as reliable and increasingly powerful prime movers, and replacement of charcoal by coal (accelerating after 1770) made it possible to build larger blast furnaces, to produce more iron, and to lower the cost of many common iron objects (Smil 2017). But even in the United Kingdom, diets were often marginal, life expectancies were short, and material possessions of most people were limited. The nineteenth century changed all that, first in the United Kingdom, parts of Western Europe and eastern US, then in most of European countries, across North America, and after 1870 Japan was the first Asian country to begin the quest for modernity. In material terms, modernization – driven by industrialization and urbanization – was marked, above all, by two processes: a greatly expanded extraction of traditional construction materials and rapidly increasing consumption of metals.

2.5.1 Construction

The first kind of material expansion is usually neglected as the historians of industrialization focus on the consumption of fossil fuels and production of metals and machines but unprecedented quantities of stone had to be cut, blasted, crushed and shaped, and even greater volumes of soil, sand, and clays had to be moved, emplaced, or incorporated in bricks and concrete in just a few decades in order to span the countries with railways and better roads to house millions of former peasants who moved every year to cities and to build the productive infrastructure of modern economies (mines, ports, and factories) that enabled the countries to switch from agrarian societies to economies dominated by manufacturing in just two or three generations.

Perhaps the most esthetically appealing paragon of a much expanded demand for the cut stone would be the modernization of Paris began during the Second Empire (1852–1870). Reshaping of Paris streets and housing – directed by Georges Eugène Haussmann between 1853 and 1870 (Carmona and Camiller 2002), and the city's subsequent growth that had nearly tripled its 1850 population by 1900 – created a new wave of demand for the characteristic creamy stone from Saint-Maximin quarries that (as already explained) was chosen by Colbert in the seventeenth century to replace locally quarried limestone. Haussmann used the stones for typical solid-looking five-stories building with angled mansard roofs lined along his straight, wide boulevards. An order of magnitude estimate could be made by assuming that 40,000 new buildings erected during Haussmann's renewal averaged 350 t of stone for the foundation and 250 t for the façade: that would imply about 25 Mt of stone and quarrying, transportation, cutting, and manipulation waste could have easily double that total.

Urbanization with it rising demand for windows had, finally, brought new methods of mass-scale plate-glass productions: in 1848 Henry Bessemer (later to become much more famous thanks to his new inexpensive way of steel production) patented the production of flat glass by drawing a slightly cooled material upward through a flat nozzle between asbestos rollers, a superior solution to the centuries-old (and severely size-limited) crown glass and to a still size-limited cylinder glass of the early nineteenth century. But it was only during the mid-1950s that Alastair Pilkington introduced molten tin bath that allows production of very large pieces of float glass with near-perfect uniformity (Pilkington 1969).

But perhaps the best illustration of a much expanded removal and emplacement of bulk construction materials, including soil, sand, and gravel is growth of railroads. Their construction began in 1830 (56 km between Liverpool and Manchester), 30 years later their worldwide length topped 100,000 km and by 1900 the railroads on five continents added up to 775,000 km, with about 250,000 km in Europe, more than 190,000 km in the United States and 53,000 km in Russia, and 30,000 km in the United Kingdom (Smil 2019). Given the wide range of terrains covered by rail tracks, it is impossible to estimate a typical volume of bulk construction materials – earth displaced and replaced to create cuts or embankments, stone cut to create tunnels or incision in mountainsides and stone quarried to produce gravel for access roads and rail beds – that had to be handled for an average kilometer of new track.

Even a highly conservative assumption of 3,000 m³/km would result in nearly 2.5 Gm³ of bulk materials associated with the world's railway construction of the second half of the nineteenth century. A similarly conservative assumption of at least 2,000 t of ballast (crushed stones packed underneath and around ties) per kilometer would translate to at least 1.5, and more likely to 2 Gt of coarse gravel applied to hold the tracks built between 1830 and 1900 in place. Mineral aggregates were also needed in unprecedented volumes for the building of new factories, for expansion of ports and for the construction of hard-top roads. But the era of asphalted roads came only after 1900 and so the most important new material in the construction of the nineteenth century was concrete produced by mixing cement with an aggregate and water. Aggregate's coarseness ranges from sand to various sizes of gravel, and hydration (an exothermic reaction) of cement hardens the mixture even under water.

2.5.2   Cement and Concrete

Suitable cement needed to produce good-quality concrete became available after 1824 when Joseph Aspdin, an English bricklayer, received a patent for a hydraulic mortar/stucco made by firing limestone and clay at high temperatures. This process vitrifies the alumina and silica materials and produces a glassy clinker whose grinding yields Portland cement, the name chosen by Aspdin because once set the color of the material resembled the limestone from the Isle of Portland (Shaeffer 1992). Hydration of this cement (its reaction with water) produces material that could be readily form shaped but that is strong in compression but weak in tension. That weakness can be overcome by reinforcing the material with iron, the combination made possible by the facts that concrete and iron form a solid bond and that hydraulic cement actually protects iron from rust.

Development and commercialization of reinforced concrete was a gradual process, with contributions made by French inventors (François Coignet in the early 1806s and Jacques Monier who patented first a reinforced concrete beam and in 1878 a general system for reinforced structures), and during the 1870s by William Ward in the United Kingdom and Thaddeus Hyatt in the United States (Newby 2001). The true beginning of modern rebar came only in 1884 when Ernest Ransome patented his system of concrete with reinforcing bars, which became widely franchised, especially to build new industrial structures. In 1886, Carl Dochring patented an ingenious idea of pre-stressing the reinforcing bars inserted in concrete: they are stretched when set in the wet material and they release the tension once the material sets (Abeles 1949). The first concrete skyscraper, Chicago's Monadnock Building, was completed in 1891, and during the 1890s the introduction of modern rotary kilns (with temperatures up to 1,500 °C) made it possible to produce low-cost, high-quality cement. All the prerequisites of concrete era were thus in place before 1900.

Even as the demand for wood was falling due to the displacement of fuelwood and charcoal by fossil fuels and coke (in France coal began to supply more than half of all energy by the mid-1870s, and in the United States the tipping point between fuelwood and coal and oil came in 1884–1885) and to the shift of ship construction from timber to steel, new markets for sawn wood were created by a large-scale expansion of coal mines and by the construction of railways. Demand for timber props in underground mining varied with the depth and thickness of seams but typical requirements in Europe and in the United States were between 0.02 and 0.03 m^3/t of coal and my best estimate is that the global need for mining timber had surpassed 20 Mm^3 (about 15 Mt) by 1900 (Smil 2013).

All ties (sleepers) installed during the nineteenth century were wooden; concrete sleepers were introduced only around 1900, but they remained uncommon until after WW II. Standard construction practice requires the placement of about 1,900 sleepers per kilometer of railroad track, and with a single tie weighing between roughly 70 kg (pine) and 100 kg (oak) every kilometer needed approximately 130–190 t of sawn (and preferably creosote-treated) wood. My calculations show that the rail tracks laid worldwide during the nineteenth century required at least 100 Mt of sawn wood for original construction and at least 60 Mt of additional timber for track repairs and replacements (Smil 2013).

2.5.3   Iron and Steel

Multiplied demand for metallic elements and their alloys included those that were known for long time but used previously in only limited quantities (cast iron, steel, and copper are the best examples in this category) as well as those that were recently isolated only during the nineteenth century and had swiftly found new industrial markets (aluminum). Improvements in iron smelting (larger blast furnaces, more efficient use of coke) reduced energy requirements of pig iron from almost 300 GJ/t in 1800 to less than 100 GJ/t by 1850 and to only about 50 GJ/t by 1900 and made the metal much cheaper (Heal 1975). Cast iron was used in some pre-1850 buildings in the United Kingdom, and afterward, once it became inexpensive, its applications became more common in the United States, particularly in the southern states, not only for columns but also for bridges. Cast iron columns work well because (as already noted) the metal is strong in compressions but its weakness in tension limited its structural use.

But the best choice to meet all those high-tension demands was steel, also an iron–carbon alloy but one whose carbon content is limited to just 0.05–1.5% and that can be tailored to many different requirements by adding variable shares (less than 2% to more than 10%) of other metals, including Co, Cr, Mn, Mo, Ni, Ti, V, and W, sometimes in combination (Smil 2016). Specialized steels have uses ranging from sheets for car bodies and cutting tools for machining to stainless steels used in medical devices, chemical syntheses, and food processing. The best varieties (tool steels) have tensile strength of an order of magnitude greater than cast iron (1,600–2,500 MPa compared to 150–400 MPa), and steel is the strongest and the hardest of all common metals: its typical tensile strength is roughly seven times that of Al and nearly four times that of Cu, and its hardness is four times that of Al and eight times that of Cu. Steel's impact resistance can be more than six times that of cast iron (130 vs. less than 20 J) and the highest temperature it can withstand before losing its structural integrity is as high as 750 °C compared to only about 350 °C for cast iron.

As the advancing industrialization needed more tensile metal, particularly for the fast growing railways, wrought iron filled the need. Low-carbon wrought iron was produced by reheating and puddling, manual turning, and pushing heavy chunks (nearly 200 kg) of pig iron in order to expose it to oxygen in shallow hearths (decarburize it) and produce a nearly pure metal with less than 0.1% C (Flemings and Ragone 2009). Wrought iron was then reheated and shaped (rolled, hammered) into final forms: all early rails laid before the late 1850s were made that way, and so was the Eiffel tower. Inexpensive steel became a possibility in 1856–1857 when Henry Bessemer (in England) and William Kelly (in US) patented their decarburation process during which molten pig iron in a tilting vessel with a refractory lining was subjected for 15–30 min to blasts of cold air that decarburized the metal (Bessemer 1905).

But the process did not remove phosphorus present in many iron ores and the solution to this problem came only during the late 1870s when Sidney Gilchrist Thomas and Percy Carlyle Gilchrist introduced basic (limestone) refractory linings and added lime to the charge: that made it possible to remove phosphorus in slag (Almond 1981). By the late 1880s, the basic Bessemer process dominated both European and American steel production – in the United States its importance peaked in 1890 at 86% of the total output (Hogan 1971) – and it produced most of the world's steel output until 1910. Its brief dominance ended with a rapid diffusion of open-hearth steelmaking furnaces built according to an 1866 patent shared by William Siemens and Emile Martin.

After a slow start the deployment of open-hearth furnaces lined with basic refractories began to soar during the late 1880s (Almond 1981) and it pushed up the alloy's share in the final output quite rapidly. In the antebellum United States only about 1% of the US pig iron was converted to steel, by 1900 the share was almost 75% by 1900, and by 1906 the conversion became virtually total (Hogan 1971). The 1880s also saw the introduction of new steels developed by Robert Mushet (self-hardening steel containing tungsten and manganese) and Robert Abbot Hadfield (with 13% Mn) made it possible to make superior metal-cutting tools as well as durable ball-bearings. And by the decade's end, American steel output surpassed British production: it rose from about 200,000 t in 1800 to roughly 2.5 Mt by 1850; then it had doubled in 20 years to almost 6 Mt in 1870 and it reached 9.5 Mt by 1900. Global production of pig iron rose from about 5 Mt in 1850 to more than 30 Mt by 1900, while steel output rose from just half a million ton in 1870 to 28 Mt by 1900 (Smil 2016).

For the first time in history, inexpensive steel could be used to build structures, devices, and machines that were previously made of wrought iron or wood, and steel products conquered new markets, thanks to their superior structural attributed and durability. In agriculture (still the dominant economic activity by 1850), steel went first into moldboard plows (first patented in 1868 by John Lane Jr.) whose mass deployment opened-up America's Great Plains and Canada's Prairies for cultivation, then into grain reapers, other implements (harrows, seeders), and the first horse-drawn combines (the first self-propelled combine was introduced in 1911). New markets for steel were created in during the late 1880s by the invention of seamless steel pipes by Reinhard and Max Mannesmann. Introduction of better rifle, inventions of machine gun and modern explosives and production of grenades and bombs and (starting in the 1880s) building of large and heavily armored battleships (but the first *Dreadnought* was built only in 1906) were other major sources of new demand.

New modes of transportation became the fastest rising markets for steel. Rapid development of railways created a huge demand for steel used in locomotives, passenger and freight cars, and for rails. Rails used during the nineteenth century weighed between 20 and 30 kg/m and assuming the average of 25 kg/m the railway construction between 1850 and 1900 would have required about 20 Mt of steel, and replacement would have more than doubled that total. Steel became a favorite material for railway bridges: its use was pioneered by the cantilevered 2,529 m long Scotland's Firth of Forth crossing designed by John Fowler and Benjamin Baker built between 1883 and 1890 with 51,000 t of the metal (Forth Bridges 2022).

The second large-scale global transportation market was created once the Lloyd's Register of Shipping accepted steel in 1877 as an insurable material for ship construction: within a generation the metal conquered the shipping market and the age of commercial wooden freight ships was over. By 1900 shipyards in the United Kingdom, Germany, and France were routinely launching passenger ships that needed more than 10,000 t steel to make. The third transportation segment that has eventually created enormous demand for steel had originated also before 1900 – but the age of mass car ownership began in the United States only in 1908 with Ford's Model T (Smil 2005).

Late nineteenth century industrialization also created enormous new markets for steel in the steelmaking industry itself (due to its massive expansion that required greater numbers of larger blast furnaces and steel mills), in the new electrical industry (starting in the early 1880s, and requiring heavy machinery including boilers and steam turbogenerators, as well as steel for transformers, transmission towers, and electrical wires), in oil and gas extraction and transportation (steel for drilling pipes, drill-bits, well-casings, pipelines and vessels, pipes and storage tanks in refineries), as well as in traditional textile and food-processing industries where advancing mechanization resulted in the adoption of a greater array of steel-based machines and other processing and storage equipment (Smil 2016). These demands led, in turn, to higher steel requirement in manufacturing of various industrial machines, tools, and components.

Inexpensive and highly tensile steel made it possible to design and to build inhabited structures of unprecedented height and functionality. Structural steel (more specifically long I-beams riveted from smaller pieces) made skyscrapers possible by doing away with thick load-bearing walls. Chicago's Home Insurance Building (designed by William Le Baron Jenney, completed in 1885, in 1931 replaced by the Field Building) was the first tall (10 stories, 42 m high) structure where steel columns and beams carried the weight, cut the total mass of the structure by two-thirds and allowed for more floor space and for the installation of larger windows (Architectuul 2022).

Five years later, the World Building in New York went up to 20 stories and 94 m, and just before the century's end Henry Gray's invention of universal beam mill capable of rolling long H-beams made steel construction easier by eliminating most of the need for laborious riveting (Gray 1903; Hogan 1971). Skyscrapers required elevators and hence more steel for machinery, cables, and cabins: Otis, still the leader today, installed its first electric elevator in 1889 (Otis 2022). The other, hidden, use of steel was its use in reinforced concrete. Joseph Monier, a Parisian gardener, patented first a version using simple metal netting in 1878 (Bosc 2001), and the 1880s saw first common applications in France, Germany, and Austria, especially in new industrial buildings.

2.5.4 Copper and Aluminum

Copper's ancient use for coins and alloys (brass, bronze) continued during the era of industrialization but two new applications emerged to dominate the market. The first one was the general adoption of indoor plumbing in growing cities, with copper as the standard choice first for water pipes, and later also in heating and cooling systems. The second one was the invention of large-scale electricity generation during the early 1880s followed by a rapid increase in demand for new power plant capacities and transmission lines. Copper is an excellent conductor of electricity, and it is the preferred metal for wires and cables (with applications ranging from massive bundled units in data centers to the circuitry in microprocessors), in electric motors; other uses include connectors, bearings, brakes, and radiators in motor vehicles, as a roofing material and (because of the metal's antimicrobial properties) in frequently touched public surfaces.

The lightest of all commonly used metals was also the latest to be produced in commercial quantities. Aluminum was discovered by Hans Christian Oersted in 1825, and for the next 60 years, it was produced only in minuscule amounts to make novelty jewelry; as late as 1884 its single largest applications was a 2.85 kg pyramidal cap topping the newly built Washington monument (Binczewski 1995). Commercial aluminum production was sought not only because of the metal's very low density (at $2.7 g/cm^3$ only a third of iron's $7.9 g/cm^3$) but also because of its excellent conductivity (surpassed only by Ag, Au, and Cu), an extraordinary combination of high malleability (making it easy to roll, extrude, or stamp) and high tensile strength (surpassed only by special steels and enhanced by alloying with Cu, Si, and Zn) and resistance to corrosion. Moreover, composite materials of aluminum and ceramics have unusual stiffness and durability, the nontoxic metal is suitable for uses where it comes in contact with food and drink and discarded items can be easily compressed for highly energy-rewarding recycling.

Henri Saint-Claire Deville was the first experimenter to produce the metal by electrolytic means during the 1850s but none of the subsequent variants of his process opened the way to mass production. Discovery of practical aluminum smelting in 1886 is among the most famous cases of a virtually identical but independent solution of a technical challenge: both inventors, Charles Martin Hall in the United States and Paul Louis Toussaint Héroult in France, were just 23 years old, and both moved rapidly to commercialize their process, with the first two plants in operation by 1888 (Borchers 1904). Hall-Héroult process still remains the only way to produce large quantities of the metal.

The process was made easier by Karl Joseph Bayer's 1888 invention of efficient alumina (Al_2O_3) from bauxite, a reddish compound particularly common in the tropics (lateritic bauxite) and named after Les Baux in southern France where Pierre Berthier first discovered it in 1821. Al_2O_3 (a non-conducting compound) is dissolved in molten, and highly conductive, cryolite (Na_3AlF_6, sodium hexafluoroaluminate), a perfect combination for efficient electrochemical separation in large electrolytic vessels; between 4 and 5 t of bauxite are needed to produce 2 t of alumina and reduction of that compound yields 1 t of pure Al. By 1900, the global aluminum output was only about 8,000 t, but the price fell rapidly to just 5% of the 1,888 level (Borchers 1904).

2.5.5 Paper, Textiles, Gases

And yet another material should be singled out besides construction aggregates and metals: paper. Its large-scale production was revolutionized at the beginning of the nineteenth century with the introduction of a continuous paper-making machine patented first in 1799 by Louis-Nicolas Robert in France but commercialized after 1801 in England by Henry Fourdrinier and subsequently called by his name (Smith 1970). This machine consists of cylinders and it forms, dewaters, and dries large continuous sheets of paper (Nuttall 1967). The pulp is laid on a continuous wire mesh at the machine's wet end, most of water is then expelled in the felt press section, and the paper-making process is completed by conveying paper over a set of heated cylinders. But the raw material remained the same: until the 1870s paper was made from recycled rags and this had obviously limited the produced mass and kept the price high.

That is why the nineteenth-century innovators experimented with scores of natural fibrous materials ranging from acacia and agave to thistles and yucca but based both on the available biomass and its properties wood provided the best solution (Smith 1970). During the 1870s making paper from wood began with mechanical pulp but its high lignin content yields inferior paper (prone to yellowing) and it is now restricted for uses including cheap newsprint, some toilet papers and, above all, cardboard and building board (Smil 2005). The first commercial operations using chemical pulp relied on the alkaline (soda) process but were soon displaced by the acid (sulfite) method that yielded a stronger pulp but a brittle paper: all printed matter using that acidic paper will eventually disintegrate. But that paper was cheap and the effects on consumption were obvious: for example, in 1872 Montgomery Ward's catalog had a single page, two decades later it had more than 600 pages (Montgomery Ward and Company 1895). The superior solution (pulping using the sulfate process) was invented in 1879 by a Swedish chemist Carl F. Dahl, but it was commercialized widely only after 1900.

As for the textiles, the nineteenth century revolution was mostly quantitative as mechanized weaving, based on Carthwright's power loom and Watt's more efficient steam engine, opened the way to a mass production of fabric and as higher incomes in urbanizing and industrializing societies created new markets for all kinds of textile products by expanding the ownership of clothes beyond what used to be commonly just a single set (or two) of outer garments (Holden 2014). Other technical innovations helped the industry's expansion: factories using water power benefitted from higher-capacity water wheels and (later) turbines, replacement of traditional wooden shafts by iron and introduction of iron of iron-frame looms lowered operating costs and improved productivity. British numbers illustrate the industry's progress: in 1800 the country had only about 2,000 looms, by the early 1830s their total surpassed 100,000, and 250,000 were operating by 1857 (Hills 1993).

Finally, scientific and engineering advances during the second half of the nineteenth century led to the first commercial extraction of gases from the atmosphere. In 1852, Thomas Joule and William Thompson (Lord Kelvin) found that a highly compressed air forced through a porous nozzle cools slightly, about 0.25 °C for every 100 kPa pressure drop (Almqvist 2003). Repetition of this expansion results in a cooling cascade that can gradually bring a gas to its liquefaction point and the first practical application of Thomson-Joule effect was patented in 1895 by Carl von Linde (Linde 1916). More than a century later, the eponymous company he founded remains the world's leader in gas liquefaction.

2.6   Materials in the Twentieth Century

I have argued that the modern, high-energy society of the twentieth century was created by the unprecedented combination of technical, scientific, and managerial advances that took place between 1865 and 1913 – but that the transformations of these advances (together with some new notable innovations) resulted not only in new quantitative but also in new economic and social arrangements and in higher quality of life (Smil 2005, 2006a). In this section I will offer brief reviews of some of the most important shifts in the production and use of materials (ranging from traditional biomass to modern electronics) that took place during the twentieth century and helped to create these new realities.

Because of their renewability, annually harvested crop residues used to be indispensable materials in all traditional agricultural societies. In many deforested regions, they were the only source of household fuel, straw-clay mixtures were made into bricks, and straw bundles were used for roof thatching, in some countries. Peasants wore straw sandals and coats, and cereal straws were used as both feed and bedding for domesticated ruminants (only they can digest cellulose that makes up as much as half of some straws). During the twentieth century, thatched roofs and straw sandals had almost completely disappeared, in affluent countries straw is now mostly (more than 60% of annual yield) recycled to improve soil quality, but in many countries crop residues have retained their importance as fuel and feed.

2.6.1   Biomass

My reconstruction of the twentieth century global crop residue production shows that in 1900 they accounted for nearly 75% of the total harvest of 1.5 Gt and that the share was still about 70% in 1950 (Smil 2013). Subsequent diffusion of modern cultivars and the rise of average harvest ratios (also known as grain/straw ratios, with cereal crops now yielding as much grain as straw) reduced the share to 65% by 1975 and to only about 58% by the year 2000. I calculated that in that year the global mass of food and feed crop residues was at least 3.75 Gt of dry weight. For comparison, Wirsenius (2000) put the global total of crop residues at 3.46 Gt for the years 1992–1994 and Haberl et al. (2007) at just 2.71 Gt for 2000. There are no reliable data about the final fate of crop residues: in many agroecosystems, they should be directly recycled to maintain soil organic matter and to prevent erosion, but often their mass is judged to be excessive and they are simply burned in fields. This undesirable practice is particularly common in rice-growing regions of Asia.

Straw continues to be burned even in some affluent countries, most notably in Denmark where it is used for house heating as well as in centralized district heating and electricity generation (Zafar 2021). Construction boards are made from shredded and compressed (or fused with resins) straw by such companies as Stramit and Agriboard Industries, and some green architects promote straw-bale buildings (King 2006; StrawBale 2022). Residues can be also used as a source of organic compounds: furfural, a selective solvent in crude oil refining and in phenolic products, can be derived from corn stover and cereal straws (Di Blasi et al. 2010). Straw can be also used as a substrate to cultivate mushrooms: wheat straw for white button mushrooms, rice straw for straw mushrooms.

The twentieth century was unmistakably the era of metals and plastics and their novelty and ubiquity made wood an underappreciated material: true, in all affluent countries its per capita consumption has declined but aggregate demand rose substantially as wood retained and actually expanded all but a few of its traditional markets. The only category of wood consumption that became relatively unimportant during the twentieth century was shipbuilding: wood is now limited to a niche market of small vessels (boats and yachts) and for interiors of more costly ships. The material's two leading uses, as lumber and pulp for making paper, will be reviewed in detail in the third chapter; here I will note the declining importance of fuelwood for railroads and coal mining.

Wooden railway tie, that quintessential nineteenth-century innovation, had maintained its high share of global market throughout the twentieth century. During the 1990s, 94% of America's ties were wooden, and even with relatively high proportion of concrete ties in parts of Europe and Asia, only 15% of the world's sleepers were made of materials other than wood (Sommath et al. 1995). Better treatment of ties prolonged their average life span from about 35 years in 1940 to 40–50 years by the year 2000 (James 2001), and wooden ties are still used to replace worn-out sleepers on routes with relatively light traffic. But new lines in China, the nation with the greatest recent railroad expansion, use only concrete sleepers as do all lines dedicated to high-speed (up to 300 km/h) trains in Europe and Asia (North America still does not have even a single such line).

Worldwide market for underground mine props kept on growing until after WW II but subsequent decline of coal-mining in Europe and a shift of extraction to surface mines in the United States and Australia weakened that demand. I calculated that the volume of timber used in coal mines had surpassed 20 Mm^3 (about 15 Mt) in 1900 and 40 Mm^3 in 1950 when it accounted for about 2% of the world's roundwood harvest (Smil 2013). Reduced reliance on coal in Europe and Japan and expansion of surface mining and widespread adoption of long-wall mining using movable steel roof supports reduced the post-WW II demand for pit props throughout the Western world.

In contrast, China's coal extraction had increased about 89-fold between 1950 and 2020 (from 43 to 3,840 Mt), but in that deforested country (recent expansion of tree cover is due to planting fast-growing poplars, pines, and eucalyptus), specific consumption of timber has been always much lower than the rate around 0.025 m^3/t that has been typical in the West. China's low specific use of mining timber (now just 0.005 m^3/t of coal) means that even with the country's large increase of raw coal extraction, the roundwood demand in Chinese coal mining is less than 5% of the country's total wood consumption. But due to its massive exports of manufactured products, China has become a leading user of wood, plywood, cardboard, and paper required for packaging.

2.6.2   Concrete

As I have already noted, all key technical breakthroughs needed to make concrete the world's most common building material were in place by 1900, but the first major initiative to make concrete a popular choice was a peculiar experiment directed by Thomas Edison who became engaged in a futile experiment to design and build cast-in-place concrete homes (Courland 2011). The inventor launched it in 1906 (after failing to develop a vastly improved battery), and five years later, as the project was failing, he tried to revive it by promising also cheap concrete furniture, including entire bedroom sets, and even a concrete phonograph. Edison's dreams of inexpensive concrete housing failed, but some architects began to choose the material for signature buildings of their careers.

In France, Auguste Perret designed elegant apartments and the Theater des Champs-Elysees already before WW I, and Frank Lloyd Wright was the material's great American proponent. After designing smaller pre-WWI structures in the United States, he built Tokyo's Imperial hotel, completed a few months before the great earthquake destroyed the city in 1923; the hotel survived with only minor damage, and it was demolished in 1968. Wright's other famous pioneering concrete structures included Johnson Wax Headquarters in Racine, WI (in 1939), the Fallingwater House in Pennsylvania (in 1935), and in 1959 the Guggenheim Museum in New York (Hess 2008).

Other famous post-WW II structures made possible by reinforced concrete include Jørn Utzon's Sydney Opera House (noted for its elegance and completed after many delays in 1973) and Burj Khalifa Tower in Dubai, the world's tallest building finished in 2010. Many new bold-looking designs took advantage of the fact that prestressed structures require much less steel and concrete to bear the same loads (reductions are, respectively, about 70% and 40%) and can be made much slender. Eugène Freyssinet pioneered this type of construction and he also introduced post-stressing that uses tensioning wires threaded through ducts in precast concrete (Grotte and Marrey 2000).

During the first half of the twentieth century, reinforced concrete also became the most important material of new bridges and dams, new offshore structures, and the foundation of modern transportation infrastructures. Swiss architect Robert Maillart pioneered the design of graceful concrete bridges already before WW I (Billington 1989), and the world's longest bridge, the 164.8-km Danyang-Kunshan Grand Bridge in China (completed in 2010) is also a reinforced concrete structure. America's largest dams build during the 1930s (Hoover Dam on the Colorado and Grand Coulee on the Columbia River) became the precursors of even larger structures built around the world after WW II; the largest one of them, 185-m tall and 2.3-km long China's Sanxia (Three Gorges) dam on the Yangzi, contains almost 28 Mm³ of concrete and 500,000 t of reinforcing steel (Ponseti and López-Pujol 2006).

In 1982, *Statfjord B* oil drilling platform became the heaviest object ever moved as 816,000 t (mostly in reinforced concrete in the structure's four massive concrete columns and storage tanks) were towed into position in the North Sea (Equinor 2022). Modern commercial flight depends on reinforced concrete that forms the runways, approaches, and aprons of airports: they must withstand repeated traffic of airplanes with mass commonly between 150 t (Boeing 737) and 277 t (Airbus 380); runway concrete is up to 1.5 m thick and the longest runways are 3,600–4,000 m long. But most of the reinforced concrete has not gone into iconic structures but into ever-increasing numbers of nondescript or outright ugly (or brutally looking) apartment buildings, high rises, factories, garages, roads, overpass bridges, and parking lots.

2.6.3 Glass

Pilkington's float-glass process made it possible to produce sizes of unprecedented size, and many architects were quick to incorporate them into their building designs. By 2022, global glass industry consisted of some 1,200 companies and produced

209 Mt of glass: about 106 Mt of float glass, 95 Mt of container glass and nearly 8 Mt domestic/tableware glass (GlassGlobal 2022). Most of this glass has sodium carbonate (Na_2CO_3), calcium oxide (CaO), magnesium oxide (MgO), and aluminum oxide (Al_2O_3) added to silicon dioxide (SiO_2), its dominant (70–74%) component, while high-transparency optical fibers are drawn from molten silica glass that contains only minute amounts of additives. By 1900, the best optical glass used in instruments was about 10,000 times more transparent than the earliest glass produced in Egypt some 5,000 years ago, and after 1970 the use of high-purity SiO_2 had improved that transparency by four more orders of magnitude making it possible to use optical fibers for long-distance, mass-dispatching of voices, and data (Agrawal 2010; Corning 2019).

2.6.4   Steel

Steel's dominance of global metal market, firmly established by 1900, had greatly increased during the twentieth century, thanks to nearly continuous advances in iron smelting and steel-making and to the emergence of new mass markets for the metal. Open-hearth furnaces remained the dominant producers of world steel for more than two-thirds of the twentieth century. The principle of open hearths remained unchanged but the unit sizes and productivities grew substantially: just before 1900 the largest US Steel furnaces had area of about 30 m², before WW I the maximum size grew to 55 m² and by 1945 it was nearly 85 m², while the typical heat (batch) capacities from roughly 40 t by 1900–200 t during WW II (Smil 2016). And electric arc furnaces, first introduced in 1902, began to be used more frequently to convert growing stocks of scrap metal into high-quality steel.

The first cars of the 1890s made only small demand for steel because they had simple wooden bodies and were made in small numbers: only once the mass production took (with Ford's Model T introduced in 1908) auto industry became the leading consumer of steel with its demand rising from about 70,000 t in 1910 to 1 Mt by 1920 (Hogan 1971). For many decades, the idea of producing non-corroding steel on a mass scale was rejected by nearly all metallurgical experts. Harry Brearley is usually credited with inventing commercial stainless steel (containing nearly 13% chromium) in 1913 (an innovation readily embraced by Sheffield's famous steel cutlery makers), but contemporaneous advances were made in Germany and in the United States (Cobb 2010).

Besides cutlery, stainless steel soon became used in surgical implants and cookware and also in many bulk food-handling applications in brewing, winemaking, bakeries, and meat industry. During the 1930s came the first deployment of stainless steel in construction: in 1930 55-m tall tower and spire of New York's Chrysler

Building was the first eye-catching application. A year later, the Empire State Building got stainless steel window trims and pilasters. Rapid passenger trains with aerodynamic steel bodies (starting with Burlington Zephyr in 1934) were also popular before they were displaced by lighter aluminum structures.

Steel output fell sharply during the Great Depression (in the United States by 75% between 1929 and 1932, down to just 12.4 Mt), but then it rose to new record highs in order to support the unprecedented wartime demand for the metal (King 1948). Inevitably, these increases brought many problems and new solutions. Higher blast rates led to faster deterioration of brick linings – and to the introduction of water cooling. Large increases in the volume of ore, coke, and limestone inputs could not be supported by manual handling of inputs – and hence the mechanized skip hoists and automated dumping. The need for clean gas to run blowing engines led to better gas-cleaning equipment, including Frederick Cottrell's electrostatic precipitators, first introduced in 1919.

Metal needed for the armaments required to win WW II reinvigorated America's steel industry, and after 1950 it pioneered the use of coke-saving smelting under pressure, use of beneficiated ores, injection of gaseous or liquid fuels, enrichment of blast air by oxygen, and automated operation controls (Gold et al. 1984). But by 1960, the innovative leadership in ferrous metallurgy shifted to Japan (the country became the world's second largest producer after the USSR) and to Europe. Larger but more efficient blast furnaces to smelt iron, basic oxygen furnaces to produce steel, and continuous casting of steel products made up the trio of technical innovations that transformed global steel industry after WW II (Smil 2016). I will cover all of them in some detail in the next chapter when dealing with advances in metal production.

Mass production of inexpensive, high-quality structural steel enabled the construction of downtown high rises in the largest cities of all inhabited continents, and many of these buildings also use steel on the outside. In 1954 New York's Socony-Mobil building was the first skyscraper completely clad with stainless steel (and cleaned for the first time only in 1995). Among the numerous steel-clad skyscrapers that have followed (usually in the form of curtain walls) are the US Steel Tower in Pittsburgh and the Petronas Twin Towers in Kuala Lumpur (temporarily the world's tallest building). Burj Khalifa uses reflective glazing and textured stainless steel spandrels.

But much more steel has gone (in the form of reinforcing rods) into cars and trucks and the new transportation infrastructures on land (ranging from multilane highways and bridges to new airports) and into the construction of large oil tankers, bulk carriers (transporting anything from grain to ores) and, starting in the 1960s, container ships and ports. Steel allows particularly captivating design of long suspension bridges with woven cables supporting lengthy road spans: Japan's Akashi

Kaikyō that links Honshū and Shikoku is nearly 2 km (1,991 m) long. Transportation sector also became the leading user of aluminum: the combination of light weight and durability made the metal, and its alloys, an ideal choice for applications ranging from cooking pots to rapid train cars and indispensable in airline industry.

Between 1900 and 1943, its output, driven primarily by an unprecedented demand for the metal in aircraft construction, expanded nearly 300 times to almost 2 Mt, with 45% of the peak demand coming from the United States engaged in the largest-ever effort to build record numbers of military planes (Yenne 2010). Many new producers were added to the global output after 1950 and by the year 2000 electrolytic production of the metal was close to 25 Mt, and by 2021, it reached a new record of 68 Mt, with China accounting for 57% of the total, with Russia the distant second (USGS 2022a; IAI 2022a). Structural aluminum spread beyond airplanes to bridge decks or superstructures (starting in the 1940s) and vessels (ranging from all-aluminum small boats to structures on offshore drilling rigs, and from large cruise ships to expensive yachts) and rail cars.

New major post-1950 markets have also included irrigation pipes (particularly for center-pivot systems now so common in many arid regions), heat exchangers (many applications in air conditioning, medical equipment, electronics), and even sheet piling. At the same time, aluminum has lost some of its traditional markets at both ends of its value range: in many demanding aerospace applications (especially for supersonic airplanes), it has been displaced by titanium, while in many simple manufactures it has been replaced by less energy-intensive and less expensive plastics. Titanium production requires about twice as much as energy (around 400 GJ/t) as the chain of operations needed to produce aluminum (for details see the third chapter), but the metal's melting point (1,667 °C) is 2.5 times higher than for aluminum (660 °C), making it well suited for the cladding of supersonic aircraft (El Khalloufi et al. 2021).

2.6.5 Color Metals

Post-WW II demand for copper has been dominated by five major final markets: copper in construction goes into electrical wiring, for plumbing, refrigeration, and air-conditioning conduits, and also for visible uses (copper sheathing and roofing); industrial machinery, fittings and wiring and heat exchangers; every category of transportation machinery; industrial electrical and electronic products, above all telecommunication and lighting; and a wide range of consumer products dominated by electronic gadgets and electrical cords and in many countries including coins. Copper has maintained the third ranking among the twentieth century metals as its global consumption rose from less than 500,000 t in 1900 to more than 13 Mt in the year 2000 and to 21 Mt by 2021 (USGS 2022a).

The fourth most important metal has been zinc, with about 14 Mt in 2021 (12.8 Mt mine production), and steadily rising demand for lead has brought this formerly more distant number five close to the zinc total: in 2021 the global refined lead supply had reached 12.3 Mt, with about 31% being primary metal and the rest coming from recycled material (ILZSG 2022). Although the use of tetraethyl-lead as the leading automotive additive (to prevent knocking in gasoline engines) was outlawed in Japan in 1986, in the United States in 1995, and in the European Union and China in the year 2000 (Smil 2023), the expansion of global car and truck market (and hence the annual installation of tens of millions new and replacement lead acid batteries) has led to record demand in automotive industry. With the total of 1.4 billion cars and light and heavy trucks and with average mass of 10 kg Pb in automobile and 13 kg Pb in truck batteries, there were nearly 15 Mt of lead on the world's roads in 2021. In addition, this heavy metal also remains an indispensable solder in electronics.

2.6.6 Silicon

Use of silicon, a light (2.3 g/cm^3) metalloid, has become synonymous with modern electronics, but the element was not present either at the industry's birth or during its pre-1950 development. Electronics' theoretical beginnings go back to the 1860s (James Clerk Maxwell and Rudolf Hertz), and practical applications got underway during the 1890s when several researchers and engineers were trying to create wireless telegraphy, that is to send signals through the atmosphere. The group's most prominent members were Nikola Tesla, David Hughes, Alexander Stepanovich Popov, William Crookes, Edouard Branly, Oliver Joseph Lodge, and Guglielmo Marconi (Smil 2005).

And it was Marconi who took the first commercial steps without relying on any new materials: his first broadcasting towers in Newfoundland were wooden, and his spark generators and wire antennas were made of common metals. Later models of high-frequency alternators (built by Reginald Aubrey Fessenden and by Ernst Frederik Werner Alexanderson, and able to produce continuous wave signal) were remarkable examples of precision engineering but superior devices that opened the way for mass-produced electronics came only with Fleming's invention diode (essentially just a light bulb with an added electrode). Its progeny – triodes, tetrodes, and pentodes – required carefully manufacturing procedures but in material terms, these vacuum tubes amounted to just lots of glass and hot (mostly tungsten) filaments.

Radios, first introduced in the early 1920s, became inexpensive and widely owned by the 1930s, and television was the first expensive electronic device to reach mass ownership. The first broadcasts took place in the United Kingdom and the United States before WW II, by 1948 still only fewer than 200,000 families had a bulky set with a small black-and-white screen, but by 1960 TVs were in 90% of all US homes. The first color broadcasts came in 1954 but affordable sets came only during the late 1960s and by 1975 two-third of American families had a color TV (Abramson 2003). By that time the conversion to solid state electronics based on silicon was virtually complete.

Silicon makes nearly 28% of the Earth's crust (second to 49% for oxygen) and while it is abundantly present as SiO_2 (silica) in sand, sandstone, and quartz and in many silicates ranging from hard feldspars (rock-forming minerals) to soft kaolinite (a layered clay mineral), it is never found in pure, unbound elementary form. But the purest crystalline silicon is the material foundation of modern electronics: intricate webs of semiconductors and connections are etched into think wafers of ultrapure Si; optical fibers (made by fusing SiO_2 and GeO_2, their deposition inside a tube to form glass and subsequent fiber drawing in a tower) carry communication among countries and continents; and photovoltaic cells power communication satellites that relay data and messages among the continents and they are increasingly used for land-based electricity generation (Smil 2006a). Silicon's importance for modern societies deserves an extended coverage.

2.6.7   Plastics

Plastics have been widely seen as quintessential twentieth century materials, with a particularly rapid post-WW II diffusion as they replaced wood, metals, and glass in many household, industrial, and transportation products (Strom and Rasmussen 2011). Although the first synthetic materials go back to the 1870s when John Wesley Hyatt patented his celluloid process and when chemists began to study reactions between phenol and formaldehyde, the real progress came only in 1907 when Leo Hendrik Baekeland, a Belgian chemist working in New York, prepared the world's first thermoset plastic formed between 150 and 160 °C (Baekeland 1909). His General Bakelite Company, established in 1910, was the first large-scale commercial producer of a plastic material and bakelite was soon turned into products ranging from telephones to electric insulators and from door knobs to parts of light weapons.

In 1912 came Jacques E. Brandenberger's cellophane (regenerated cellulose, first in France, since 1924 in the US) and styrene (its polymer is now a leading insulator and packaging material) and cellulose acetate (Brydson 1975). But the 1930s became the still unsurpassed era of major plastic discoveries as a result of systematic institutionalized research by large chemical firms, above all the US DuPont, Germany's IG Farben, and the UK's Imperial Chemical Industries. First, in 1930, Wallace Hume Carothers led a Du Pont team that produced neoprene (synthetic rubber), and

IG Farben synthesized polystyrene. In the early 1930s, the ICI began to research organic reactions under very high pressure and by 1935 this led to polyethylene (Smil 2006a).

The company also began its synthesis of methyl methacrylate in 1933, the year when Ralph Wiley at Dow Chemical made an accidental discovery of polyvinylidene chloride (Saran wrap for cling-packaging of food). In 1935 came the first Plexiglas, a year later polyurethanes (Otto Bayer at IG Farben), and in 1937 Carothers he got the patent for his polymer 66, commercially known as Nylon (Carothers 1937; Hermes 1996). Toothbrush bristles were the first nylon products on 1938, soon followed by stockings. And also in 1938, DuPont's Roy Plunkett made a serendipitous discovery of polytetrafluoroethylene, branded as Teflon. This was followed by alkyd polyesters and polyethylene terephthalate (PET) during the 1940s. PET has been made into fibers (Terylene and Dacron), a film (Mylar), and since 1973 it has been the leading plastic for bottling water (Smil 2006a). These ubiquitous PET bottles and containers are also among the most recycled plastics and are transformed into polyester carpets, fiberfill for sleeping bags and coats, and car bumpers and door panels.

The greatest discoveries of the 1950s included polyimides (for bearings and washers and for heat- and chemicals-resistant applications in electronics) and polycarbonates (for optical lenses and windows, later for CD covers), but the most consequential advance was Karl Ziegler's method of synthesizing polyethylene at normal temperatures and low pressures by using new organometallic catalysts. Post-1960 additions have included polysulfone (a flame retardant), polybutylene (a flexible polyolefin used for pipes and plastic packaging), liquid crystal polymers (aromatic polyesters used in electronics), and Du Pont's plastics marketed under such well-known names as Lycra (Spandex, for athletic wear and clothes), Kevlar (bullet-proof para-aramid used for body armor), flame-resistant Nomex (used in firefighting equipment and pilot suits), and Tyvek (a form of high-density polyethylene used to wrap houses in an insulating but water vapor-permeable barrier).

Global production of all plastics remained below 50,000 t until the early 1930s did not reach 1 Mt until 1949 and surpassed 6 Mt in 1960. Then, with a greater availability of hydrocarbon feedstocks, the worldwide synthesis rose rapidly by an order of magnitude, reaching 100 Mt in 1989, 200 Mt in 2002, and its pre-Covid high was 368 Mt in 2019 (PSP 2022). The latter total was nearly six times the worldwide output of aluminum and about 20% of steel production in that year. But if the comparison is done in terms of volume rather than mass, then plastic production surpassed the output of steel: assuming average density of 1 g/cm^3 it amounted to 368 Mm^3 compared to 239 Mm^3 of steel (averaging 7.8 g/cm^3).

Increased availability of plastic waste and many of its undesirable environmental impacts have led to intensifying efforts to recycle at least several major plastic varieties, and concerns about the long-term supply of hydrocarbon feedstocks rekindled interest in renewable (plant) sources of raw materials needed for plastic syntheses. Plastics industry now offers more than 50 kinds of materials but only a small number of them accounts for most of the global output. I will take closer looks at the three dominant products – polyethylene, polypropylene, and polyvinylchloride – in the next chapter.

2.6.8   Fertilizers

I saved the fertilizer advances for the last entry in this brief review of material innovations of the twentieth century – but if the order of presentation were to be determined by the existential importance for the survival of our species, then the Haber-Bosch ammonia synthesis should have to come first. After all, we could have (and until the 1950s we had) a prosperous civilization without any Si-based electronics, and we could have done (with more expense and sometime with less comfort) with much less steel and without scores of plastics and metal alloys – but we could not support the twentieth century global increment of 4.5 billion people consuming increasingly better diets without a huge increase in nitrogen applications (Smil 2022).

Availability of nitrogen was a key factor in limiting the yields of traditional cultivars; assiduous recycling of organic matter (crop residues, animal manures, human urine and feces), rotations including leguminous species capable of fixing atmospheric nitrogen (alfalfa, clover, vetch), and regular fallowing were the only options to maintain soil fertility. In many affluent countries higher yields were the only way to expand the needed food supply and higher applications of nitrogen were the key (the other two macronutrients, P and K, are more readily secured by extracting P- and K-containing minerals). In 1900, there were only three limited options to get fertilizer nitrogen: declining imports of guano, nitrates from Chile, and ammonia as a by-product from coke ovens. Global consumption from these three sources amounted to less than 350,000 t N, only about 2% of N needed by that year's crops (Smil 2001).

Continuing reliance on recycling of organic matter and planting N-fixing legumes could not guarantee the high yields required to provide better nutrition to more urban population with higher disposable incomes. The solution came in 1909 when Fritz Haber, chemistry professor at the University of Karlsruhe, demonstrated the practicality of catalytic synthesis of ammonia from nitrogen and hydrogen; a no less remarkable achievement was that of Carl Bosch, at the BASF corporation in Ludwigshafen, whose team was able to transform Haber's laboratory demonstration into full-scale commercial synthesis in just four years (Smil 2001). Most of Germany's ammonia synthesis was then diverted to make munitions and explosives for WW I and the interwar years saw only a limited progress in increased nitrogen fertilization. But the post-1950 change was rapid, and after 1970 most of the gains came from nitrogen applications in populous, low-income nations.

Global output of synthetic fertilizers (in terms of pure N) rose from just 150,000 t in 1920 to 3.7 Mt in 1950 and to 85.13 Mt in 2000, nearly a 60-fold increase in 80 years. And that was not an exceptionally large gain as the global production of other new materials saw even greater increases during the course of the twentieth century: three orders of magnitude for aluminum (roughly 3,600 times, from just 6,800 t in 1900 to 24.3 Mt in 2000) and nearly four orders of magnitude for plastics, from about 20,000 t in 1925 to 150 Mt in 2000 (Smil 2019, 2022). And even traditional materials had very large global production multiples, roughly 30 times for paper, 30 times for steel (from 28.3 to 850 Mt), and 27 times for copper (from 495,000 t to 13.2 Mt). In comparison, between 1900 and 2000 the global population increased 3.8 times and the gross world product (in constant monies) rose about 20-fold, which means that during the twentieth century, material use had greatly intensified both in per capita terms and per unit of economic output.

Before examining in detail some long-term trends in material consumption, presenting a systematic review of their energy requirements (energy costs), highlighting some of the environmental burdens resulting from their production and use, and noting some successes and challenges of material recycling (all in the fourth chapter), I will take closer, systematic looks at the most important components of the modern world's material consumption – ranging from bulk construction materials to extremely pure silicon – and I will do so in all cases at generic and global levels, and in many instances I will also focus on specific national developments.

3

What Matters Most

Alexey Rezvykh/Adobe Stock

Materials and Dematerialization: Making the Modern World, Second Edition. Vaclav Smil.
© 2023 John Wiley & Sons Ltd. Published 2023 by John Wiley & Sons Ltd.

Given the enormous variety of materials used in modern civilization, any systematic, but still far from comprehensive, review of their production, their energy costs, their most common uses and associated environmental, social and economic impacts would require either a multi-volume coverage or it would fit into a mid-size book only when restricted, encyclopedically, to a sequence of very brief entries. I am adopting what is perhaps the most efficacious solution of this challenge: in this chapter, I will note the diversity of every major category of materials whose incessant supply enables the functioning of modern civilization – but I will take a closer look only at a limited number of products that are either quantitatively dominant or qualitatively indispensable. I will survey basic production methods of every one of these key materials and note some of the important industrial advances of recent decades.

My reviews start with the two leading biomaterials, lumber used in construction and furniture-making and wood pulped to make paper. Then I will turn to the world's leading construction materials, that is to all varieties of sand and stone and then to cement, a fairly energy-intensive product whose combination with aggregates and water yields the world's most important construction material, concrete. Modalities and consequences of iron and steel production will dominate the third section that will also take a closer look at the two other leading metals, aluminum, and copper. Afterward I will deal with the most important plastic materials, with their derivatives and, necessarily, with the hydrocarbon feedstocks required to produce them, and with many worrisome environmental consequences of their use and challenges of their recycling.

This will be followed by a section devoted to three leading industrial gases whose commercial supply has become indispensable for industrial sectors ranging from high-quality steel to nitrogen fertilizers whose synthesis will be covered alongside the less complicated efforts to produce the two key mineral fertilizers, phosphates and potash. As we have entered the second century of electronics – I am dating the field's birth to Fleming's invention of diode in 1904 (Fleming 1934) – I will take a closer look at the material that is at the very core of modern computing, as well as imaging and telecommunication. Hence, the last section will explain the ways of producing polysilicon, the growing ever-larger, ultra-pure silicon crystals and wafers, and slicing them to provide substrate for ever-more powerful microchips.

3.1   Biomaterials

Category of biomaterials is a large and diverse one: it includes products as different as construction lumber, straw, cotton, wool, beeswax, birds down, and animal hides – but in mass terms, as well as in the overall importance and indispensability of uses, it is dominated by wood. As already noted, crop residues, above all cereal straws, come next, and they remain an important source of fuel in rural areas of some low-income countries. But as most of them are either burned in fields (particularly rice straw in Asia), recycled to renew organic matter and to prevent soil erosion or removed as a feed or bedding for animals, their use as construction materials in high-income countries is now marginal (Barreveld 1989; Unger 1994; Smil 1999; King 2006).

3.1.1   Fibers

In mass terms the third largest category of biomaterials are fiber crops whose aggregate harvest was about 31 Mt in 2020 (FAO 2022). Cotton lint is dominant, with global harvests rising from 7 Mt in 1950 to more than 12 Mt in 1975, nearly 19 Mt in 2000 and more than 25 Mt in 2020. Jute (3.4 Mt in 2019) comes second, coir (coconut husk fiber, 1.2 Mt in 2020) is the distant third. Wild flax (*Linum usitatissimum*) fibers (dated to 30,000 years BCE and used to make cords) were found in a Paleolithic cave in Georgia (Kvavadze et al. 2009), and the plant is the oldest cultivated fiber crop whose harvest surpassed 1 Mt in 2020, followed by hemp (about 250,000 t), sisal (fiber from stiff *Agave sisalana* leaves, about 200,000 t in 2020), and ramie (fibers from a bushy perennial *Boehmeria nivea*, nearly 10,000 t in 2010). Worldwide harvests of each of two less important crops – kapok (fiber surrounding seed pods of a large tropical tree *Ceiba pentandra*) and abaca (fibers from leaves of banana species *Musa textilis*) – are around 100,000 t a year.

Two species of ancient plant fibers that used to be woven into fabrics – coarse hemp made into sailcloth, sack, and canvas, fine linen that furnished many societies with excellent cloth fabric and crisp handkerchiefs – have been in retreat, in many countries to the point of near disappearance. How many men are now regularly wearing puff-folded pocket squares in their jackets, and how many ships are tied to bollards with hempen ropes in a new age when cannabis is grown mostly as a drug? But cotton is more ubiquitous than ever as it clothes more people than any other natural fabric. The next most massive category of biomaterials is natural rubber, latex extracted from the Amazonia rubber tree *Hevea brasiliensis* (Sethuraj and Mathew 1992). Transplanted throughout the tropics, the species now produces natural rubber mainly in Asia, with Thailand, Malaysia, Vietnam, and Indonesia being the largest exporters. Annual production was below 2 Mt until the early 1960s, it surpassed 3 Mt a decade later, reached nearly 7 Mt in 2000 and 15 Mt by 2020 (FAO 2022), just slightly ahead of annual synthetic rubber production.

Animal hides and skins come next, and their production has been rising along with the expanding demand for meat: by 2020 annual output of bovine (cattle and water buffalo) hides surpassed 8 Mt, and that of sheep, lambs, and goats is more than 3 Mt Production of wool, the most important animal fiber, rose from about 960,000 t in 1950 to 2.9 Mt in 1970, fluctuated afterward (peaking at 3.3 Mt in 1990) and a long-term decline brought it to less than 1.8 Mt by 2020 (FAO 2022). In contrast, production of silkworm cocoons has grown to about 660,000 t by 2020, and (just for the sake of completeness) here are recent annual outputs of minor biomaterials: about 70,000 t of beeswax; 20,000–30,000 t of lac, resinous secretions of insects belonging to genus *Kerria*, with 70–80% of the output coming from India (Singh 2006); and just above 100,000 t of gum arabic (exudates from *Acacia seyal* and *Acacia senegal*) mostly from Sudan (UNCTAD 2018).

Consumption of many of these biomaterials has been in long-term decline and some of them will get even more marginalized; aggregate demand for other plant and animal products may diminish, but they will remain highly valued by some buyers due to their intrinsic qualities or, at least, due to the perceived status conveyed by their acquisition (be they beeswax rather than tallow candles, and silk rather than synthetic saris or purses). And there can be no doubt that the demand for the three biomaterial leaders – lumber, paper made from wood pulp, and cotton clothes – will remain strong. The only major wood-based industry that has seen significant post-1950 changes and inroads by substitutes has been furniture making.

3.1.2 Wood

Quality furniture has seen a great shift to veneer-covered particle board or plywood and those two products can be sourced from many species; low-end furniture uses laminate materials (most commonly melamine-clad panels of particle board) and many smaller household items formerly made of wood have been replaced by plastics and metals, and solid wood pieces remain favorites only in some markets and for some specific applications. Similarly, wood use in wine-, spirit-, and beer-making has declined with the adoption of stainless steel tanks and aluminum kegs, but cumulatively significant and irreplaceable (or at least preferable) uses of wood remain: for many products oak barrels and casks will remain in demand, as will select woods for musical instruments and some sport implements.

Scores of tree varieties are still harvested (both for lumber and pulp), but the preference of modern, large-scale processing for fairly uniform inputs has led to an increasing dominance of a relatively small number of coniferous (softwood, above all pines, spruces, firs) and deciduous (hardwood, mainly ash, eucalyptus, poplar) species, with an increasing share of them harvested from newly planted and replanted monocultures: in parts of Europe coniferous monocultures of firs and spruces are already in their third, even fourth, generation. Some of these managed forests have also seen remarkable increases in productivity but climate change and drought and wild fires may have the opposite effect (Kauppi et al. 2010; Lindner and Verkerk 2022).

In forestry statistics, industrial roundwood is the composite category that includes all sawlogs, veneer logs, as well as wood harvested for pulp to be made into paper. Reliability of wood harvest statistics ranges from excellent in Western countries to chronically unreliable in many nations in Latin America, Africa, and Asia where illegal cutting has been common. FAO's data show the worldwide roundwood harvest at about $1.7\,Gm^3$ in 1950, $2.88\,Gm^3$ in 1975, $3.43\,Gm^3$ in the year 2000, and a slightly higher total of $3.91\,Gm^3$ in 2020 (FAO 2022). The 2020 total was split 49 : 51 between fuelwood and industrial roundwood whose harvest was just a bit below $2\,Gm^3$. Assuming average dry weight of 420 per kg of green industrial roundwood, the total 2020 harvest would have been about 830 Mt.

Actual annual cut of woody phytomass is considerably higher not only because the standard statistics exclude illegal harvests, but above all because they leave out all phytomass that does not become a part woody of commercial products. A very conservative adjustment for illegal cutting would raise the global industrial roundwood harvest by 15% to about 950 Mt in 2020. A much larger adjustments must be done for the cut phytomass that is not commercially harvested including small trees (with breast-height diameter less than 12 cm) and non-merchantable woody biomass, a category including trees of poor form, stumps, bark, branches, and tapering tree tops.

Birdsey (1996) and Penner et al. (1997) published formulas for converting merchantable stem volumes to the totals of cut above-ground phytomass. US multiplier averaged about 2.1, with the species-specific values ranging from about 1.7 (southeastern loblolly pine) to more than 2.5 (spruce and fir). Consequently, even a very conservative assumption would double the harvested total to about 1.9 Gt in 2020, and adding at least 20% in order to include the root phytomass that remain place after wood harvests would yield nearly 2.3 Gt, with half of it assignable to the category of hidden material flows. Is this an excessive harvest or could this extraction continue without further massive deforestation? To answer this question on the global level is complicated by unreliable forest inventories, but it is fairly easy for affluent nations with good forestry data.

In the United States, the latest data (for 2016) show net annual growth (708 Mm³) being nearly twice the average annual removals (368 Mm³): in the Norther forests, growth was 2.4 times higher, in Southern woodlands 1.8 times that of average annual removals, and twice as high in the Pacific Coast; only in Rocky Mountains removals outpaced growth because of high tree mortality rates (Oswalt 2021). Clearly, modern management makes it possible to exploit forests in non-destructive manner. Going back to the global roundwood harvest, those 1.9 Gm³ cut in 2020 ended up as about 1.1 Gm³ of sawlogs and veneer logs and nearly 700 Mm³ of pulpwood for making paper. Before I will follow each of these material streams, I will first describe the main handling and processing methods that lead to commercial lumber (sawn pieces) as well as other structural wood-based products, above all particle board and plywood, and then I will look at a variety of paper products and the current changes in their consumption.

As I will show in the next chapter, substitutions by concrete, metals, and plastics have greatly reduced wood intensities of modern societies (that is mass of wood used per unit of GDP), but wood remains an important structural material and its annual production surpasses the combined output of all metals and plastics. Rising post-WW II wood demand has been due to great surges in wood-based house construction, use of wood for concrete forms, packaging of goods for long-distance shipments and intercontinental trade, and in paper-based information flows. Similarly, progressing adoption of fossil fuels and hydroelectricity in modernizing countries of Latin America, Africa, and Asia reduced average per capita demand for wood and charcoal, but rapid population growth on those three continents resulted in large increases in aggregate demand for fuelwood.

Hundreds of tree species are used to make building materials, with their properties (density, resistance to moisture, durability, homogeneity, etc.) and cost determining many specific applications ranging from structural timber, cladding, internal paneling and flooring to roofing, windows, doors, veneers, plywood, and engineered components. Wood properties (chemical, mechanical), moisture relations, drying, durability, fire safety considerations, and methods increase the material's strength and lifespan have been studied for generations in great detail: centennial edition of *Wood Handbook* is perhaps the best single-volume source of this information (Forest Products Laboratory 2010).

Turning logs into sawn wood generates inevitable milling losses and modern automated computer-guided operations make accurate cuts and minimize waste. Typical yields (averages for coniferous and non-coniferous species) are 54% of logs ending as sawn wood (range of 45–62%), 29% as chips and slabs, 11% as sawdust, and 1% as shavings with the shrinkage loss averaging 4% (UNECE 2010). Nothing gets wasted, chips, shaving, and sawdust are used for making particle board and pulp, for mulching, cellulose extraction, or to produce steam and electricity. Using a rounded average of 0.55 for logs-to-sawn wood conversion would mean that the global 2020 production of about $472\,Mm^3$ of sawn wood required about $860\,Mm^3$ of roundwood.

Durability of lumber is obviously a key consideration in all external wood uses. Durability of untreated wood is determined by its location: extremes range from close to a millennium (when dry elm, oak, and beech as well as pine and spruce can last 800 years, sometimes even longer) to less than five years for wood in contact with earth (Berge 2009). Treated timber can last significantly longer. Creosote derived from coal tar remains the first choice for treating railway ties and electricity poles but various copper-based compounds and new nonmetallic preparations have replaced chromated copper arsenate used to pressure-treated construction lumber (Richardson 1978). Biodeterioration (wood attacked by fungi and insects) is a universal problem, but it is particularly rapid in the tropics; borates offer the best, and the safest, protection against termites, the main natural destroyers of wooden structures in warm climates.

New housing construction in industrializing economics shows a clear divide among a minority where wood has remained dominant (USA, Canada, Japan) and a majority where traditional brick structure have been increasingly augmented by buildings using concrete and some structural steel (above all EU and China). Thanks to North America's abundance of low-cost lumber, wood remains often the only structural component of single-family houses. Boards with a rectangular cross-section (two-by-fours, now actually planed down short of these measurements) are used for framing studs, plywood for outer walls, wooden trusses for roofs, and wooden joists, subfloors (plywood) and often floors as well, staircases, doors, and door and window frames. A new North American house of $200\,m^2$ requires about 14 t of lumber (typically yellow pine) and another 14 t of panel products (mostly plywood) for the total mass of 28 t. Greater use of plywood and engineered lumber has reduced the relative demand for dimensional sawn wood, but in aggregate terms this trend has been countered by a steadily increasing size of an average American house (from $110\,m^2$ in the early 1950s to $210\,m^2$ by 2020).

Wood-based composite products – including fiberboard, particle board, plywood (layered wood veneers whose commercial use began during the 1850s), laminated veneer lumber, glued-laminated lumber, manufactured trusses, and finger-jointed lumber – have been displacing traditional sawn wood in many applications. Glue-laminated lumber (glulam) is a particularly outstanding material (APA 2022). Because it is made from thin flexible lamellas it is easy to machine, it can be used not only for columns, rafters, and trusses but can be turned into optimal shapes including arches and domes for large clear spans and curved portals.

Glulam is lightweight (its density is $0.55–0.7\,g/cm^3$ compared to $7.8\,g/cm^3$ for steel), which means that even though it must be fastened with steel bolts or dowels and steel plates, the overall weight of its structures is much lighter than those made of steel. But the material is strong enough (more than twice as much as the sawn timber, per unit of mass comparable to structural steel) to be shaped into major load-bearing structures, including large bridges. Glulam is also fairly resistant to aggressive chemicals, it burns poorly, it insulates well, and the same piece can be used both inside and outside of a building without any cold bridging problems.

Industrial construction has been also a major wood consumer because of the need for concrete forms. Plywood makers produce framework panels with moisture-resistant adhesive of specified strength and stiffness. Wood requirements vary with the thickness and shapes of the poured concrete; forming a typical 20 cm-thick basement wall of an American house will require $10\,m^2$ of plywood (weighing about $20\,kg/m^2$) for every m^3 of emplaced concrete. But the forms, particularly the standard-sized pieces, can be reused and those properly engineered (with smooth, abrasion-resistant surfaces) can be deployed up to 200 times (APA 2012). And while in the Western countries, standard-sized tubes or pipes are now the norm for vertical scaffolding components; wooden boards still provide horizontal pieces and bamboo scaffoldings (often on buildings in excess of 30 stories, built with standard lengths of 6 m) with wooden boards continue to be common in East Asia.

3.1.3 Pulp and Paper

In contrast to wood's stagnating or declining fortunes, paper has been one of the signature materials of the modern world. By the late 1930s, Dahl's sulfate process, producing stronger (kraft) paper, became the leading chemical method in large-scale papermaking. Pulp treatment with inexpensive sodium sulfate and net energy production were its two main advantages, offensive emissions of hydrogen sulfide its leading drawback. By the year 2000, sulfate process was producing roughly two-thirds of the world's pulp, semi-chemical, and mechanical processes accounted for nearly all of the rest, with sulfite method reduced to less than 2% of the total. Demand for raw material varies with the processing method and with the kinds of paper (Bajpai 2010).

Mechanical pulping needs $2.5\,m^3$ of roundwood for a ton of pulp and the rates are nearly $2.7\,m^3$ for semi-chemical process and almost $4.5\,m^3$ for the kraft process, with the overall global mean of about $3.8\,m^3/t$ (UNECE 2010). This would mean that the 2010 global output of some 185 Mt of pulp required just over $700\,Mm^3$ of roundwood, slightly less than the total used for sawn wood. Analogical conversion rates (all in m^3 of roundwood per ton of product) are about 2.9 for newsprint, 3.25 for packing paper, and 4.35 for sanitary and household paper, with the mean for all paper and paperboard products at about $3.6\,m^3/t$.

Worldwide expansion of papermaking since the middle of the twentieth century was driven by a combination of several trends: growing book, newspaper, and journal publishing; increasing demand for photographic paper (due to expansion of both professional and amateur photography, the latter use promoted by inexpensive pocket cameras); much higher need for typing paper in the offices of the pre-computer era; new demand created by the introduction of plain-paper copiers (first in the United States in 1959 by Xerox, soon a worldwide office necessity); massive diffusion of office printers (first limited to line printers using perforated paper and serving large computers, since the late 1970s smaller machines attached to desk computers using standard-size office paper); post-1980 acquisition of small printers attached to personal computers; and a much increased demand for paper used in industrial and retail packaging. Ironically, rapid increase in the consumption of printing paper coincided with the diffusion of what was to be paperless office of a new electronic age.

In the United States, additional paper demand came from the post-1950 housing expansion. By the mid-1950s, drywall had replaced traditional gypsum plaster in half of all new homes, and a decade later it had all but a small share of the household market for internal partitions and ceilings. American drywall is a thin (the standard thickness for interior partitions is half an inch or $1.27\,cm$) layer of gypsum that is sandwiched between two heavy liner paper boards; the mineral was formerly all mined, now nearly 50% of it originates as the by-product of flue-gas desulfurization in coal-fired power plants (Gypsum Association 2022). Paper used in drywall production amounts to about $400\,g/m^2$ (on two sides) of a panel. In 2021, US gypsum board consumption reached about $2.6\,Gm^2$ and that means that the industry used about one million tons of paper, virtually all of it being a secondary (recycled) material (USGS 2022a). Sheathing paper has been replaced by Tyvek, but tar paper is needed for the waterproofing of roofs as underlayment for asphalt or wood shingles.

By the turn of the century, paper consumption became saturated in many affluent countries, and wood pulp for domestic production began to decline because of more intensive paper recycling. In the United States, wood pulp production peaked in 1994 (at nearly 65 Mt/year), was in decline until 2010, and since that time it has stagnated at around 50 Mt/year, a trend roughly matched by the output of paper and paperboard whose decrease (more than 10% since 2000) was accelerated after 2007 due to the economic downturn, and by 2014, it declined to less than 80 Mt a year. This was accompanied by a massive decline in pulp, paper, and board mill employment (from 200,000 in 1999 to less than 120,000 a decade later and to just 96,000 by 2019) and hence by a substantial rise in output per mill employee, from 450 t/year in 2000 to about 820 t/year by 2019 (Data USA 2022).

Among the trends that have contributed to the decline of paper consumption, none have been relatively as abrupt as the virtual demise of the market for photographic paper and the shift of disseminating news and images from printed to electronic forms. This formerly important paper-consuming sector was first affected by a widespread adoption of color slides, but its near demise came when printed photographs became gradually displaced by electronic images, first taken by digital cameras and viewed on screens, later by a mass ownership of camera-equipped cellphones with billions of images increasingly stored in the cloud. More importantly, the publishing sector has seen closures, reductions, and retrenchment as many newspapers (several hundred in the United States since 2008) and magazines have folded, including such well-known periodicals as *Life* (1883–2000), *Mademoiselle* (1935–2001), *Home* (1951–2008), and *Gourmet* (1941–2009) or stopped printing and launched only e-editions (perhaps most notably America's second oldest weekly, *Newsweek*, published between 1933 and 2012).

But global consumption of paper keeps rising, driven by large increases of demand in populous Asian countries, above all in China. Worldwide production of paper and paperboard rose from 127 Mt in 1975 to nearly 325 Mt in the year 2000, in 2011 it surpassed 400 Mt, rose to 415 Mt in 2017 and by 2020 it was just above 400 Mt (FAO 2022). China's 2020 use of 117 Mt accounted for 29% of the global supply, up from just 11% in 2000 and 3% in 1975. Not surprisingly, given China's limited forest resources, the country has become the world's largest importer of waste paper (accounting for about 18% of global trade), with the US being its largest supplier with 37% of global sales (OEC 2022). In mass terms, sales of waste paper peaked at 56 Mt in 2018 and fell to 49.3 Mt in 2019 because of China's stricter regulations: about 11 Mt were sent to Asia from the United States and 7.4 Mt from Europe, while about 12 Mt of recovered fiber moved between different EU members (BIR 2020).

FAO (2022) data also show that packaging and paper paperboard accounted for about 62% of the output in 2020, followed by printing and writing paper (20%), newsprint (about 8%), and household and sanitary paper (about 9%). Per capita differences in average national consumption rates (calculated using FAO's production and trade figures) have been reduced during the last two generations when the global mean rose from about 25 kg/year in 1960 to about 90 kg in 2020 (FAO 2022). During the same period, the US rate fell slightly from 195 to 187 kg and the Chinese rate soared from only about 3 to nearly 90 kg. Even so, large gaps persist, particularly in the consumption of household and sanitary paper with the 2020 rates ranging from about 22 kg in the United States to less than 8 kg in China and to less than 200 g in India.

3.2   Construction Materials

Before I review production methods and uses of construction materials that dominate the world of the early twenty-first century, I must point out that hundreds of millions of people – Berge (2009) put the total at more than 30% of humanity – continue to live in structures whose material, locally available clay, has not undergone any elaborate processing and that can be made without any modern energy inputs. Although the construction of earthen houses requires only few simple tools, it is rather labor intensive. Walls are usually built by ramming the manually dug-out earth between shuttering or by pressing earth blocks and drying them in the sun before emplacement. These adobe bricks are made by adding fibrous plant materials: straw is used most often, but stalks of wild plants and cattle manure are also common ingredients.

3.2.1   Bricks and Aggregates

Adobe buildings remain common in arid regions, and while some massive structures have been surprisingly durable, many earthen houses have poor structural integrity and their collapses have been a major reason for exceptionally high death tolls in areas that experience repeated earthquakes. In contrast, production of all durable soil- or earth-based materials requires firing in kilns, with temperatures ranging from less than 500 °C for low-quality bricks to as much as 1,100 °C for ceramic tiles, 1,300 °C for vitrified bricks, 1,400 °C for glass, and 1,400–1,450 °C for the pyro-processing of Portland cement (Berge 2009). Damp clay used to make quarry tiles and terracotta is the same raw material that is used for bricks, and its firing in kilns sinters the minerals. Dry pressed clay, often with added kaolin and finely ground waste glass, is used for vitrified ceramic tiles and the firing produces smoother and finer pieces. The main problem with attractive glazing is that many of the pigments used (oxides of Cd, Co, Cr, Ni, V) are toxins that require careful disposal.

The classification of aggregates, bulk construction materials whose extraction dominates the mass of annually mined minerals, is based not on their chemical composition but on their basic physical attribute, the grain size (Reeves et al. 2006; The Constructor 2022). Clays contain particle smaller than 0.002 mm in diameter and result from disintegration of rocks; their composition is dominated by the two oxides most common in the earth's crust, SiO_2 and Al_2O_3. Firing of clays produces hard and durable materials ranging from building bricks (with clays making up 20–30% of the initial mixture that contains 50–6% of sand) to highly heat-resistant tiles. Silt – be it in such massive aeolian deposits as China's Loess Plateau or as river-borne material left behind floods – is coarser than clay, with diameter going from 0.004 up to 0.06 mm.

Construction aggregates include sand, gravel, and crushed stone. Sand grains have diameter up to 2 mm, gravel's irregular dimensions are between 2 and 6 cm, and pieces larger than that belong to various kinds of crushed stone. Sand and gravel are unconsolidated results of rock disintegration whose composition is dominated by siliceous and calcareous minerals. They are widely distributed in alluvial deposits and sedimentary beds and hence they can be usually produced from nearby deposits for local and regional consumption, but there is some longer-distance and even international trade. Their modern extraction is done either as pit excavation or by barge-based dredging in rivers, lakes, or along coastlines. These extractions produce raw, moist materials that could be used directly as a fill or a base layer, but most excavated sand and gravel is processed to meet market specifications.

Sequential washing, screening, crushing, and dewatering eliminate any organic matter and clay and produce specific coarseness of material with low moisture. The best estimates for 2020 show about 46% of construction sand and gravel used as concrete aggregates, 21% going into road base, coverings, and road stabilization, 13% for construction fill and 12% for asphaltic concrete aggregate and other bituminous mixtures, with the remaining uses ranging from filtration and golf course maintenance to plaster sands, railroad ballast, roofing granules, and snow and ice control (USGS 2022a). Replenishment of eroding beaches (Hawaii's Waikiki has been perhaps the most famous example, with the sand imported from California's Manhattan Beach during the 1920s and 1930s, now dredged from the near offshore deposits) is yet another minor use. Sand and gravel production correlates highly with the fortunes of construction industry, and during building slumps it can contract rapidly and significantly: in the United States, it fell from 1.25 Gt in 2007 to just 760 Mt in 2010, a 40% decline in just three years, and it rose from 888 Mt in 2017 to 1 Gt in 2021 (USGS 2022a).

Crushed stone used in construction (mostly in road building and maintenance and for cement manufacturing) is produced mostly from sedimentary rocks (above all limestone and dolomite), while granite is the most commonly quarried igneous rock.

Dimension stone can be used in construction in both rough and dressed forms. Dressed stone is used in traditional ashlar buildings for curbing, and thinner pieces for flagging and cladding, and because durability is important most of it is granite and harder varieties of limestone (including different marbles) and slate (in addition to flooring and cladding and also for roofing). Laborious stone cutting has been fully mechanized by using diamond blades on circular or frame saws, as well as water and flame jets. Finishing (grinding and polishing) can be done to exacting specifications.

Brick-making, one of the world's oldest industries, has reached new global highs, thanks to China's extraordinary post-1990 construction effort. During the late 2010s, the world was producing about 1.5 trillion bricks annually, with China accounting for 67% and India for 17% (Li 2012; Eil et al. 2020). When assuming, conservatively, average mass of 2.8 kg/brick, then the global output in 2020 had the total mass of about 4.2 Gt, the same total as the global production of cement in that year.

3.2.2   Cement

Cement is prepared by high-temperature heating of limestone, shale, clay, slate, slag, and silica and iron ore (providing Ca, Si, Al, and Fe) to make the key ingredient of concrete, by the far the most massive anthropogenic material. Depending on the desired qualities of the materials, concrete contains 7–15% of cement and 14–21% of water; this paste binds with fine and coarse aggregates (sand, gravel, crushed stone) that make up 60–75% of the finished material, with the remainder (up to 8%) being air enclosed in the porous material (PCA 2019). Cement hydration produces durable solids even when used under water, finer aggregates produce stronger concrete but require more water in the mix, coarser aggregates need less water, but within the commonly used ranges the size of aggregate matters less than its other qualities. Additives may be used in order to accelerate the rate of setting at a worksite to delay setting during extended transport in mixer trucks or to allow entrainment of a small volume of air and to reduce or repel water content.

Final disposition of cement varies by country: in the United States about 75% of the material goes to make ready mixed concrete, less than 15% for the production of concrete masonry units and precast concrete, in Germany the analogical ratios are about 55% and 30% (Sika 2020). Because of its near-total dependence on construction, cement consumption fluctuates with economic booms and downturns. Construction of highways, dams, factories, and houses pushed the US cement consumption by an order of magnitude in just 28 years: about 3 Mt of cement (87% of it produced domestically) were mixed with aggregates in 1900 and the total surpassed 30 Mt in 1928 (Kelly and Matos 2016).

The Great Depression cut it to just less than 11 Mt in 1933, and then it rose once again to 31 Mt in 1942, the peak year of wartime plant construction. After a brief post-war dip, the domestic output, as well imports, kept growing almost without any interruption until 1973 when the consumption reached nearly 82 Mt. Construction of the US Interstate Highway System was a major component of this rising demand (USGS 2006). About 60% of these multilane highways are paved in concrete whose standard thickness is 28 cm and hence 1 km of the four-lane highway (each lane is 3.7 m wide) requires about 4,150 m³. This adds up to roughly 10,000 t of concrete for every kilometer and the entire system of 73,000 km embodies about 730 Mt of concrete in driving lanes, with more emplaced in shoulders, medians, approaches, and overpasses.

The 1973 US cement production of 77.5 Mt was surpassed only in 1994, a new peak was reached in 2005 with 128.25 Mt, and by 2010 the total was down by 45% to just 71.2 Mt (reached for the first time in 1968). But by 2021, the output rose to 92 Mt (USGS 2022a). Until the mid-1980s, imports were less than 10% of domestic consumption, but then they rose to 22% of apparent consumption in the year 2010 and were about 17% in 2021. Since 1986 the global cement production has been dominated by China; during the 1980s its cement output rose by 20%, during the 1990s it had more than doubled and during the first decade of the twentieth century it had more than tripled to nearly 1.88 Gt, and by 2021, with 2.5 Gt, it accounted for nearly 57% of the global output of 4.4 Gt (USGS 2022a). India was the distant second with 330 Mt, Vietnam the unexpected third with 100 Mt ahead of the United States (92 Mt) and Turkey (76 Mt).

Cement industry used to be a large source of uncontrolled particulate emissions (as much as 1 kg/t of the product), but modern fabric filters capture 99.6% of these particulates that are returned into the kiln. This leaves relatively high CO_2 emissions as the industry's most serious environmental impact: in mass terms cement production emits roughly a unit of gas for every unit of material, with half of the amount originating from the decomposition of $CaCO_3$, the rest from heat and electricity required to supple thermal and mechanical requirement of a rotating kiln. Recent emissions average about 600 kg CO_2/t of cement, and global compilations of CO_2 emissions from cement industry show its contribution rising from 1% of all carbon released from fossil fuels in 1950 to 2% in 1975 to about 4% by 2020 when the total surpassed 1.6 Gt CO_2 – and when China's CO_2 emissions from cement production were equal to nearly 75% of Japan's total emissions of the gas (Andrew 2019).

There has been a great deal of interest to improve the industry's energy efficiency and to reduce its CO_2 emissions (Madlool et al. 2011; Barcelo et al. 2014; IEA 2021a). Options include using cement and construction materials based on MgO, geopolymer cement, and cement using slag and fly ash. Use of fly ash is particularly appealing: not only does it reduce energy cost of producing more clinker but it also avoids landfilling the captured fly ash and extracting new raw minerals, and it is the most effective way to curb the CO_2 emissions. By 2010, China's cement industry used all available slag (223 Mt) as well as 66.3% of 395 Mt of fly ash (Lei 2011). And in the US Ceratech, a cement company in Virginia, produced blends containing 95% fly ash and 5% liquid ingredients (compared to standard 15% fly ash blends), making a stronger concrete that also reduces the required mass of steel reinforcing bars (Amato 2013).

3.2.3   Concrete

Assuming 11% average cement content, the world has used some 38 Gt of concrete in 2020 and with density ranging between 2.2 and 2.4 t/m^3 that would amount to about 16.5 Gm^3, which is an equivalent of a concrete cube about 2.5 km high. Concrete has poor tensile strength of just 2–5 MPa but excellent compressive strength that has been steadily increasing. In 1900, the best mixtures had compressive strength of about 25 MPa, by 1970s 40 MPa were standard, and 50 MPa became common by the century's end. Today's best ultrahigh performance concrete (UHPC) has a minimum compressive strength of 120 MPa, and it can be formulated to bear in excess of 200 MPa (Ulm 2012).

Recycling of concrete from demolished buildings or deteriorated roads and runways recovers the aggregate through crushing and sieving, and the material can be used as a bulk fill (a cheaper option) or incorporated in new concrete (Wilburn and Goonan 2013). Energy savings can be also realized by better ways of transporting and emplacing concrete, including improvements in pneumatic conveying, cold weather concreting, curing, and reusing returned material (Kermeli et al. 2011). Not surprisingly, even national data on concrete recycling are incomplete, with the US estimates ranging as high as 80% of annual demolitions and as low as 50%, but US Portland Cement Association put the total mass at 140 Mt and it has not provided an update (PCA 2022).

Affordability of concrete and its ubiquity in modern civilization is not (even after leaving the contribution to anthropogenic CO_2 emissions aside) without major long-term costs. Some 100 Gt of cement were produced worldwide between 1945 and 2020, which means that on the order of one trillion tons of concrete were emplaced in structures, with 40% of the total put in place during the second decade of the twenty-first century. Concrete (particularly its reinforced form) is now by far the most important man-made material both in terms of global annual production and cumulatively emplaced mass. While this material provides shelter and enables transportation and energy and industrial production, its accumulation also presents considerable risks and immense future burdens.

These problems arise from the material vulnerability to premature deterioration that results in unsightly appearance, loss of strength and unsafe conditions that sometimes lead to catastrophic failures and whose prevention requires expensive periodic renovations and eventually costly dismantling. Concrete, both exposed and buried, is not a highly durable material and it deteriorates for many reasons (AWWS 2004; Cwalina 2008; Stuart 2012). Exposed surfaces are attacked by moisture and freezing in cold climates, bacterial and algal growth in warm humid regions (biofouling recognizable by blackened surfaces), acid deposition in polluted (that is now in most) urban areas, and vibration. Buried concrete structures (water and sewage pipes, storage tanks, missile silos) are subjected to gradual or instant overloading that creates cracks and to reactions with carbonates, chlorides, and sulfates filtering from above. Poor-quality concrete can show excessive wear and develop visible cracks and surficial staining due to efflorescence in a matter of months.

Alterations of freezing and thawing damage both the horizontal surfaces (roads, parking) that collect standing water as well as vertical layers that collect water in pores and cracks. While concrete's high alkalinity (pH of about 12.5) limits the corrosion of reinforcing steel embedded in the material as soon as that cover is compromised (due to cracks or defoliation of external layers), the expansive corrosion process begins and tends to accelerate. Chloride attack (on structures submerged in seawater, from deicing of roads, in coastal areas from NaCl present in the air in much higher concentrations than inland) and damage by acid deposition (sulfate attack in polluted regions) are other common causes of deterioration, while some concretes exhibit alkali-silica and alkali-carbonate reaction that leads to cracking. Unsightly concrete blackened by growing algae embedded in the material's pores is a common sight in all humid (especially when also warm) environments. Given the unprecedented rate of the post-1990 global concretization, it is inevitable that the post-2030 world will be facing an unprecedented burden of concrete deterioration.

This challenge will be particularly daunting in China, the country with by far the highest rate of new concrete emplacement where the combination of poor concrete quality, damaging natural environment, intensive industrial pollutants, and heavy use of concrete structures will lead to premature deterioration of tens of billions of tons of the material poured into buildings, roads, bridges, dams, ports, and other structures during the past generation. Because the maintenance and repair of deteriorating concrete have been inadequate, future replacement costs of the material will run into trillions of dollars. To this should be added the disposal costs of the removed concrete: some concrete structures have been recycled, but the separation of concrete and reinforcing metal is expensive.

The latest report card on the quality of American infrastructure gives poor to very poor grades to all sectors where concrete is the dominant structural material: bridges fared relatively best with C; aviation, schools, and inland waterways got D+; dams, roads, and levees just D, and there was an estimated 10-year investment gap of at least \$2.59 trillion needed to prevent further deterioration (ASCE 2021). Transposed to post-2030 China, this reality implies the need for an unprecedented rehabilitation and replacement of hundreds of billion tons of concrete emplaced during the first decade of the twentieth century, at a cost of many tens of trillions of dollars.

Finally, here is concrete (or, more accurately, all man-made nonmetallic construction materials) seen from two unusual global perspectives: one comparing areas, and the other one volumes. Elvidge et al. (2007) combined satellite observations of night-time lights and population numbers to estimate the world's impervious surface area (built-up, paved) at about 580,000 km^2: that is 0.43% of ice-free surface but an area equal to Kenya. In per capita terms, high-income countries in northern latitudes had the largest areas of impervious surfaces (Canada 350 m^2, USA 300 m^2, Sweden 220 m^2), while in low-income countries the rates were below 100 or even below 50 m^2 per capita.

Two much more accurate studies were published in 2022. The first one, based on remote sensing with 30-m resolution, found the global impervious area doubling from 511,000 km^2 in 1985 to 1.08 million km^2 in 2020 (Zhang et al. 2022). The second one based on 10-m resolution data found a significantly higher total increasing from 1.27 million km^2 in 2015 to 1.20 million km^2 in 2018 (Sun et al. 2022). As expected, United States has the largest share of all impervious surfaces (about 20%), followed by China (about 12%) and Russia (nearly 10%). Inevitably, further gains, with the highest growth rates in South America, Africa, and South Asia, are inevitable.

Of course, not all impervious surfaces are concrete but the material accounts for their largest share – and there are good reasons why there should be even more of it on the ground. Replacing mud floors by concrete floors in the hundreds of millions of the world's poorest dwellings would cut down the incidence of parasitic diseases by nearly 80%, while paving the streets boosts land and rental values, school enrollment, and overall economic activity, and it also improves the access to credit (Kenny 2012). The case for more concrete is thus persuasive – and so is the argument, made first by J.R. Underwood in the early 1970s, that the enormous mass of concrete, bricks, tiles, and glass deserves to be put into a new category of anthropic rocks to complement the standard division into igneous, sedimentary, and metamorphic materials (Cathcart 2011); three decades later he published the idea (Underwood 2001) and more than two decades later further huge increases in concrete production strengthen his suggestion.

At the same time, we should keep the mass of the anthropic rocks in perspective. In 2020, humanity put in place more than 40 Gt of them (dominated by about 38 Gt of concrete and 4.2 Gt of bricks), an equivalent of at least 18 km^3. For comparison, the volume of one of the world's best known mountains, Japan's Fuji, is about

$400\,km^3$ when approximated by a cone with the radius of about $13\,km$ rising about $2,800\,m$ above the surrounding countryside: the mountain is $3,776\,m$ and the Lake Yamanaka is $980\,m$ above the sea level (Kaneko et al. 2022). This means that we are emplacing worldwide anthropic rocks equivalent to Fuji's volume roughly every two decades. In mass terms, it would be more like every three decades because basalt that forms the volcano has density of about $3\,g/cm^3$ compared to, respectively, 2.4 and $1.8\,g/cm^3$ for concrete and bricks. You can judge this achievement to be both impressive and quite modest: after all, the volume of the world's most massive volcano, Mauna Loa (including the mountain's entire $17\,km$ elevation from its depressed sea floor foundation to its peak at $4,170\,m$ above the sea level), is, at about $75,000\,km^3$, nearly 200 times larger than Fuji's volume (Kaye 2002).

3.3   Metals

The combination of advancing industrialization, intensified private and public transportation, mechanization of agriculture, and the emergence of mass consumption created rising demand for every metal during the twentieth century. In some affluent countries specific metal consumption (per capita, per unit of GDP) has leveled off but, as I will show later, there were few absolute declines. Iron, used overwhelmingly as steel, remained the dominant metal of the twentieth century and by the year 2000 the global output of iron ore, pig iron, and steel reached new global records: at 1 Gt/year iron ore extraction was surpassed only by the output of fossil fuels and bulk construction materials, pig (cast) iron production rose to nearly 600 Mt, and at roughly 850 Mt/year steel output was about 30 times higher than in 1900. That total was also almost 20 times larger than as the aggregate smelting of aluminum, copper, zinc, lead, and tin, and in per capita terms it rose from less than 20 to about 140 kg/year. Demand for copper increased by a similar rate (27-fold, to 13.2 Mt) and zinc production rose almost 20-fold, from about 480,000 t to 8.77 Mt (Kelly and Matos 2016).

Silver had seen the smallest relative production increase during the twentieth century (only about 3.4-fold, from 5,400 t in 1900), followed by lead (4.3-fold, from 750,000 t to 3.2 Mt) and gold output rose nearly sevenfold, but in absolute terms, it amounted only to about 2,600 t in the year 2000, compared to 18,100 t for silver, and to millions of tons for Pb, Zn, and Cu. Aluminum emerged as the century's new signature metal, with global output rising from less than 7,000 t to nearly 25 Mt, surpassing copper extraction nearly twofold. In demanding (high strength-to-weight) application, the metal is often substituted by titanium: about two-thirds of the annual production of titanium sponge metal are used by the aerospace industry, the rest goes for armaments and to chemical industry, while TiO_2 is added as a whitener to paints, paper, and plastics (USGS 2022a).

3.3.1 Iron and Steel

Advances in iron and steel industry had changed every aspect of ferrous metallurgy changed during the twentieth century. Although most of the primary metal still comes from blast furnaces, their sizes, capacities, and energy efficiencies have been transformed by continuous quest for lower cost and higher productivity (Kawaoka et al. 2006; Smil 2016). The largest furnaces now have volumes in excess of 5,000 m^3: 5,500 m^3 Shougang Jing Tang's furnace in Caofedian (new in 2009), 5,513 m^3 Schwelgern 2 in Duisburg (ThyssenKrupp since 1993), 5,775 m^3 Japan's Ōita 2 (enlarged from 5,245 m^3 in 2004) and (since 2013) the world's largest, South Korea's (POSCO) Gwangyang with 6,000 m^3. These furnaces have hearth diameter of about 15 m and their maximum daily output is on the order of 12,000 t of hot metal (Hoffmann 2012). Larger furnaces need relatively less coke: in the United States, gradual decreases in coke charging and, later, supplementary injection of pulverized coal, oil, or natural gas reduced the specific consumption of coke in blast furnace from about 1.3 t/t of hot metal at the beginning of the twentieth century to less than 0.5 t at its end (de Beer et al. 1998).

By 1914, nearly 75% of the US steel output was coming out of open hearths and nearly half a century later their share of steel production peaked at 88%. By that time, Japan and Europe were rapidly converting to top-blown basic oxygen furnaces (BOFs). They are charged with molten pig and scrap iron that is subjected to blasts of supersonic oxygen, and the adjective does not imply a simplicity of design, it refers to basic pH level of furnace's magnesium oxide lining. The first designs of these new furnaces were developed by Robert Durrer in Switzerland, and in 1948 he demonstrated the practicality of the process that used scrap as more than half of the total furnace charge (Durrer 1948). The subsequent commercialization of the process was not done by a major established steel-making company but by two Austrian steelmakers, *Vereinigte Österreichische Eisen- und Stahlwerke* (VÖEST) in Linz and Alpine Montan in Donawitz, with the first production starting before the end of 1952 (Geschichte-Club VÖEST 1991).

As a result, this new way of steelmaking became also known as the Linz-Donawitz process, and it was almost instantly adopted in Japan as the country was rebuilding its steel industry destroyed in WW II. In contrast, major US steelmakers were reluctant innovators, adopting their first BOF only in 1964. By 1970 BOF smelted half of the world's steel output and 80% of Japanese production. Oxygen was initially blown only from the top but later also from the furnace's bottom at rates of 50–60 m^3/t of hot metal. BOF capacities range mostly between 150 and 300 t/heat can complete decarburization (to less than 0.1% C) in only 35–45 min (in contrast, in open-hearth furnaces it took 9–10 h), and they boosted labor productivity 1,000-fold (Berry 1999; Smil 2016; González et al. 2021).

By the century's end, BOFs produced slightly more than 70% of the world's steel and by 2020 the share reached 73%, with the rest coming from electric arc furnaces (EAFs) using scrap metal (WSA 2022a). Commercial use of EAF began with aluminum smelting in the 1880s, but their adoption by major steelmaking companies came only after WW II once the electricity prices declined and once there was an abundant supply of scrap metal. Their spreading adoption cut the traditional connection between iron- and steel-making, as new smaller steel mills (mini mills) could operate by using only cold scrap metal as their raw material. As a result, the steel/pig iron ratio in the US output stood at 2.1 by the end of the twentieth century and the country's steel output was split between EAFs and BOFs. Two decades later, the United States became an even better user of its obsolete metal: steel/pig iron ratio rose further to 3.97 and despite the fact that the country has been a major exporter of scrap its EAFs produced nearly 71% of all steel, compared just 42% in the EU, 25% in Japan, and just 9.2% in China (WSA 2021).

Post-1950 changes have also transformed the processing of newly smelted metal. Standard practice was to cast steel into ingots (oblong chunks of 50–100 t) and reheat them in order to form semi-finished products (thick slabs, square-profile billets, rectangular blooms) that were eventually rolled into final shapes as bars, beams, coils, plates, rods, rails, sheets, or wire. During the 1950s, these energy-inefficient practices began to be replaced by continuous casting of hot metal pioneered by a German metallurgist Siegfried Junghans and promoted by Irving Rossi, an American engineer (Morita and Emi 2003; Tanner 1998; Fruehan 1998). Continuous casting speeds up production (from more than a day to less than an hour), cuts the metal losses (from about 10% to 1% or so) and saves 50–75% of energy compared to the traditional ingot-reheating-rolling sequence. Japanese steelmakers were the early adopters of the process, by the century's end nearly 90% of the global steel output was cast that way and by 2020 even their global mean reached 97%, with the United States at 99.8% and China at 98.6% (WSA 2021).

As the steelmaking expanded and matured, both the primacy in production and the technical leadership shifted from the United Kingdom to the United States; but starting in the 1950s, the center of innovation moved to Europe and above all to Japan, while the USSR became the world's largest, but technically inferior, steel producer (Smil 2016). After 1974 (coincident with the first round of large oil price rises), the industry's almost uninterrupted post-WW II rise turned into a sequence of advances and retreats, with the global production dipping in 10 of the next 25 years. The next dip came in 2008 and 2009 as a result of global recession, but the output remained above 1 Gt/year (first surpassed in 2004), mainly because of China's huge, and rising, demand, and by 2021 the global output (1.953 Gt) came close to 2 Gt of crude steel (WSA 2022a).

In 1900, the US Steel Corporation produced nearly 30% of the world's steel and the US output was 36% of the global total. At the end of WW II (with Japan and Germany in ruins), the US share was nearly 80% of the global output, but the post-1955 expansion of the Japanese and Soviet steelmaking decline began to reduce the industry's global share. By 1975, United States still had three companies among the world's top 10 producers (US Steel as number 2), by 1990 the US Steel was alone in the top group (number 5), and a year later USX, its parent company, was dumped from the Dow 30 to make place for Disney. By the year 2000, there was no American company among the top 10 (the US Steel placed 14th); by 2010, 6 out of 10 largest steelmakers were in China; and in 2012, the CEO of Rio Tinto, the world's second largest metals and mining corporation predicted that China's steel production will reach 1 Gt by 2030 (Albanese 2012). His forecast was badly off, China reached one billion total already in 2020, and in 2021 it produced nearly 57% of the world's crude steel, compared to less than 4% in the United States (WSA 2022a).

Fairly reliable worldwide statistics make it possible to outline the global flow of materials in steelmaking in 2020. Integrated steelmaking (blast furnace-basic oxygen furnace) remains the dominant route; and in 2020, it averaged 1,600 kg of iron ore, about 450 kg of metallurgical coke and 250 kg of limestone (or dolomite) to produce 1,000 kg of iron (WSA 2022b). Globally, this meant mining 2.3 Gt of iron-bearing ores that were (often after being concentrated) charged into blast furnaces together with coke supplemented by pulverized coal injection (almost exactly one billion tons of coal were used for these two inputs) and about 340 Mt of raw flux (limestone, dolomite) to produce 1.34 Gt of crude steel.

This route depends on large-scale intercontinental shipments of iron ore: in 2019 nearly 79% of its production (1.6 Gt) was traded, with Australia (836 Mt) and Brazil (340 Mt) being the largest exporters and China the largest buyer, taking 81% of the world's exports and deriving about 82% of ore it uses from imports (WSA 2021). Larger ships have been used to carry this ore, with Brazil's Vale acquiring 35 362-m long Valemax carriers rated at 400,000 deadweight tons (Vale 2022). In 2020, the second route of steel production, using recycled (scrap) metal and a smaller amount of directly reduced iron (without coke, using natural gas) in EAFs required 710 kg of recycled steel, 586 kg of iron ore, 150 kg of coal, 88 kg of limestone, and 2.3 GJ of electricity to make 1,000 kg of crude steel. Global output of 493 Mt of EAF steel needed nearly 500 Mt of scrap and 1.1 EJ of electricity.

As already noted, global output of steel rose 30-fold between 1900 and 2000 (to nearly 850 Mt), it surpassed 1 Gt in 2004, reached 1.43 Gt in 2010 and nearly 1.88 Gt in 2020, with China producing 56% of the total. In per capita terms, the worldwide consumption of finished products prorated to about 228 kg, with the national means of about 415 kg in Japan, 691 kg in China, 274 kg in the EU, 242 kg in the United States, and only 64 kg in India, with all western rates much affected by the post-2008 economic downturn (in 2006, the US rate was at 400 kg). Exceptionally high per capita rates in South Korea and Taiwan (respectively, 955 and 777 kg) are not due to domestic consumption primarily but to large steel demand by ship-building industries producing vessels for export.

Fairly accurate output statistics make it possible to calculate that the global pro-duction of steel amounted to nearly 31 Gt during the twentieth century, with half of the mass produced after 1980. Nearly, 12 Gt were added during the first decade of the twenty-first century and almost 17 Gt between 2010 and 2020, bringing the 1900–2020 total to about 60 Gt. An obvious question to ask is what happened to all of this metal? Obviously, much of it has been lost (in shipwrecks, mines, shafts) and much more has deteriorated (mostly due to relentless corrosion) and is not available for reuse. I estimated that at the beginning of the twentieth century, the accumulated steel stock that could be potentially turned into new metal was about 15 Gt, or roughly 2.5 t/capita (Smil 2005).

My estimate was not that different from a later Japanese attempt to quantify the global steel stock (Hatayama et al. 2010). According to that analysis for 42 coun-tries, the total reached 12.7 Gt in 2005 after it had doubled since 1980. Most of the metal is in structures (about 60%) and 10% is in vehicles; they also forecast that the structural and vehicular stock will reach 55 Gt by 2050, mostly due to 10-fold increase of steel consumption in Asia. For comparison, Müller et al. (2006) esti-mated the total anthropogenic iron stocks at 25–30 Gt, another Japanese analysis (Takamatsu et al. 2014) calculated the changes in accumulated global steel stock since 1870, and it put it at about 12.5 Gt in the year 2000, about 15% lower than my approximation, but according to Fujitsuka et al. (2013) they reached only 16 Gt by 2010. What is not in doubt that the stock increased rapidly during the second decade of the twenty-first century when additions topped 1.5 Gt/year.

Missing steel consumption statistics led some researchers to rely on nighttime satellite light images as proxies of steel used in engineering infrastructures and build-ings. Hsu et al. (2015) ended up with 11 Gt, almost equally split between infrastruc-tures and buildings. National estimates do not appear to be more reliable: the US stocks of steel rose from 200 Mt in 1920 to 1.25 Gt in 1950 and to 3 Gt in the year 2000 and 3.2 Gt in 2004 (Müller et al. 2006). The latter total would be second only to the mass of iron present in the ores in the American shares of the lithosphere. But Sullivan (2005) put the country's steel stocks in use at 4.13 Gt in 2002, and Buckingham's (2006) total was 4.09 Gt in 2001.

Even the lower totals would put America's 2010 per capita stocks of steel at no less than 12 t, as much as in Germany and more than in Russia (9 t) and France (8 t) but less than Japan where the stocks were about 16 t/capita (Watari and Yokoi 2021). As expected, per capita steel stocks are still low in India and very low in Africa where there are measured in tens to hundreds of kilograms rather than in tons. China's post-1990 economic expansion brought an unprecedented accumulation of steel stock in new infrastructures that now includes the world's longest freeway, railway, and high-voltage transmission networks, the largest number of major container ports and hydroelectric dams, as well as an unmatched speed of building new factories and houses and acquiring new cars and household appliances.

China became the first nation whose accumulated in-use stock reached 5 Gt (by 2010) and then went on to soar to 15 Gt as the country added 11 Gt during the second decade of the twenty-second century and as the per capita rate, less than 1 t/capita until the year 2000, has approached 10 t/capita. This, of course, means that in the coming decades – as the economy matures, population ages and steel production will begin to decline – China will be relying less on imports of iron ore as it will be "mining" the accumulated stocks of the metal and using it in its rising EAF steel production.

3.3.2 Aluminum

During the twentieth century, aluminum became the second most important metal of modern civilization, but the fundamentals of its production remain as they were invented in the 1880s, combining Bayer's production of alumina from bauxite (extraction with caustic soda followed by calcination) and Hall-Hérault's electrolysis of cryolite and alumina (Smil 2005). But the entire process (from bauxite mining to electrolytic smelting and casting of ingots or rolling of sheets) consumes a great deal of raw materials and it remains highly energy-intensive. Material requirements for the primary production of one ton of the metal include 4–6 t of bauxite as well as about 425 kg of carbon anode (it is about 80% calcined petroleum coke and 20% coal tar pitch), and in 2020 the global average of electricity use in Al smelting was 14,250 kWh (Djukanovic 2018; IAI 2022b).

Many industries, led by automaking, have turned to aluminum in order to reduce the weight of their products without compromising structural integrity: as already noted, the metal's density is only a third of steel's density but tensile strength of many aluminum alloys is 400–500 MPa, overlapping with that of the most common structural steels (high-tensile steels rate well above 1 GPa). Aluminum can reduce weight of many key car components by up to 40% compared to only 11% for high-strength steel, and that is why the global car industry, required to meet more stringent emission standards, has been the metal's leading and steadily growing market. In 2010, average American vehicle contained 154 kg of aluminum (making up 9% of the average car mass but more than a third of the material scrap value), in 2020 it was up to 208 kg with more than 30 heavier vehicle platforms having more than

500 pounds (nearly 227 kg), and by 2026 the mean is expected to rise to 233 kg of aluminum peer vehicle (DuckerFrontier 2020). Aluminum is mostly in engine blocks and wheels, and increasingly in hoods, trunks, and doors.

A new twentieth century manufacturing sector has been particularly aluminum-dependent. The earliest airplanes had wooden and fabric covers but aluminum bodies appeared by the late 1920s and monocoque construction were all-aluminum, with the demand spurred above all by America's unprecedented program of new fighter and bomber construction during WW II when the country's aluminum consumption more than quintupled in just four years. Military demand remained strong after WW II and by the mid-1950s came the era of commercial jetliners (starting with Boeing 707 in 1957) whose bodies are 70–80% high-strength aluminum alloys, with a single wide-bodied jet containing more than 100 t of the metal – but the latest designs are dominated by composite materials. The Boeing 787 is 80% composite by volume and 50% composite by weight, with aluminum at 20% (wing and tail leading edges), titanium at 15% and steel at 10% (Hale 2006).

Other post-WW II markets with a high reliance on aluminum alloys have ranged from house construction (window and door frames, siding, eavestroughs) to communication and spy satellites cars and beverage cans (Hosford and Duncan 1994). And aluminum wires became the most important long-distance conductors of electricity. Between the beginning of its commercial production during the 1880s and 2012 almost 1.5 Gt of the metal (about 1,430 Mt) were extracted from bauxite, and the best calculations show that nearly 75% of that mass is still in productive use, with about a third in buildings, close to 30% in transport and a similar share in electrical applications and about 10% in machinery (IAI 2021). This means that a significant share of the extracted metal has been recycled many times – but there is a limit to this reuse due to rising levels of other elements added to make aluminum alloys (use of pure, 99.8%, aluminum has declined as alloys were selected to meet a variety of specific demands).

3.4 Plastics

Many important types of plastics were first synthesized before WW II, but by 1950 the global output of these new, light, and durable materials was just 2 Mt but continued exponential growth (interrupted by only two small annual declines, in 1975 and 2008) brought the production to 35 Mt by 1970, 213 Mt by the year 2000, and 367 Mt by 2020, the total that does not include the production of recycled plastics (Plastics Europe 2021). This means that more than half (56%) of all plastics ever made were produced during the first two decades of the twentieth century, and further expansion is expected.

Not only has the production of plastic grown at a faster rate than of any other mass-produced material, but about half of the annual output has been for packaging and shopping, to be thrown away after a single, often very brief, use. As expected, North America and Europe lead in per capita plastic waste, but in aggregate terms Asia dominates. As with most other mass-produced materials, China is now the world's leading synthesizer of plastics: in 2020, it delivers 32% of the global output, followed by 19% from North America and 15% from Europe.

Of course, in per capita terms, China, with just above 80 kg/year (including a large share incorporated in exported consumer goods), is still far behind the European Union and the United States where the average consumption rate of plastics was, respectively, about 110 and 150 kg in 2020. The final use of plastic products is well illustrated by the detailed breakdown available for the European market (Plastics Europe 2021): in 2020, packaging consumed 40% of the total (mostly as various kinds of PE and PP), construction 20% (mostly for plastic sheets used as vapor barriers in wall and ceiling insulation), auto industry claimed nearly 9% (interior trim, exterior parts), and electrical and electronic industry more than 6% (mostly for insulation of wires and cables).

Most of the now ubiquitous plastic materials belong to two categories of manmade polymers: thermoplastics account for nearly 80% of the total, thermosets for most of the rest. Thermoplastics are either linear or branched molecules that lack any chemical bonds and are softened by moderate heating and hence are easy to process; they harden again once cooled and retain fairly high impact strength. Polyethylene (PE) is by far the most important thermoplastic (it accounted for 28% of the world's aggregate plastic output, or roughly 104 Mt in 2020), polypropylene (PP) comes next (with about 20% or 74 Mt in 2020), followed by polyvinyl chloride (PVC, about 6% or 21 Mt in 2020). In contrast to thermoplastics, thermoset plastics have bonds among molecules (and hence a greater dimensional stability) and cannot be softened by moderate heating (Goodman 1998). Their output is dominated by polyurethanes and epoxy resins, followed by polyimides, melamines, and urea-formaldehyde.

The three dominant polymers that now make up about three-fifths of the global output of synthetics – PE, PP, and PVC – are structurally very similar: they all share the backbone of carbon atoms, but it carries only hydrogen atoms in PE, while in PP every other carbon atom carries a methyl (CH_3) group and in PVC chlorine alternates with hydrogen. There are other similarities: ethylene (IUPAC name ethane, C_2H_4) and propylene (correct IUPAC name is propene, C_3H_6) are polymerized with the help of the same class of catalysts, they have many overlapping physical characteristics, and they share many more everyday uses for packaging, construction, and a variety of household and industrial items. But there is a notable difference in terms of consumer perception and environmental impact: PE and PP do not get much negative coverage, but PVC has been widely portrayed as environmentally dangerous – and its synthesis does require hazardous chemicals, it is not easy to recycle, and its disposal has undesirable consequences, no matter if the material gets landfilled or burned.

3.4.1 Polyethylene

PE has been the world's most important plastic for four generations (Malpass 2010; Spalding and Chatterjee 2017). ICI patented high-volume polymerization of PE in 1936 and began its production of low-density polyethylene (LDPE, a branched ethylene homopolymer) using a chromium catalyst in September 1939 on the eve of Britain's entry into WW II. During the war, the new material was used as a cable insulator and to make light radar domes, but its production required very high pressure and high temperature (respectively, in excess of 100 MPa and 200 °C). This changed with the discovery of mixed organometallic catalysts in 1953 by a group led by Karl Ziegler at the Max Planck Institute: it found that nickel in combination with triethylaluminum dimerized olefins (Ziegler 1963). Further research discovered that other catalysts containing heavy metals, above all titanium in the form of $TiCl_4$, could polymerize ethylene to produce high-density polyethylene (HDPE, a linear, semicrystalline homopolymer of C_2H_4).

In first trials, this was done at a relatively moderate pressure of 20 MPa, later at ambient conditions (Ziegler 1963). Italy's Giulio Natta used these catalysts to produce polymers from propylene, and in 1963, the two chemists shared the Nobel Prize in Chemistry (Natta 1963). HDPE is a more resilient material: its density is only slightly higher than that of LDPE (0.95–0.97 g/cm³ compared to 0.91–0.94 g/cm³) but its melting temperature is considerably higher (135 vs. 115 °C) and its tensile strength is superior, 32 MPa compared to 9 MPa for LDPE. LDPE is more flexible and more transparent and both polymers are inert. Propensity of early HDPE products to crack with aging was solved by synthesizing a material with a small share of branches in the linear chain.

Later, during the 1970s, the efficacy of titanium-based catalysts was greatly enhanced by the addition of MgCl, and this opened the way to the commercial synthesis of linear low-density polyethylene (LLDPE, a random copolymer of ethylene and α-olefins with density between 0.915 and 0.925 g/cm³). The material is stronger than LDPE and can be used to make thinner films. The range of ethylene-based materials now also includes ultra-high-molecular-weight polyethylene (UHMWPE, whose molecular weight is typically between 3.5 and 5.5 million and density ranges between 0.93 and 0.935 g/cm³), medium-density polyethylene (MDPE, 0.94 g/cm³), and cross-linked polyethylene (PEX or XLPE, an elastomer due to its cross-link bonds).

All of these products begin as ethane. In North America and the Middle East, ethane is separated from natural gas and low gas prices and abundant supply (in the United States from hydraulic fracturing of shales) led to surplus production for export and favored further construction of new capacities. Dominant feedstock for ethane in Europe, where prices of imported natural gas are high, is naphtha derived by the distillation of crude oil. Ethane is cracked to make ethylene ($C_2H_6 \rightarrow C_2H_4 + H_2$) and the subsequent multistage ethylene purification yields a compound with 99.9% purity whose exothermic catalytic conversion (about 97% efficient) produces powdered PE.

Different kinds of PE are shaped into final products by techniques best suited to particular specifications: by extrusion and molding (through impact, blowing, or rotation) of objects and by casting or blowing of sheets (single layer or laminated) and insulation foams. Besides its low stiffness, PE's other disadvantages include flammability, poor weathering resistance, limited range of temperature tolerance, and proneness to stress cracking. Final PE products are used for a still-expanding range of visible and hidden applications. Virtually unavoidable are thin LLDPE, LDPE, and MDPE films made into transparent or opaque bags (sandwich, grocery, or garbage), sheets (for covering crops and temporary greenhouses), wraps (Saran, Cling) and squeeze bottles (for honey), HDPE garbage cans, containers (for milk, detergents, motor oil) and toys (including Lego bricks). Among a myriad of hidden PE applications are HDPE for house wraps (Tyvek) and water pipes; PEX for water pipes and as insulation for electrical cables; and UHMPWE for knee and hip replacements.

Flexibility, strength, durability, and resistance to UV radiation makes LLDPE an excellent choice for heavy-duty uses including pool liners and geomembranes used in construction, and the material is also preferred for ice and frozen food bags. Sizewise, the PE applications range from massive LDPE water tanks (50,000 l or 50 m³) to small machine parts (bearings, gears) made of wear-resistant UHMPWE. Another extreme range is indicated by soft LDPE cling and bubble warps and UHMPWE (Spectra) used in bulletproof clothing as an alternative to aramid (Kevlar). PE products have been increasingly recycled. HDPE (numbered 2 inside the recycling triangle molded at an item's bottom) returns again as garbage cans or detergent containers or is made into flower pots. LDPE (number 4) comes back as new bags.

3.4.2   Polypropylene

Propene, a colorless but highly flammable and asphyxiating gas, is produced by hydrocarbon cracking, and it was first polymerized to make a crystalline isotactic compound (with all methyl groups attached to the same side of the carbon backbone) in 1954 by Giulio Natta in Italy and Karl Rehn in Germany. Much like PE, isotactic PP continues to be made by using metallic Ziegler-Natta catalysts, while the production of syndiotactic PP (with regular alternation of opposing monomers) is catalyzed by soluble metallocenes, organometallic complexes whose deployment allows the control of molecular structure in order to produce polymers with specified qualities. Its commercial varieties include copolymers (homopolymer, block, and random) and rubber-modified blends. PP's disadvantages include flammability, degradation by UV radiation (both can be minimized by additives), susceptibility to chlorinated solvents, and poor impact strength at low temperature.

PP shares many market niches with PE – it is made into many kinds of food containers (yogurt, sour cream, mayonnaise) and their caps, crates, totes, garbage cans, storage bins, bottles, baskets, pails, and pipes – as well as with PVC (as a wire and cable insulator or a film). But its combination of low density (range 0.89 to 1,06, typical rate just $0.905 \, g/cm^3$), substantial strength (25–40 MPa, more than MPa for a film) and high melting point (171 °C for perfectly isotactic polymer, 160–166 °C for commercial material), and resistance to acids and solvents makes it an ideal raw material for high-temperature uses (laboratory and hospital items requiring sterilization), heavy-duty applications (industrial pipes for hot or cold liquids, ropes, parts tolerating bending, such as hinges, and impact, such as container lids), nonwoven materials (including diapers and liquid and air filters) and fibers. The fibers range from indoor–outdoor carpeting to lightweight fabrics woven from PP yarn and used particularly for outdoor apparel as the material insulates while staying dry.

3.4.3   Polyvinylchloride

PVC's history began unusually early for a plastic: vinyl chloride was first prepared by Henri Victor Regnault in 1835 and in 1872 Eugen Baumann succeeded in polymerizing the compound in laboratory setting. Real commercial breakthrough came only in 1926 when Waldo L. Semon (working for BF Goodrich Company in Akron, Ohio) dissolved heated polymerized vinyl halide in a nonvolatile organic solvent and, after cooling, produced what he called a stiff rubbery gel (Semon 1933). PVC production begins by combining ethylene and chlorine and converting ethylene dichloride into vinyl chloride that is polymerized (mostly in suspension but also in emulsion) inside reaction vessels to yield white PVC powder.

In order to illustrate the material's ubiquity in modern society, I wrote a brief narrative of a city woman's day pointing out PVC's presence in objects she uses in the morning and she leaves for work (ranging from insulated wires, water, and sewage pipes to food wraps and her car's interior and body undercoating), and in conclusion I noted that an even greater PVC presence would surround her if she had an accident and ended in a hospital:

There she would be enveloped by objects made of different kinds of PVC: disposable and surgical gloves, flexible tubing for feeding, breathing and pressure monitoring, catheters, blood bags, IV containers, sterile packaging, trays, basins, bed pans and rails, thermal blankets, labware. (Smil 2006a, 131)

PVC is also used in construction (house siding, window frames) for outdoor furniture, water hoses, office gadgets, toys, and credit cards, but new plastic banknotes (now in circulation in Australia and Canada) are made from PP. But there are concerns. Greenpeace called the material "one of the most toxic substances saturating our planet and its inhabitants" and "the single most environmentally damaging of all plastics" that contaminates people and the environment not only during its production but also during its use and incineration or landfilling – and called for substitution by other materials in order to create a PVC-free world (Greenpeace 2003). Plastic makers have been rebutting these claims and the compound has become so ubiquitous that creating a PVC-free world would be very challenging.

Ubiquitous exposure to phthalates, used as plasticizers to soften normally rigid PVC and as additives in cosmetics, has been even more controversial. These compounds are classified as endocrine disrupting chemicals that can impair the human endocrine system, induce fertility problems, cause respiratory diseases, and contribute to childhood obesity and neuropsychological disorders. Children are particularly affected as they are relatively more exposed to phthalates than adults. Many uncertainties remain despite extensive, and sometime contradictory, inquiries but with the adoption of new strict regulations human exposure to phthalates has been decreasing (Wilkinson et al. 1999; Tickner et al. 2001; Katsikantami et al. 2016).

Plastics have a limited life span in terms of functional integrity: even the materials that are not in contact with earth of water do not remain in excellent shape for decades. Service spans are no more than 2–15 years for PE, 3–8 years for PP, and 7–10 years for polyurethane; among the common plastics, only PVC can last two or three decades, and thick PVC cold water pipes can last even longer (Berge 2009). But complete degradation may take decades or more than a century, and as torn films, broken pieces and degraded bits of plastics accumulate in the environment. I will provide more detailed data on the extent of this problem in the next chapter.

3.5 Industrial Gases

When asked to make a short list of materials indispensable for the functioning of modern civilization, few people will include industrial gases; moreover, these gases are also usually excluded from the totals of material flows monitored in the EU and North America, and they are also commonly neglected by authors writing histories of modern inventions. Those readers who have not skipped any previous sections already appreciate that the three most important elements – oxygen, hydrogen, and nitrogen – deserve such ranking because without them we could not produce steel in the most efficient way and could not have modern petrochemical and nitrogen fertilizer industries. Other elements and compounds classified as industrial gases include acetylene, argon, carbon dioxide, helium, neon, and nitrous oxide.

As already noted, much of the early development to liquefy gases was done by Carl von Linde who used combination of Thomson-Joule effect and countercurrent heat exchange. In 1902, he patented separation of principal gases from the liquefied air (at temperatures lower than $-140.7\,°C$) through purification by rectification (countercurrent distillation), and by 1910, he perfected this process for large-scale commercial applications by developing a double-column rectification that made it possible to produce pure oxygen and nitrogen simultaneously (Linde 2022). Also in 1902, Georges Claude, after working with acetylene, introduced his variant of cryogenic air separation (Almqvist 2003) and a century later his and Linde's designs remain the foundations of modern production of high-purity streams of nitrogen, oxygen, and argon from the atmosphere where their concentrations are, respectively, 78.08%, 20.95%, and 0.93%.

But Linde and Claude might be even more surprised to see how consequential their inventions have become. Most importantly, without the synthesis of ammonia (predicated on large-scale supply of pure nitrogen), we would not be able to feed billions of people and without oxygen we could not produce most of the world's most important alloy. Ammonia synthesis is the world's largest consumer of nitrogen: in 2020, it required 145 Mt of the gas (about 116 Gm^3 of N_2). Nitrogen's other key uses as a feedstock include ammonia for the synthesis of nitric acid, hydrazines, and amines.

Liquid N_2 is used not only to preserve vaccines and tissues but also to freeze soils for easier drilling and tunneling, in tire and solvent recycling, enhanced oil recovery and production of plastic moldings and to achieve material shrinkage: nitrogen cooling of metal parts enables tight assembly fits and, in reverse, it allows to take apart closely fitted parts. With the expansion of modern electronics nitrogen found a new market, particularly during soldering when it is necessary to reduce the presence of oxygen and to maintain clean atmosphere. And nitrogen is also used as an inert blanket for flammable or explosive compounds or as protector of wine in opened bottles.

Ferrous metallurgy is by far the largest user of oxygen: the gas is blown into blast furnaces; EAFs and BOFs need about 50 m^3/t of hot metal. Chemical syntheses (above all ethylene oxidation) are the second largest market, and oxygen is also used in smelting color metals (lead, copper, and zinc furnaces) and in the construction material industries (producing more intense flame and reduced fuel use in the firing of glass, mineral wool, lime, and cement); for delignification and bleaching (a superior choice to ClO_2) in pulp and paper industry. Liquid O_2 is an excellent rocket propellant (combined with kerosene it powered the first stage of Saturn rockets used for all Apollo lunar missions); welding and cutting metal is done with oxygen and oxygen-acetylene flames; waste water treatment uses it for oxidation, incineration, and vitrification; incineration of hazardous wastes is another growing market; and aquacultural operations use it to aerate their ponds. Global market for liquid oxygen was nearly 400 Mt in 2020, with industrial uses dominating but with medical demand rising sharply during the first year of the pandemic (Path 2022).

Argon, the cheapest truly inert gas goes into incandescent and fluorescent lights (typical mixture is 93% Ar, 7% N_2) and between the panes of high-efficiency windows; it is used in stainless steel production as a shield gas in casting and when blown into a converter it reduces Cr losses and prevents the formation of nitrides; it removes hydrogen and particulates from molten aluminum; it provides a shield for tungsten arc welding as well as for the production of virtually pure silicon and germanium crystals and for the protection of wine in opened bottles; and it is used in micro-cryosurgery to destroy tiny bits of tissue.

There is too little hydrogen in the atmosphere (0.00005% compared to about 0.04% for CO_2) to separate it economically from air and the standard pre-WW II choice was to produce by steam reforming of coal and since the 1950s by steam reforming of hydrocarbons with methane (CH_4) being obviously the best feedstock. By the end of the second decade of the twenty-first century, the worldwide production of hydrogen (intentionally separated gas as well as by-product gas generated by petrochemical industry for its own consumption) was nearly 90 Mt a year (IEA 2021b). Crude oil refineries require hydrogen for cracking, dearomatization, and desulfurization.

Hydrotreating, hydrodesulfurization, and hydrocracking used to process roughly 4.2 Gt of oil in 2020 claimed (assuming that H_2 demand averaged 0.5% of the total crude input, or roughly 60 m^3/t) about 23 Mt of the gas. Synthesis of 176 Mt of ammonia (NH_3) required about 31 Mt of hydrogen and the remainder was divided between other chemical syntheses (methanol, polymers, solvents, vitamins, pharmaceuticals), and industrial uses including glass-making, semiconductors, and food processing (to hydrogenate unsaturated fatty acids in order to produce solid fats) and rocket fuel. By 2022, the industry appeared to be at the beginning of unprecedented expansion as the long-standing dreams of developing a new hydrogen economy began to receive record amounts of private funding and government subsidies (IEA 2021b).

Nitrous oxide (N_2O, prepared by heating of ammonium nitrate) is a common anesthetic and aerosol spray propellant. Acetylene (traditionally produced by hydrolyzing calcium carbide made by reducing lime by coke) was widely used for welding and cutting before it was largely displaced by arc process. Helium (extracted from natural gas) is used mainly in cryogenic applications (including basic scientific research and magnetic resonance imaging) to purge and pressurize rockets and in chromatography to provide lift for weather (and party) balloons. Finally, CO_2 has many uses ranging from enrichment of greenhouse atmospheres in order to accelerate growth of vegetables to enhanced oil recovery, and liquefied and pressurized CO_2 can also replace petrochemical-based solvents in dry cleaning.

Although they receive only a fraction of attention given to metals and nonmetallic minerals, industrial gases are now used in sectors that account for more than half of the world economic output and the value of their production has been growing faster than the growth rate of global economy. In 2000, their global market was worth about $34 billion, a decade later it had nearly doubled as it exceeded $60 billion and by 2020 it reached $92 billion (Grand View Research 2021). Inclusion of industrial gases in a comprehensive appraisal of material flows is thus justified both because of the aggregate magnitude of annual use and because of the indispensability of those elements and compounds in virtually every important sector of modern economy. I will take a closer look at the processes producing the three most important (and most voluminous) streams, those of oxygen, nitrogen, and hydrogen. The main reasons for the continuing expansion of these industries have been technical innovations that cut costs and raised the efficiency of gas extraction, shift of production to consumption sites, and greater economies of scale.

The most important pre-WW II innovations were Fränkl's regenerators (heat exchangers, patented in 1928) and expansion turbines replacing piston engines used for refrigeration that were introduced during the 1930s. The first Linde-Fränkl oxygen plant began to operate in 1950 and the diffusion of this process cut the oxygen prices by an order of magnitude. A major progress in air separation took place during the mid-1980s when Linde introduced structured packings in rectification columns: compared to previously used tray sieves packed columns make it possible to enrich the downward-flowing liquid in oxygen and the ascending stream to get increasingly richer in nitrogen with reduced energy consumption (Smil 2006a).

Cryogenic extraction of the two dominant atmospheric constituents starts with the compression of ambient air to about 600 kPa (normal pressure is 101.3 kPa) and the removal of H_2O and CO_2 by molecular sieves. These are made of synthetic zeolites, aluminosilicates whose molecular-size pores and diverse chemical compositions make it possible to tailor them for highly specific uses; these materials are able to adsorb selectively H_2O, CO_2, and N_2 and produce a stream of 90–95% pure O_2 (Ruthven et al. 1993). After removal of water and carbon dioxide, elements of cooled liquid air are separated by distillation, with N_2 boiling away first (at 77.4 K), followed by Ar (at 87.3 K) and O_2 (at 90.2 K). Cold gases then flow back to heat exchangers where they cool the incoming air while warming to near-ambient temperature and pressure, an exchange that reduces the plant's energy consumption and boosts its overall efficiency to about 35%.

A competing technique for gas separation that emerged during the1950s is pressure swing adsorption (PSA), a process using a concomitant pressurization/adsorption and depressurization/desorption in two identical adsorber vessels filled carbon molecular sieves and taking advantage of the fact that O_2 diffuses faster into their pore structure than does N_2 (Linde 2019). PSA popularity is due to its simplicity and low operating costs and the largest plants delivered by Linde have capacity up to 5,000 Nm^3/h and yield N_2 and Ar with 97–99.0000% purity. Vacuum swing absorption (VSA) is another non-cryogenic process that uses zeolites and produces oxygen with lower energy cost than PSA (as it operates at near-normal pressure and temperature) and it has been widely adopted by plants delivering at more than 20–30 t/day. Synthetic zeolites are thus a perfect example of a relatively obscure but very important material; by 2010, their global output rose to 2.8 Mt (USGS 2022a).

When BASF began commercialization of Haber's ammonia synthesis process, it adopted a process of hydrogen production that was newly patented by Wilhelm Wild in 1912. This steam-reforming *Wasserstoffkontaktverfahren*, using catalysis to shift CO to CO_2, remained the standard process for large-scale H_2 generation until the late 1940s (Smil 2001). This was a solution based on a resource constraint: German chemists knew that it would be best to reform pure methane, the compound with the highest H : C ratio, or naphtha, a mixture of light hydrocarbon liquids, but these feedstocks were not available in large quantities in the interwar Germany, one of the world's leading coal producers.

Not surprisingly, the first steam reforming operation using CH_4 was built in hydrocarbon-rich America in 1939, for an ammonia plant of Hercules Powder Company in California. After WW II, the US production of hydrogen became entirely based on the steam reforming of methane, while in Europe and Asia coal and liquid hydrocarbons continued to dominate for several more decades. By the century's end, about half of the world's hydrogen output was based on methane. Reaction of CH_4 and steam at temperatures between 700 and 1,100 °C and pressures of 0.3–2.5 MPa – $CH_4 + H_2O \rightarrow CO + 3H_2$ – is followed by shifting CO to CO_2 – $CO + H_2O \rightarrow CO_2 + H_2$ – and producing additional hydrogen, and the process is completed by removing CO_2 and unreacted CO. PSA, a standard choice since the1960s, yields hydrogen that is 99.9999% pure and Linde's plants with the largest capacities can produce more than 400,000 Nm^3 of the gas every hour (Linde 2019).

Liquid hydrocarbons (principally naphtha) are the feedstock for hydrogen production in crude oil refineries where the gas is needed for the catalytic conversion of heavier fractions to lighter fuels and also to comply with ever stricter environmental regulation and to desulfurize the refined products. Steam reforming of coal is still done in China and India, and only small fractions of the global hydrogen output come from highly energy-intensive electrolysis of water (needs $4.8\,kWh/m^3$ H_2) and from methanol cracking. Synthesis of ammonia remains the leading user of hydrogen, followed by refinery needs (above all in processing heavier oil and in desulfurization).

Consumers requiring a steady supply of large volumes of industrial gases are best served by pipelines that carry highly pressurized gas. Major producers maintain their own pipelines and maps show expected high densities of O_2, N_2, and H_2 lines along the Gulf of Mexico to serve the region's exceptional concentration of refineries and petrochemical plants, and in the most industrialized regions of the United States, Germany, and Japan. At the same time, there has been a significant expansion of decentralized gas production. Leonard Parker Pool of America's Air Products pioneered this option during the 1940s: on-site custom-designed cryogenic units are much less expensive than distribution of gases in cylinders from central plants (Air Products 2022). As with most modern industries geared toward mass-scale industrial consumption, there is now a very high degree of concentration. In 2020, only four companies accounted for more than 80% of the global industrial gas market. Linde, and old German company, now a multinational organization headquartered in Dublin but with principal executive offices in Surrey, is the global leader. Air Liquide (headquartered in Paris) claims the second place, Air Products (head offices in Allentown, PA) is third, and Japan's Tayio Nippon Sanso the fourth.

3.6   Fertilizers

As already described in the previous chapter, the real breakthrough in satisfying the rising demand for nitrogenous fertilizers came only in 1909 with Fritz Haber's demonstration of ammonia synthesis from its elements; this was followed by an uncommonly rapid progress of turning that lab-bench reaction into an industrial commercial synthesis. BASF, Germany's leading chemical company, had done so, under Carl Bosch's leadership, by 1913 and I have argued that the resulting impact on global food production justifies the labeling of Haber-Bosch synthesis as perhaps the most consequential technical innovation of the twentieth century (Smil 2001). But that impact remained limited for decades, first due to the diversion of ammonia for the production of explosives during WW I, and then to the Great Depression and WW II.

Post-1950 expansion was rapid, with the global ammonia synthesis rising from less than 6Mt in 1950 to about 120Mt in 1989 before a stagnation of the early 1990s caused by the saturation of nitrogen use in many affluent countries and by the collapse of the USSR, at that time ammonia's largest producer. By the year 2000, the synthesis surpassed 130 Mt and it was 159 Mt in 2010, and 147 Mt in 2020 (USGS 2022a). In that year, the world consumed about 257 Mt of sulfuric acid but because ammonia has a much lower molecular weight (17) than H_2SO_4 (98), NH_3 remains the world's most synthesized compound in terms of moles (more than three times as much as H_2SO_4).

About two-thirds (65–57%) of all synthesized NH_3 have been recently used as fertilizer, with the total global usage more than tripling since 1970, from 33 Mt N to about 113 Mt N in 2020. Because ammonia is a gas under ambient pressure, it can be applied to crops only by using special equipment (hollow steel knives), a practice that has been limited to North America. The compound has been traditionally converted into a variety of fertilizers (nitrate, sulfate), but urea (containing 45% N) has emerged as the leading choice, especially in rice-growing Asia, now the world's largest consumer of nitrogenous fertilizers; ammonium nitrate (35% N) comes second.

Worldwide pattern of nitrogen fertilizer use has changed during the past two generations. Consumption reached plateaus in many affluent countries (in the United States it was 11.5 Mt in 1996, 10.5 Mt N in 2000, 11.5 Mt N in 2010, and 11.62 Mt in 2020), and in some of them it had appreciably declined: in Japan from nearly 800,000 t N in 1979 to only about 369,000 t N in 2020 (FAO 2022). Populous modernizing countries, led by China, emerged as the largest users. China surpassed the United States to become the world's largest consumer of reactive nitrogen in 1979 and a decade later it also became its largest producer: by the year 2000, it produced 22.1 Mt N, in 2020 32 Mt N and it used nearly 26 Mt N.

Large-scale availability of inexpensive nitrogen fertilizer made it possible to realize the yield potential of new crop varieties. Hybrid corn was introduced first (during the 1930s in the US), high-yielding, short-stalked wheat, and rice cultivars were made available by the CIMMYT in Mexico and the IRRI in the Philippines during the 1960s (Smil 2000). Compared to traditional harvests, the best national yields of these three most important grain crops have risen to about 10 t/ha for the US corn (from 2 t/ha before WW II), 8–10 t/ha for European wheat (from about 2 t/ha during the 1930s), and 6 t/ha for East Asian rice (from around 2 t/ha). High-yielding US corn now receives, on the average, about 160 kg N/ha, European winter wheat more than 200 kg N/ha, and China's rice gets 260 kg N/ha, which means that in double-cropping regions, annual applications are about 500 kg N/ha. According to my calculations in the year 2020, at least 40% of nitrogen present in the world's food proteins came from fertilizers that originated in the Haber-Bosch synthesis of ammonia.

Efficient use of fertilizers depends not only on appropriate rates and modes of application but also on matching the crop needs for the three macronutrients (N, P, and K) and, if need be, on an adequate supply of micronutrients (above all boron, copper, iron, manganese, molybdenum, and zinc). Rising use of nitrogen had to be accompanied by rising use of the other two essential macronutrients and, as with nitrogen, this rise was temporarily interrupted after 1989, mainly due to the demise of the Soviet bloc economies (where the overuse of nutrients was common) and to higher efficiencies of fertilizer use in the Western agriculture.

In mass terms, phosphorus is the second most important fertilizing material used overwhelmingly in the form of beneficiated phosphate rock that is extracted mostly in surface mines. Mining of phosphates predates the synthesis of ammonia: global production reached 1 Mt in 1885, 10 Mt in 1928, and 100 Mt in 1974. In the year 2000, the output was 133 Mt and 10 years later it reached 181 Mt, and in 2020 it was 219 Mt with China (88 Mt), Morocco, and the Western Sahara (37 Mt) and the US (23.5 Mt, mined mostly in Florida and North Carolina) dominating the world output (USGS 2022a). Nearly all (>95%) of the US phosphate rock is used to produce phosphoric and superphosphoric acid, the two intermediate feedstocks for the production of solid and liquid phosphate fertilizer and also feed additives. Treatment with H_2SO_4 produces single superphosphate (containing 8–9% P), phosphoric acid (H_3PO_4) is used to make triple superphosphate (with about 20% P), and H_3PO_4 and NH_3 are used to make ammonium phosphates (up to 23% P).

Agricultural phosphate consumption is usually expressed in terms of P_2O_5 but I prefer to make the comparisons in terms of actual nutrient (multiply P_2O_5 by 0.436 to get P): worldwide agriculture consumed increasingly higher amounts of phosphates until 1989 when the annual rate peaked at 16.5 Mt, and after the growth resumed (it was only 14.3 Mt in 2000) the total reached 20.3 Mt P in 2010 and 19.5 Mt P in 2020 (FAO 2022). Potassium is applied to fields mostly as ground potash (KCl), obtained mostly by underground mining of sylvinite, a mixture of about a third of KCl and two-thirds of NaCl; Saskatchewan has its largest reserves of the rock and is the leading global producer and Canada and Russia are the largest exporters. Worldwide extraction (expressed in terms of K_2O equivalent) rose to nearly 45 Mt by 2020. About 85% of all KCl ends up as fertilizer, the rest is used by chemical industry: crop applications rose from less than 18 Mt K in 1975 to about 22.5 Mt K in 2010 and 37 Mt K in 2020 (FAO 2022). Global nutrient (N : P : K) ratio in 2020 was thus 1 : 0.17 : 0.32 compared to 1 : 0.25 : 0.40 in 1975.

Increase in crop yields was also made possible by the introduction of synthetic pesticides (mainly insecticides) and herbicides during the 1940s. Dichloro-diphenyl-trichloroethane (DDT) was the first important insecticide (Müller 1948), but its excessive application led to resistance against the chemical and its bioaccumulation in body lipids caused a great deal of environmental damage. After some 6,000 t of it were applied in the United States since the mid-1940s its use was banned in the United States in 1972, but many new, more acceptable, insecticides are now in use (Smil 2023). Herbicide applications began in 1945 with 2,4-D and while many new compounds followed none became as important glyphosate synthesized in 1971 by John E. Franz at Monsanto Company (Franz 1974).

This broad-spectrum herbicide (better known under its brand name Roundup™) inhibits the production of a key growth enzyme and destroys roots of weed plants, but it does not travel through soil to affect other plants and to contaminate water (Wesseler 2005). Rapid adoption of transgenic varieties of glyphosate-resistant crops introduced during the 1990s changed corn, soybean, and rapeseed farming in the United States and Canada, and later also in Brazil, Argentina, and China, but the diffusion of transgenic crop has met with, so far, persistent resistance in Europe as well as with obstacles in India (ISAAA 2021).

3.7 Materials in Electronics

If the classical designation of metal-specific eras, first proposed by Christian Thomsen in 1836, were to be extended to all materials, a strong case could be made to label the post-1954 world (after Texas Instruments released the first silicon transistor) – or perhaps the post-1971 world, after Intel released the first universal microprocessor – as the Silicon Age. This was an unexpected development and, fortunately, one based on a superabundant material, the most important mineral element in the Earth's crust ubiquitously available in silica (SiO_2). Elaborate exploitation of this material is nothing new: with about 70% of the mass SiO_2 (as quartz sand) dominates the raw materials used to make glass, with the rest evenly divided between Na_2CO_3 (soda ash) or K_2CO_3 and $CaCO_3$ or (in heavy leaded glass) with Pb.

In 1900, nobody could foresee that SiO_2, the commonest of all surface minerals, will acquire new importance as the quintessential material of a new electronic era. The element has been used for decades in metallurgy, above all in ferrosilicon alloys (containing 15–90% Si) for deoxidation in steelmaking. This market accounts for about 80% of the element's annual global production that surpassed 1 Mt during the late 1950s and 2 Mt by 1975; in the year 2000, the total reached 3.5 Mt, rose to 8 Mt by 2011, and then it stagnated and during the last pre-Covid year it was 8.4 Mt (USGS 2022a). China has been by far the largest producer of both ferrosilicon and silicon metal. The two main uses of silicon metal (polycrystalline silicon) have been in aluminum alloying and chemical syntheses, above all for the production of silicones, Si–O polymers with attached methyl or phenyl groups. Their commercial production began during the 1940s and that make excellent lubricants and resins as well as insulators and water repellents.

The raw material for producing silicon is abundant, but an energy-intensive high-temperature deoxidization with carbon – $SiO_2 + 2C \rightarrow Si + 2CO$ (using graphite electrodes in electric furnaces) – is required to yield the element that is 99% pure. But even 99% purity is quite unacceptable for solar and electronic industries, and hence the metallurgical-grade Si has to undergo elaborate and costly processing that makes it many orders of magnitude purer to meet the specifications for producing semiconductors, solar cells, and optical fibers (Föll 2000). Solar-grade polysilicon used to make photovoltaic cells has purity of 99.9999–99.999999% (6–8 N), while electronic-grade silicon used to make wafers (microchips) has purity levels of 9–11 N, that is up to 99.999999999% pure.

3.7.1   Electronic-Grade Silicon

The first step of converting metallurgical-grade to electronic-grade Si is the catalytic production of trichlorosilane ($Si + 3HCl \rightarrow SiHCl_3 + H_2$) developed by Siemens during the 1960s (McWhan 2012). This is followed by the trichlorosilane decomposition with hydrogen ($SiHCl_3 + H_2 \rightarrow H_2 + Si$) and concurrent addition of tiny amounts of impurities (dopants) whose presence will increase the element's conductivity. This is a demanding step given the high toxicity of common dopants (AsH_3 and PH_3), high combustibility of H_2 and $Si HCl_3$, and high corrosiveness of HCl vapors. After this exacting process, the extremely pure polycrystalline silicon must be transformed into a crystal that can be cut into thin wafers (Duffar 2010). A serendipitous discovery opened the way to this transformation when in 1916 Jan Czochralski, a Polish metallurgist working in Berlin, dipped his pen into a small crucible of molten tin and lifted a thin thread of metal that turned out to be a single crystal.

When Czochralski used a small seed crystal to pull a molten metal from the melt and combined the lifting with a slow rotation and gradual lowering of the temperature he could produce crystals of tin, lead, and zinc whose diameter was a few mm across and whose length was up to 1.5 cm (Czochralski 1918). For three decades, this remained just a useful way to study the growth of metallic crystals, but the invention of transistors made it the obvious candidate for pulling the crystals of semiconducting materials. Germanium, rather than silicon, was the first semiconducting material and in 1948 Gordon K. Teal and J.B. Little at Bell Labs deployed Czochralski's method to produce first pure Ge crystals. By 1950, shortly after the appearance of first Si-based transistors, Teal and Ernest Buehler could grow pure silicon crystals and by 1956 with their improved crystal-pulling and crystal-doping methods, their largest crystal was eight times as heavy and had twice the diameter (2.5 cm) of their first products (Buehler and Teal 1956).

And the growth still continues: diameter of 7.5 cm and crystal mass 12 kg were reached by 1973, 10-cm, 14-kg crystals followed soon afterward, and by the year

2001 the largest crystals were 30 cm across and had a mass of 200 kg (Hahn 2001; Zulehner 2003). In contrast to dark brown powdery amorphous Si, the crystals are black or black-gray and lustrous, with impurities being less than one part per billion. These perfect crystals are cut into layers of about 1 mm thick (and deviating less than 1 μm from an ideal flat surface) and the polished wafers are ready to undergo the final steps in the production of microprocessors. Cutting of the crystals has defined the successive wafer eras, starting with 20 mm-wafers in the early 1960s, 100-mm wafers by the mid-1970s, 200-mm by 1991, and the first 300-mm wafers in 2001 (Hattori 2015).

The 300-mm era was to be followed by crystals and wafers with diameters of 450 mm. Plans announced by the consortium of the largest semiconductor producers (including Intel, Samsung, and TSMC) in 2008 had soon unraveled, as did the initiative launched in New York in 2011 that hoped for the first deliveries by 2015–2016 and for volume production by 2018. In the early 2020s, we are still in 300-mm era, with some people asking if 450-mm wafers will ever come, while others see an eventual progress. In the past, semiconductor industry had steadily reduced its manufacturing costs per chip by increased integration, better production yield, higher throughput, and larger wafer diameters, but high costs (and technical difficulties) involved in stepping up from 300 to 450 mm do not appear to favor the step.

Williams (2003) traced the global silicon production chain for the year 1998, and Takiguchi and Morita (2011) reconstructed worldwide material flows of silicon used in electronics during the years 1997 and 2009. In 2009, purification of metallurgical-grade Si by the Siemens process produced about 23,000 t of electronic-grade Si and that mass was converted to 16,100 t of single-crystal Si that yielded about 7,500 t of wafers sent for microchip fabrication. By 2020, the mass of electronic-grade polycrystalline Si increased to about 32,000 t and the aggregate weight of cutting-ready crystals surpassed 10,000 t. Crowding of more transistors on a single wafer, the process subject to much-publicized Moore's law, has been the key driver of advances in computing.

In 1965, when the number of transistors on a microchip doubled to 64 from 32 in 1964, Gordon Moore predicted that this rate of doubling will continue, and forecast a chip with 65,000 transistors in 1975 (Moore 1965). In 1975, he relaxed the pace of his already famous Moore's law to a doubling every two years (Moore 1975). And that tempo has prevailed ever since a chip-making progressed with Intel's pioneering designs and with advances introduced by its competitors selling chips of ever-increasing complexity. The world's first universal microprocessor, Intel 4004 released in November 1971, had 2,250 metal-oxide semiconductor transistors and its operating speed was equivalent to that of the room-size ENIAC computer built in 1945 (Intel 2022).

In 1979, the company's 8088 microprocessor had 29,000 transistors, soon the fabrication of microprocessors moved from large-scale integration (up to 100,000 transistors on a microchip) to very large-scale integration (up to 10 million transistors) and then to ultra-large-scale integration (up to a billion transistors) by 1990. By 2000 Intel's Pentium 4 had 42 million transistors, by 2012 the count climbed to 5 billion (Xeon Phi Coprocessor), 10 billion mark was reached in 2015 (SPARC M7); in 2021 Apple M1 Max reached 57 billion. Mass deployment of these increasingly powerful microprocessors in conjunction with increasingly capacious memory devices has transformed every sector of modern economies, thanks to unprecedented capacities for communication, control, storage, and retrieval of information.

Not surprisingly, the entire process of turning quartz into the substrate for microprocessors is one of the best examples of steeply adding value along a production sequence. Detailed analysis by Williams (2003) showed that by the end of the twentieth century, pure quartz wholesaled at less than $0.02/kg, metallurgical-quality Si was available in bulk at $1.10/kg, trichlorosilane cost about $3/kg, polycrystalline Si was between $50 and 100/kg, monocrystalline Si was at least $500/kg, polished Si wafers were at least $1,500, and epitaxial slices as much as $14,000/kg. By 2020, electronic-grade polycrystalline silicon sold for about $50/kg and the global market for the highly purified metal was worth about $3.8 billion (Fact MR 2022). Wafer shipments for semiconductor applications rose from just over $3.5\,mm^2$ in the year 2000 to $5.8\,mm^2$ in 2010 and $8\,mm^2$ in 2020 (SEMI 2022), and global accounts show the value of these shipments rising from less than $4 billion in 1977 (the first years for which worldwide data are available) to $50 billion by 1990, surpassing $200 billion in the year 2000, and reaching $440 billion in 2020 (SIA 2021).

During the first decade of the twenty-first century, electronics ceased to be the major consumer of high-grade silicon as most of that material now ends up in photovoltaic (PV) cells. Using silicon to make PV cells is an application nearly as old as the making of transistors. Bell Labs had the first prototype in 1954 and in 1958 *Vanguard I* was the first satellite receiving PV electricity, just $0.1\,W$ from about $100\,cm^2$ of cells, more than enough to power a 5-mW transmitter. By 1962, Telstar, the first commercial telecommunications satellite, had PV cells delivering $14\,W$ and in 1964 weather-monitoring *Nimbus* drew $470\,W$ (Smil 2006a). Thousands of launches have followed, with the numbers of new satellites increasing as their size and cost keep declining. According to the United Nations Office for Outer Space Affairs, more than 1,100 satellites were launched by the end of April 2021, with 7,389 remaining in orbit, but by the end of January 2022 the total rose to 8,261 (UNOOSA 2022).

But PV cells found a far larger market on the Earth where they now supply a small, but steadily increasing share of renewably generated electricity. Global share of PV generation rose from negligible 0.007% in the year 2000 to 3.6% by 2021. This ascendance was helped not only by various government subsidies but even more so by steadily declining cost of modules (since 2010 by 18–20% a year) and by their increasing efficiency. These land-based PV applications began with less-expensive (but still very costly) amorphous silicon cells in 1976; by 2012, their best conversion efficiencies of thin films rose to 20% in laboratory settings; and by 2022 the best amorphous silicon cells were about 14% efficient and crystalline cells reached the top performance around 27%, with the best commercially available models rated at 15–19% (NREL 2022a; Svarc 2022).

For decades, PV cells were made with off-grade polycrystalline material that was not good enough for electronic applications, but as the heavily subsidized market for PV installation rose from less than 100 MW/year in 1995 to more than 10 GW/year in 2009 and surpassed 100 GW a year in 2017, it was necessary to divert increasing amounts of purified polycrystalline metal into solar cell industry: poly-silicon is melted in order to grow monocrystalline silicon ingots that are sliced into silicon wafers processed into solar cells, connected, sandwiched between glass and plastic sheets, and framed to make rigid PV modules, mounted on steel supports and connected to the grid via inverters.

In 1997, the industry used only 800 t of such solar-grade silicon, by 2009 it required 69,100 t, three times as much as consumed by electronics, to produce about 44,500 t of solar cells, mostly by the casting of polycrystalline metal (Takiguchi and Morita 2011). During the second decade of the twenty-first century, China took over the global PV market, and by 2022, 97% of the world's production of silicon wafers used in electricity generation were made in China. Incredibly, the United States, the world's second PV market, "has no active ingot, wafer, or silicon cell manufacturing capacity, and polysilicon production capacity is not being used for solar applications" (US DOE 2022). By 2020, global polysilicon output rose to 520,500 t – with PV industry using 93.4% (486,000 t) and semiconductor industry processing 34,600 t – and in 2021, the production capacity rose to 627,000 t a year (IEA 2021c).

4

How the Materials Flow

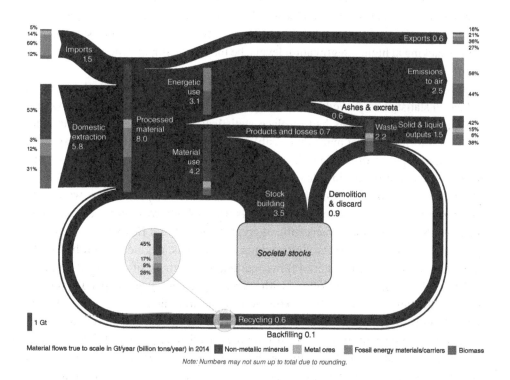

Material flows true to scale in Gt/year (billion tons/year) in 2014 ■ Non-metallic minerals ■ Metal ores ■ Fossil energy materials/carriers ■ Biomass

Note: Numbers may not sum up to total due to rounding.

Materials move through societies and environments in ways ranging from simple, fairly linear, and easy-to-trace flows to complex and difficult-to-follow disappear-ances; and while some materials provide ephemeral services (plastic discarded immediately after purchasing an item in a blister or clamshell packaging), others are

Materials and Dematerialization: Making the Modern World, Second Edition. Vaclav Smil.
© 2023 John Wiley & Sons Ltd. Published 2023 by John Wiley & Sons Ltd.

long-lived with life spans measured not only in decades but even in centuries and millennia; and while some metals can be reused indefinitely (albeit with some mass losses) recycling of most materials entails often considerable loss of quality and functionality. There is nothing new about keeping detailed records of material flows required for particular operations, factories, and companies: without them artisans could not fill their orders, and engineers and managers could not control production costs or strive to improve productivity and profitability of their enterprises.

But after setting aside the exceptional case of the world's earliest modernizer (in the UK data on basic material flows were both fairly comprehensive and fairly accurate already by 1800), reconstruction of even a few basic material flows for the early decades of the nineteenth century entails a great deal of approximations and assumptions for individual industries and faces even more difficulties when trying to assemble aggregates at national level. Pre-1850 data on production, imports, and exports of individual raw materials are commonly questionable, unsystematic, or simply lacking at higher aggregation levels, be they those for specific industrial uses or for aggregate demand on national levels.

Between 1850 and 1870, availability and reliability of European and American industrial statistics improved considerably, and before the century's end, many rapidly modernizing countries had in place statistical services collecting a widening range of information that makes it possible to reconstruct material flows of specific sectors (steel, paper, or textile industry) and input and output aggregates for principal material requirements in at least half a dozen major European economies, United States, Canada, and Japan, and those nations also had good data on crop, wood, and fish harvests and on extraction of fossil fuels. But nobody was interested in adding up all of these items and constructing aggregates of material flows required for the growth of national economies. Statisticians, economists, engineers, and historians paid attention to sectoral material flows, particularly to those of dominant inputs.

This makes it possible to follow, with increasing accuracy and expanding coverage, the evolution of fossil fuel extraction or the mining of many minerals and harvesting of wood in a number of industrializing economies for more than a century, and for some countries we can also reconstruct detailed material flows in some industries. Beginnings of a new approach to the studies of material requirements in modern societies had their origins in concerns about increasing burdens of environmental pollution and the critique of economic thinking that tended to ignore such matters. Ayres and Kneese (1969, 283–284), describing the reality in clear physical terms, noted that such omission "may result in viewing the production and consumption processes in a manner that is somewhat at variance with the fundamental law of the conservation of mass," and pointed out the obvious consequences for the environment, namely that in the absence of trade and net stock accumulation "the amount of residuals inserted into the natural environment must be approximately equal to the weight of basic fuels, food, and raw materials entering the processing and production system, plus oxygen from the atmosphere."

But it took nearly two decades before this admonition was transformed into first fairly comprehensive studies of material requirements on national level as several research teams began to reconstruct direct material inputs (DMIs) as well as outflows, and total material requirements (TMRs) of the world's leading affluent economies only during the late 1990s. Fischer-Kowalski et al. (2011) provide a concise history of this material flow accounting as well as a survey of methodological foundations of these studies (system boundaries, compartments, stocks), major material flow indicators, and data reliability and uncertainties across various data sets. I will take a closer look at these material flow accounts, first in general analytical terms, and then I will review the results of global and national studies tracing the inputs, as well as outputs, in the world's largest economy, in European countries and (inevitably, given the country's recent economic rise) in China.

There are other approaches to the investigation of material flows: one that attempts to trace life cycles of individual commodities on a national, regional, or global level; another one that looks at the energy costs of commodities and products; and one that traces the environmental impacts of their production, use, and abandonment (or recycling). Life-cycle assessments (or analyses, in either case the acronym is LCA) follow materials from extraction through incorporation into products to final disposal (cradle-to-grave studies), and by the early 2020s, they have been performed at different scales for many elements and compounds (from hydrogen to titanium, from alcohols to zeolites) and for manufactured items ranging from aluminum cans to steel truck wheels. Many of these analyses are available in open literature, most of them are accessible only after paying licenses to LCA databases (openLCA Nexus 2022).

Not surprisingly, flows of leading metals in general, and iron and steel in particular, have received especially close attention. Because of their depth and complexity some of these analyses are far more revealing, and fundamentally far more informative, than the standard tracing of annual production, trade and consumption flows. Similarly, a comprehensive assessment of environmental impacts associated with the extraction, processing, use and disposal of a specific commodity or manufactured product may be much more consequential than a simple tracing of its flows through a national or global economy. While there are many concerns specific to a particular material or a group of commodities there is now an important and truly universal consideration – carbon burden of materials or, more accurately, greenhouse gas emissions associated with their consumption. These emissions are dominated by releases of CO_2 with CH_4 and N_2O as lesser contributors and with other greenhouse gases making noticeable difference only in the case of a few materials.

4.1   Material Flow Accounts

Inquiries into total material requirements can follow a hierarchy of scales – starting with accounts for a specific production process; progressing to those for an entire industrial sector; and culminating in national, continental, and global totals – and even a brief reflection about the challenges inherent in preparing these summaries and about the utility of such aggregate measures will reveal their dual qualities: they are useful and revealing, particularly in order to quantify the extent of human acquisitiveness and its environmental consequences, but also inadequate and misleading, particularly due to their non-discriminative preoccupation with quantity and inability to gauge many fundamental qualitative differences.

Four kinds of these accounts appear to be most useful: detailed analyses on national scale that could be compared within a group of economically similar countries or with states at different stages of modernization; longer-term historical perspectives that illustrate impacts of technical innovation, efficiency improvements, managerial advances, and shifting market choices; combinations of these two approaches that make it possible to compare the differences in long-term national trajectories; and global accounts of total material requirements as well as the resulting outputs into the environment that reveal the extent of human extractive and productive activities on a civilizational scale and that make us aware of the burdens that these processes put on the biosphere's finite (and still poorly understood) capacity to deal with the changes to the atmosphere, waters, and ecosystems.

4.1.1   Accounts and Boundaries

I would argue that setting of the system boundaries, rather than coping with the inherent complexity and heterogeneity of these accounts, is the greatest challenge in constructing the national and global summaries of total material flows. This challenge is reflected by different choices made by the pioneering studies of national material balances – they began during the late 1990s, and between 1997 and 2001 they included comparisons restricted to short-term accounts for a handful of affluent economies (Adriaanse et al. 1997; Matthews et al. 2000; Bringezu and Schütz 2001), long-term assessments for the United States (Matos and Wagner 1998) – as well as by the first deliberate annual effort by a statistical agency to provide data for DMIs as well as outflows, and total material requirements (TMRs, sums of DMI and hidden flows, extractions that are not used in further processing) of European economies (EC 2001).

There are problems with all of these indicators. Limiting the account to DMI would greatly underestimate the overall resource demand in all modern economies engaged in intensive international trade, and particularly in such major powers as the United States, Germany, or Japan that rely on imports for large shares of many materials. Correcting this by the inclusion of net imports of all raw materials is only a partial (and increasingly deficient) solution because increasing shares of many metals and other minerals are not imported in form of ores or concentrates or bulk shipments but embodied in finished products. Identifying specific material contents of these products (even their limited inventory would run to many hundreds of individual machines, tools, components and consumer items) presents a major challenge – but the adjustment should not end there as many items imported from a particular country contain components made of materials in a number of other countries that, in turn, imported parts or raw materials from yet another country or, more likely, a set of countries.

A popular consumer item offers a perfect illustration of these, now truly global, accounting challenges. The tear-down of Apple's iPhone 6 shows that its major components are made by nearly 20 companies, most of them headquartered in the United States, Japan, South Korea, China, Germany, Switzerland, and Taiwan (Costello 2021). For example, US-headquartered Corning and Qualcomm make, respectively, the glass screen and microchips for cellular networking, Japan's Sony and Toshiba make camera and flash memory, and South Korea's Samsung and LG make batteries and LCD screens – but all of them have their networks of their own subcontractors and maintain manufacturing facilities in many (some in scores of) countries around in Asia, Europe, and Americas. The label "Made in" is thus entirely inappropriate for products of this kind containing parts and materials that originated in dozens of countries, and only the final assembly can be unambiguously located: with iPhones it primarily Foxconn's (Hon Hai Precision Industry's) giant manufactory in Shenzhen in China's Guangdong province.

I have already introduced the major material flow data series published by the Eurostat and by the USGS. Starting in 2004, Eurostat expanded the coverage to all of 27 member states as well as to Norway, Switzerland, Montenegro, Croatia, Macedonia, and Turkey, compiled according to detailed methodological guides for economy-wide material flow accounts (EW-MFA, or simply MFA). USGS has been releasing periodic summaries of the country's material flows (Matos and Wagner 1998; Matos 2009; Kelly and Matos 2016) and its annual *Minerals Yearbook* contains the latest production, trade, and consumption data for more than 80 minerals and materials both for the United States and on the global level.

The most comprehensive material flows database is offered by the UN's Environment Programme: it reports extraction and direct trade of raw materials, indirect trade flows and intensities derived from these material measures for more than 200 countries, and in 2022, it covered the period from 1970 to 2019 (UNEP 2022a). An online portal materialflows.net, hosted by Vienna University of

Economics and Business, is based on the UN's database (Material Flows Net 2018). Papers on global material flows include, chronologically, Schandl and Eisenmenger (2006), Giljum et al. (2008), Krausmann et al. (2009), Steinberger et al. (2010), Krausmann et al. (2017), and Schandl et al. (2016, 2017).

Notable long-term national MFA studies include Schandl and Schulz (2002) for the United Kingdom between 1850 and 1997, Krausmann et al. (2011) for Japan between 1887 and 2005, Kovanda and Hak (2011) for Czechoslovakia between 1855 and 2007, and Infante-Amate et al. (2015) for Spain between 1860 and 2010. Shorter-term national MFAs are available for Australia (Schandl et al. 2008; Wood et al. 2009), China (Xu and Zhang 2007), and India (Singh et al. 2012). I should also note that the global on-line database and most of the global and national studies of material flows have been, overwhelmingly, products of a small number of individuals from Austria and Germany, most of them associated with the Institute of Social Ecology in Vienna.

In conclusion, I question the utility of constructing those all-encompassing national or global flow accounts – those aggregates of oxygen, biomass, all fuels, all minerals, and all hidden and dissipative mobilizations where critical differences in qualities and no less important differences in environmental consequences are submerged in indiscriminating totals dominated by oxygen and hidden material flows – because I am not sure what other revealing conclusions to derive from these summations of disparate input and output categories besides the obvious confirmations of substantial differences in national aggregates and in the rates of long-term growth. Of course, the maximalist aggregates of the all-encompassing variety also have an undoubted heuristic and curiosity value and they do convey the truly massive scale of global mobilization of raw materials.

4.1.2 Global Material Flows

Half a dozen studies of global material extraction at the beginning of the twenty-first century that include all harvested biomass, all fossil fuels, ores and industrial minerals, and all bulk construction materials (but exclude waste flows, water, or oxygen) cluster fairly tightly around 50 Gt/year. This is hardly surprising given the fact that these studies derive the flows from the same sets of data: from the Food and Agriculture Organization's agricultural and forestry statistics for biomass, from USGS complemented by some national statistics for ores and industrial minerals, and from the International Energy Agency and the UN databases for fossil energy. The lowest value for the year 2000 was 48.8 Gt (Schandl and Eisenmenger 2006), the highest one was 58.7 Gt (Krausmann et al. 2008); a year later Krausmann et al. (2009), in their summation of flows for 1900, 1925, 1950, 1975, and 2005, put the last total at 59.5 Gt. Other published values for the year 2000 ranged from 48.5 Gt (Steinberger et al. 2010) to about 53 Gt (Behrens et al. 2007).

Leaving the high outlier aside, the average of five estimates for the year 2000 was almost exactly 50 Gt/year, with roughly 18 Gt coming from biomass, 10 Gt from fossil fuels, about 5 Gt from ores and other minerals, and more than 17 Gt from nonmetallic minerals (mostly bulk construction materials). *Sensu stricto* global material flows in the year 2000 (direct inputs of raw materials excluding all food, feed, energy, and hidden flows) were on the order of 25 Gt: about 22 Gt from minerals and 3 Gt from wood and other biomaterials. An estimate conforming to the USGS list should be increased by about 500 Mt in order to include all non-renewable organics destined for non-energy uses (the total dominated by naphtha and methane feedstocks and bitumens for paving).

Given the uncertainties in estimating the mass of bulk construction minerals (above all for the extraction of sand and gravel) that account for at least two-thirds of the all material flows, the mass of 0.5 Gt is well within the minimal range of estimation error (±2 Gt), and the total of roughly 25 Gt thus remains my preferred aggregate of directly used global materials in the year 2000. That total prorated to just over 4 t of materials per person (global population was 6.08 billion in the year 2000), with at least 2.5 t (and perhaps as much as 3 t) accounted for bulk construction materials and only about 0.8 attributed to all metals and nonmetallic minerals. These rates compare to nearly 1 t of food and feed crops (fresh weight), close to 0.5 t of wood (excluding fuelwood), and about 1.7 t of fossil fuels (roughly 0.8 t of coal, 0.6 t of crude oil, and 0.3 t of natural gas) extracted for every inhabitant of the world in the year 2000.

Raw material inputs into the global economy, and hence the eventual worldwide output of finished products have soared in absolute terms during the first two decades of the twenty-first century, mostly thanks to China's continued economic rise, with substantial new demand coming also from India, Indonesia, and Brazil, and the global material use (again *sensu stricto*) for 2019 (before the demand was depressed by the pandemic) rose to about 45 Gt or about 5.7 t/capita: for comparison, by 2019 per capita combustion of fossil fuels fell a bit to about 1.65 t/capita. Clearly, in the new century, the global economy has not shown any signs of absolute dematerialization.

While Asia accounted for most of the consumption increase during the first two decades of the twenty-first century, growth of raw material inputs has been more subdued in affluent societies, partly because they have already put in place material-intensive infrastructures, partly because of the ongoing outsourcing of material-intensive (and often polluting) industries to foreign low-cost producers. But even in EU, this did not mean translate into any notable declines in material demand: in 10 years between 2012 and 2021, aggregate demand for mineral and organic materials (leaving all fossil fuels aside) rose by about 12% (Eurostat 2022a).

America's most massive material use, that of construction sand and gravel, peaked in 2006 at 1.32 Gt; by 2010, it was barely above 800 Mt; and it reached 1 Gt once again only in 2021. All other material-use categories had also experienced (less pronounced) aggregate declines as a result of the Great Recession of 2008. By 2019, gradual recovery brought the value of industrial minerals (dominated by sand, gravel, and crushed stone) some 30% above the previous (2007) peak, but apparent consumption of all major metals remained below the highs reached before the 2008 recession, with steel about 20%, aluminum about 17%, and copper about a third lower.

A naive interpretation would see this as a strong proof of advancing dematerialization of the world's largest economy – but these apparent domestic consumption rates (production plus net imports of metals or scrap) exclude significant amounts of materials embedded in imports of finished products. Cars are the most obvious major example. In 2019, the United States imported 7.8 million passenger vehicles, and average contents of 900 kg of steel and 200 kg of aluminum per vehicle translate into about 7 Mt of steel and 1.5 Mt of aluminum embedded in the machines made in Japan, South Korea, Europe, and Canada. Those totals alone equaled nearly 9% and 34% of apparent domestic consumption of the respective metals! I will take a closer look at these misleading dematerialization claims in the fifth chapter.

All of this is interesting, and if we had reliable historical series, we could also trace the rise of global material consumption for each of these major categories: all pre-1950 global totals are nothing but questionable estimates, and even the recent aggregates depend critically on what gets included. For example, Krausmann et al. (2009) put the worldwide biomass extraction (crops, their residues, roughages, and wood) at 19.061 Gt in 2005, while in my detailed account of phytomass harvest (Smil 2013), I showed that in the year 2000 the total for woody phytomass alone could be anywhere between 2 and 13.4 Gt depending on the boundaries chosen for the analysis. Even greater absolute differences arise when accounting for hidden flows because of different assumptions regarding typical thicknesses of overburden that needs to be removed before reaching metal-bearing ores and typical (and generally declining) metal content of those ores. Similarly, all estimates of accumulated stocks have considerable margins of uncertainty: Krausmann et al. (2017) estimated that their global aggregate increased 23-fold during the twentieth century to 792 Gt in 2010.

Consequently, there can be no single accurate totals as the search for global totals will be always determined by assumptions, and if everybody agrees on common boundaries, the basic results will be largely predictable. Physical realities dictate that the mass of sand and gravel used to emplace and maintain modern concrete-based infrastructures must be substantially greater than the mass of metallic ores; or that the mass of iron, a metal of outstanding properties produced from abundant ores with a moderate energy intensity must be orders of magnitude higher than the mass of titanium, an even more remarkable metal but one derived from relatively rare ores with a great expense of energy. At the same time, it must be kept in mind that data for inexpensive, readily available bulk construction materials (particularly for sand and gravel) that are usually sold not far from their points of extraction are generally much less reliable than the statistics for metal ores and industrial nonmetallic minerals that are traded globally.

Economic realities dictate that the diffusion of successive changes of human organization – urbanization, industrialization, post-industrial societies characterized by mass consumption – promote, widen, and often also accelerate, use of materials. Although the resulting outcomes had to be expected some of the contrasts still amaze: for example, the world now consumes in one year more than it did during the first post-WWII decade, and (even more incredibly) more cement than it consumed during the first half of the twentieth century. At the same time, these admittedly fascinating but highly aggregate global-scale data series that amalgamate quantities and ignore qualities offer little guidance for future decision-making (beyond the obvious point that recent high growth rates cannot continue for many decades).

Useful insights can come from two kinds of finer focus: through closer examinations of material flows on national level and by putting more restrictive analytical boundaries on the set of examined materials and tracing the flows of individual commodities with some clear goals in mind. That can be done by detailing their uses, dispersal, and persistence in a society, and by attempting life-cycle analyses of those materials that circulate on human time-scale, that is by quantifying their direct and indirect requirements for energy or by identifying and assessing environmental impacts of their production and use.

4.2   US and European Material Flows

The world's largest economy (nearly $21 trillion in current monies in 2020 or still about 40% ahead of China) is obviously the leading candidate for a closer examination of national material flows. Fortunately, the task of assembling the primary data is made easy, thanks to a ready availability and high accuracy of the country's production, trade, and consumption statistics – but even those simple additive chores are unnecessary thanks to the USGS researchers who have collated the primary information from more than 100 individual material streams into clearly defined categories and, moreover, have done so in fully comparable manner, originally for the period from 1900 through 1995 (Matos and Wagner 1998) with subsequent extension to 2006 (Matos 2009) and with the latest disaggregated statistics available in the regularly updated historical series (Kelly and Matos 2016).

4.2.1 Materials in the US Economy

As already explained, the USGS series contains data for metals and minerals (sub-divided into primary and recycled metals, and industrial and construction materials) and non-renewable organics (asphalt, waxes, oils and lubricants, and fossil fuels used as feedstocks) that were used to make products in the United States: it leaves out food and the bulk of the fossil fuels that are used as source of energy (as large shares of these materials are consumed directly), but it includes all phytomass that is used in manufacturing (ranging from lumber to new and recycled paper, and from cotton to wool). A centennial perspective traces material demands from the peak of the first phase of industrial revolution (energized by coal and steam engines and characterized by growing but still limited mass consumption) through the rise and maturation of the second phase (increasingly energized by refined oil products and natural gas, with steam turbogenerators, internal combustion engines and gas turbines as the dominant prime movers and resulting in unprecedented extent of mass consumption) to the post-1950 rise of service economy.

This perspective makes it possible to make several important generalizations that are undoubtedly valid for other major affluent economies: rise of absolute consumption in every material category; substantial decline in relative importance of biomaterials; dominance of the total mass by bulk construction materials; expanded demand for metals; rising importance of recycling; emergence and a huge increase in consumption of non-renewable organics; and material-specific shifts (fluctuating stagnation, steady rise or long-term decline) in per capita consumption. During the twentieth century, natural growth potentiated by immigration had increased the US population nearly fourfold (3.7), and the country's GDP (expressed in constant monies) was 26.5 times higher in 2000 than in 1900: not surprisingly, combination of these two key factors had driven absolute consumption increases in all material categories, with the multipliers ranging from 1.7 for materials produced by agriculture to more than 90 for non-renewable organics (and 8 for primary metals, 34 for industrial minerals, and 47 for construction materials).

Importance of renewable materials (wood, fibers, leather) fell from about 46% of the total mass (when bulk construction materials are included) or 74% (with stone, sand, and gravel excluded) in 1900 to just 5% (or 22%) for analogical rates in the year 2000, an expected trend given the rising reliance on light metals and plastics. Aggregate wood demand rose less than 1.4-fold during the twentieth century, but consumption of primary paper and paperboard had multiplied about 19 times, and it has been supplemented by rising quantities of recycled paper: when the data collection in the latter category began in 1960, recycled paper accounted for about 24% of all paper and paperboard use, but by the year 2000, its share was up to 46% even as large quantities of waste paper were exported (in 2000 this amounted to about 22% of all domestic collections) primarily to China (FAO 2022).

The fact that bulk minerals used in construction (crushed stone, sand and gravel) have increasingly dominated America's annual flows during the twentieth century – in 1900, they accounted for 38% all materials; by 2006, their share reached 77%; by 2015, it fell slightly to just above 70% – is not surprising given the enormous expansion of material-intensive transportation infrastructures after WWII. Construction of the Interstate system began in 1956 and it also required building of many new bridges while the introduction of commercial jetliners led to a rapid expansion of airports. Rising demands for bulk construction materials also came from the building of new container ports, stream regulation (above in the Mississippi basin), electricity generation (hydroelectric dams, nuclear power plants), new factories, commercial real estate (warehouses, shopping centers), and housing. In the next section, I will show how all of these infrastructural programs have been recently repeated, on even larger scales, in China.

Mass of construction materials used in the United States rose about sevenfold between 1900 and 1940 and then had doubled between 1945 and 1951, doubled again by 1959 to 1.1 Gt, but the next doubling, to 2.26 Gt, took until 1997. Use of these materials shows notable declines during economic recessions: during the Great Depression the peak-to-bottom decline between 1929 and 1933 was 50%, between 1979 and 1982 the total demand fell by a third, and the peak of more than 2.8 Gt in 2007 was followed by the greatest retreat on record as the US consumption of sand and gravel fell by 44% between 2007 and 2010 and demand for crushed stone declined by 29% during the same period (Matos 2009).

End-use data indicate that the largest identifiable category of sand and gravel consumption (nearly half of the total of about 1 Gt in 2020) is to produce concrete, followed by road base and coverings (about a fifth of the total), construction fill (about 13%), and as aggregates added to asphaltic and bituminous mixtures with minor shares used in golf course maintenance, plaster sands, roofing granules, and snow and ice control. Differently sized aggregates used in the production of concrete are also the leading final uses for crushed stone, and railroad ballast is another indispensable application. With the ballast minimum depth of 15 cm and up to 50 cm for high-speed lines and the overall width of roughly 4.5 m, this amounts commonly to more than $1,000 \, m^3/km$ or (with density of $2.6 \, t/m^3$) to around $3,000 \, t$ of crushed stone per kilometer. Dimension stone (rough or dressed) used in construction (for facing, curbing, flagging, and monuments) is a smaller category and some four-fifths of the quarried mass actually never leaves the sites (constituting a rare instance of a hidden flow quantified by the USGS data).

Annual use of industrial sand is only about 10% as large as the consumption of construction sand and gravel, but it is qualitatively very important. Glassmaking and foundry sand were dominant before the large-scale adoption of hydraulic fracturing in oil and gas production. Nearly three-quarters of the annual use of industrial sand are now used as ingredients of fracking liquids as well as for packing and cementing wells. Glassmaking is now distant second, and smaller but functionally irreplaceable uses include abrasives used in blasting and sanding, sands for water filtration, and for creating artificial beaches and sporting areas. Other materials aggregated by the USGS into the heterogeneous group of industrial minerals include elements (carbon in the form of graphite and diamonds, boron, bromine, hafnium, helium, lithium, nitrogen, sulfur, strontium, zirconium) as well as both abundant (phosphate, potash, salt) and relatively rare (gemstones, industrial garnet, quartz crystal) compounds.

In total mass terms, this is the second largest group of materials in the US economy: its mass equals only about 12% of all construction materials, but it has been roughly 2.5 times the mass of all (primary and recycled) metals. Its total rose only about 10 Mt in 1900 to 150 Mt in 1950; it surpassed the 300 Mt mark in 1972; and then, after two decades of stagnation, it rose to 370 Mt in 2000; and by 2015, it was about 400 Mt. Its largest constituents include salt (about 40 Mt in 2020), phosphate rock (about 30 Mt), nitrogen (about 14 Mt), and sulfur (about 11 Mt). America's salt consumption is remarkably high (about 15% of the world total in 2020); the two dominant uses (each about 40% of the total in 2020) are for road deicing and in chemical industry (for production of alkaline compounds and chlorine). Small shares go into food production, animal feed, and water treatment (in water-softening to remove mineral ions).

Rising consumption of metals, and the changing shares of individual elements and alloys in this broad material category have been other expected developments. Consumption of 10.3 Mt of primary metals in 1900 doubled in just six years and then it doubled again to 42 Mt in 1929 before it was cut drastically by the Great Depression to just 11.3 Mt in 1932. The pre-depression record was surpassed in 1940 and the annual rate remained around 59 Mt until 1944; and after years of decline and fluctuations, it moved above 60 Mt only by 1963. By that time recycled metals provided almost a third of the total supply of roughly 87 Mt, and by the year 2000, with the total metal consumption at nearly 144 Mt, they supplied 44%.

But while metal recycling has shown a generally upward trend (albeit with some notable recession-induced downturns), domestic consumption of primary metals reached a fluctuating plateau during the late 1960s and the early 1970s followed by a general decline for the next 30 years ascribable to the first wave of deindustrialization of America, that is to the country's loss of global dominance in iron and steel-making and declining output of color metals. Crude steel production peaked in 1978 at 137 Mt; by 1990, it was 35% lower (89.7 Mt) and subsequently it continued to stagnate, falling to 80.5 Mt by 2010 and ending the second decade of the twenty-first century at 87.8 Mt in 2019 (USGS 2022a). Peak years of domestic output were 1969 for zinc, 1972 for copper and nickel, and 1973 for lead, and only the copper production was eventually surpassed for eight years starting in 1992, before it fell back again.

The second most important metal in mass terms is aluminum whose domestic consumption rose from less than 900,000 t in 1950 to more than 7 Mt by the year 2000, thanks to aircraft production and to widespread adoption of aluminum and its alloys in automotive industry and in many consumer products (from cookware to computers). Afterward, the domestic use fell back to about 3.5 Mt by 2010 and it was nearly 5 Mt in 2019 before Covid-induced decline – but all recent data are made of a complex combination of production methods and trade. In 2019, output of the primary metal was only about 1.1 Mt but secondary production was 3.4 Mt, and imports were about 5.9 Mt and exports were 3 Mt), leaving year-end stocks of about 1.6 Mt.

Copper has seen similarly large shifts, with 2010 domestic output nearly halved compared to the late 1990s and with apparent consumption of just 1.74 Mt compared to 3.1 Mt in 1999 – and the rate was unchanged (1.8 Mt) by 2020. The only other metal whose recent annual US consumption has surpassed 1 Mt is lead (1.65 Mt in 2019). Consumption of manganese, whose dominant use is to alloy steel, peaked at 1.35 Mt in 1974, but as the American production of specialty steel declined so did the consumption of the metal: by 2019, it fell to less than 800,000 Mt, all of it imported (its small domestic production stopped in 1990). In contrast, molybdenum consumption, whose annual rate had fluctuated mostly between 15,000 and 25,000 t for decades prior to the year 2000, has risen to record levels since 2005 (24,000–34,000 in 2017–2019), the only metal to have done so. The metal is another important ingredient to steel alloys (used for hardening and heat resistance; catalysts are another important Mo market) and the United States is its third largest producer (after China and Chile) and a significant net exporter of the metal.

I should reiterate that actual domestic US consumption of virtually all metals is, often significantly, higher than shown by the USGS balances because substantial amounts of various metals are reaching the country not only as imported raw materials (all of those are included in the totals of domestic consumption) but also embedded in products, and that presence is excluded from nationwide aggregates of apparent domestic consumption. Major components of these unaccounted flows include not only such leading metals as steel and aluminum in cars, airplanes, machinery, and appliances but also such toxic heavy metals as lead in automotive lead-acid ($PbSO_4$–H_2SO_4) batteries and cadmium in rechargeable Ni–Cd batteries.

In 1900, total consumption of nonrenewable organics (mostly paving materials and lubricating oils) was less than 2 Mt but subsequent extension of paved highways, mass ownership of cars, rise of the trucking industry, and, above all, rapid expansion of crude oil- and natural gas-based synthetic materials made this the fastest growing material category in the United States: by 1950, the flow surpassed 30 Mt, and by 1999, it reached 150 Mt, with nearly two thirds being hydrocarbon feedstocks (naphtha and natural gas) used to make ammonia, the starting compound for all synthetic nitrogen fertilizers. The second largest input by mass is asphalt and road oil; consumption of these paving and surfacing materials rose from less than 10,000 t in 1900 (when few paved roads existed outside cities) to 100 times that mass in less than two decades, it reached more than 10 Mt by 1950 and its recent use has been about 20 Mt a year.

I will address the levels of America's material intensities – that is material flows per unit of economic product – and their long-term trends in the next chapter dealing with apparent dematerialization of modern economies but before leaving this section I will note some notable per capita consumption levels. In aggregate terms, the USGS accounts translate to domestic consumption of about 1.9 t/capita in 1900, 5.6 t in 1950, 12 t/capita in the year 2000; after leaving out bulk construction materials these rates are reduced, respectively, to 1.2, 2.3, and 3 t/capita, which means that the use of construction materials rose from about 0.7 t/capita in 1900 to 3.3 t in 1950 and to 9 t in the year 2000. Partial data sets extending beyond the year 2000 indicate mostly stable or slightly declining rates of use.

Wood is the only material category showing a century-long decline of per capita consumption, from about 800 kg in 1900 to about 400 kg by 1950, about 300 kg/capita in 2000. Materials produced by agriculture rose slightly from 40 to 47 kg/capita during the first half of the twentieth century but afterward they declined to just 18 kg (but cotton and wool in imported finished apparel, now the dominant source of clothes worn in the United States, is not included in these totals). Paper consumption rose from 35 to about 175 kg/capita between 1900 and 1950; it stood at 179 kg/capita in 2000 and at about 200 kg/capita in 2020. Consumption of all metals has shown a similar pattern, rising from 135 kg in 1900 to 515 kg/capita in 1950, but by the year 2000 it was essentially the same at 510 kg/capita (once more a somewhat misleading rate given the country's large post-1970 net imports of cars, airplanes, and machineries).

As far as the three leading metals are concerned, per capita consumption of steel kept on rising, with fluctuations, from 125 kg in 1900 until 1974 when it just surpassed 500 kg, but the subsequent decline brought it to 425 kg in the year 2000, to only 260 kg/capita in 2010 and to just below 300 kg/capita in 2020. Copper reached domestic consumption peak in 1972 at 10.5 kg/capita (compared to 2.2 kg in 1900), after a temporary decline it rose to 11 kg in 2000, but then it fell to less than 6 kg by 2010 and it was about 5.5 kg/capita in 2020. Aluminum, too, has experienced a recent drastic fall, from the peak of more than 26 kg/capita in 1999 to less than 11 kg in 2010 and about 10 kg/capita in 2020. Are these declines characteristic of a new phase of material consumption in affluent, material-rich economies? Obviously, those who see the signs of permanent dematerialization everywhere would argue so, but they should consider material requirements of the unfolding decarbonization (gasoline fueled car average less than 15 kg of copper, electric vehicle need 80 kg), as hundreds of millions of tons of steel would be needed to lift America's decaying infrastructure from ASCEs (2021) common sectoral D marks at least B levels. I will return to these matters in the next chapter on myths and realities of dematerialization.

4.2.2   European Material Flows

Although several European countries have excellent historical statistics, none of their national institutions has produced a study of long-term material flows comparable to the USGS series. Instead (as already noted), we have three distinct kinds of material accounts: long-term reconstructions done by groups of researchers for a few economies, including Schandl and Schulz (2002) for the United Kingdom, Kovanda and Hak (2011) for the Czech Republic and Infante-Amate et al. (2015) for Spain; a set of economy-wide material flow accounts for EU-15 between 1990 and 2004 (Weisz et al. 2006); and Eurostat's series for more than 30 countries that begins only in the year 2000.

As different as these studies are, they share most of the underlying statistical sources as well as their maximalist approach to material accounting: they include all food and all fuels besides the materials used in production and services. As in all modern economies, bulk construction materials are the single largest category of Europe's material flows: at 7.1 t/capita in 2020, they were an order of magnitude greater than the consumption of metallic ores (0.7 t/capita in 2015). In comparison with the United States, EU-27 has similar metal consumption but lower demand for construction minerals, a difference that is due mostly to the continent's much higher population density and more compact transportation infrastructure.

Some European accounts have also compared the nations based on the spatial intensity of their material use (DMC/km^2), but that is a dubious measure because the extraction of both construction materials (dominant category by aggregate mass) and metallic ores are highly localized affairs, especially compared to much more extensive crop cultivation or forestry harvests whose density is a valid indicator of prevailing intensity of harvests. As expected, large countries with relatively small populations will have (even if they have considerable mining sectors) low rate of DMC per unit of area and the opposite will be true about small densely populated nations.

EU's material flows include all food, and this inclusivity is carried to excessive lengths by counting wild berries and mushrooms and all hunted animals. Indeed, EU's compilation guide and the materials flow handbook (Eurostat 2018) contain lists of scores of mammals and birds with recommended live weights to be used in aggregating their consumed biomass but I have to wonder how many marmots, brown bears, or pink-footed geese are actually killed for food every year to make any difference to the mass of available food, especially given the well-known impossibilities to account for millions of tons of staple grain harvests with uncertainty lower than ±3%. Even more importantly, in contrast to these irrelevant game-hunting details, the guidelines contain single conversion factors for translating gross ores into metal content or ore concentrate, assuming (quite unrealistically) that all iron ores have 43.32% of Fe and all copper ores have 1.09% of Cu (far above the global mean for 2020s).

In any case, given the brevity of the Eurostat series, its main value lies in allowing us to compare national levels of material use based on uniform accounting methods and in revealing some notable national peculiarities (Eurostat 2022a). These include expected high per capita consumption of forest phytomass in Finland and Sweden (4.2 t/year), more than twice the EU mean, and high consumption of metal ores in Sweden and nonmetallic minerals in Ireland (due to still considerable extraction of peat). Per capita consumption of construction materials tends to be higher in less populous countries, mostly because of higher demands for building more extensive transportation infrastructure.

Aggregate data make no stunning revelations. EU is nearly self-sufficient in biomass, sand, gravel, and many other nonmetallic minerals and extremely dependent on direct and indirect imports of fossil fuels and metal ores; and the findings of the highest raw material equivalents associated with the imports of crude oil and the largest export-generated raw material equivalents due to machinery and motor vehicles are as expected. European data also make it possible to contrast national differences in resource productivities expressed as GDP (in current prices at purchasing power standards) per unit mass (€/kg) of all materials used. In 2021, the all-EU mean was €2.3/kg, with the extremes (leaving Luxembourg out) running from €0.8/kg for Bulgaria and Romania to €5.7/kg for the Netherlands, with Sweden turning out just €1.6, Germany €2.7, Spain €3.0, France €3.1, and Italy €3.4/kg (Eurostat 2022c). But this comparison also reveals the limits of the aggregate approach where all materials are included without any regard to their fundamental qualitative differences. Is Italy really so much better than Germany and Sweden much worse than Spain?

4.3   Materials in China's Modernization

In October 1949, when Mao Zedong proclaimed the founding of the People's Republic of China (after the Communist Party won the country's long civil war with Kuomintang), China, the world's most populous country, was an impoverished nation, an overwhelmingly agricultural economy with rudimentary industrialization, with barely enough to eat and with hardly any foreign trade. Its alliance with the Soviet Union helped it, following the Stalinist economic model, to develop some heavy industries during the 1950s, but Mao's delusionary goal of rapid industrialization (The Great Leap Forward that was to accomplish in years what other nations took decades to achieve) led directly to the world's worst famines: between 1959 and 1961 at least 30 million people died, and after the famine's end significant resources were diverted into the development of nuclear weapons.

A brief post-famine respite was followed by another, this time more protracted, catastrophic policy known under the most incongruous term of the Great Cultural Revolution launched by Mao in 1966. This violent social upheaval extended China's economy misery for another 15 years: its acute phase ended only with Mao's death on September 9 1976, but it was only just before the end of 1979 when Deng Xiaoping emerged victorious from the brief power struggle following Mao's death, and in 1980, he began to turn China onto a long-postponed road to economic modernization combined with reduced repression. The first important step was a de facto privatization of farming that had rapidly improved food supply and ended food rationing in 1984.

4.3.1   China's Delayed Economic Rise

Economic reform gathered pace during the 1980s, trade increased as modern industrialization began to advance with the help of foreign companies (Japan and Hong Kong were prominent in the earliest stages) - but on June 4 1989, the future became uncertain after the Communist Party decided to end the popular movement calling for democratic rights by killing students protesting at Tian'anmen. At that time, few people would have predicted that this will have no effect on China's economic progress. Most notably, foreign direct investment did not falter even for a single year, it was higher in 1990 than in 1989; 10 years later it was 12 times higher ($42.1 billion); in 2013, it reached the peak of $290.93 billion; and in 2020, it was not far behind with $253.1 billion. The annual total of foreign direct investment surpassed $50 billion in 2003, stayed above $150 billion since 2010, and amounted to about $3.5 trillion during the first two decades of the twenty-first century (World Bank 2022b).

This enormous money transfer was accompanied by an equally unprecedented acquisition of modern extraction, industrial and transportation methods, and machines brought by the world's leading multination companies, a priceless advantage enjoyed by a late-starter to modernization that has been also accompanied by extensive industrial espionage (O'Connor 2019). But two others factors were even more important for sustaining rapid rates of economic modernization. The first one was tremendous pent-up desire of the world's largest population, kept for decades in Maoist economic misery, to improve their quality of life. The second was China's huge demographic dividend (accelerated economic growth that accompanies falling birthrate caused by one-child policy and increasing numbers of economically active people) together with delayed urbanization that released hundreds of millions of young productive workers from the countryside to cities, powering their expansion and new manufacturing capacities that made the country the world's largest exporter.

The results have been impressive. According to official statistics, between 1980 and 2020 China's annual rate of economic growth was only three times below 5% (in 1989 and 1990 and, Covid-induced, in 2020) and 15 times above 10%, and this performance lifted China from contributing just 2.3% of the global economic product in 1980 to creating 18.2% of it by 2020 (IMF 2022). In 1980, China's economy was (in PPP terms) only about half the size of Italy's; by 2020, it was the world's second largest, about three times as large as that of Japan (IMF 2022). In per capita terms, this growth lifted China's average GDP from $288 in 1980 to $10,530 in 2022, a 36-fold increase. Actual rates were undoubtedly lower than the officially exaggerated statistics, but still as high or higher than the fastest pace achieved during their peak-growth spells by Japan or South Korea). Looking back, it is now clear that China's four post-1980 decades added up not only to the fastest but also to the largest economic advance in history.

Given a relatively low starting point of post-Maoist expansion (in 1980, China's per capita GDP, at just $250, averaged less than that of Pakistan), it had to be expected that all material flows had to multiply in order to enable four successive doublings of the country's economy in three decades. And it was also clear that China's domestic endowment would not be able to satisfy all of these needs and that the country will have to resort to massive imports: it has done so and the country has become the world's largest importer of commodities ranging from iron ore to potash fertilizer.

At the same time, the expansion of domestic mining surpassed most expectations, and by 2020, China was the world's largest producer (alphabetically) of aluminum, bismuth, copper, gold, mercury, tin, titanium, vanadium, and zinc as well as of rare earths, the group of 17 metallic elements ranging cerium to yttrium with a wide range of high-tech applications (USGS 2022a). In some of these cases, China not only leads, its extraction dominates the global market: nearly 90% for mercury, about 80% for bismuth, and about 60% for vanadium and titanium and rare earths. And while it is only the fourth largest producer of copper ore, it is the largest producer of refined metal (nearly 40% of the global output in 2021). This all is in addition to China becoming by far the largest producer of coal (in mass terms nearly 4 Gt/year, half of the global extraction, in energy terms nearly 60%), the fourth largest producer of natural gas and the world's sixth largest producer of crude oil.

Chen et al. (2022) provided the most detailed account of this material expansion by constructing a database covering hundreds of materials from thousands of data sources covering the three decades (1990–2020) of unprecedented growth. These accounts include not only all biomass, minerals, and fossil fuels but also the rising demand for oxygen and water, and they indicate overall tripling of aggregate material demands during the three decades that was accompanied by more than quintupling of CO_2 emissions. Here I will take a closer look at the most notable material shifts that made China and unrivalled leader in several production categories.

4.3.2   Concrete, Steel, Waste

Not surprisingly, China's huge post-1980 construction boom - by any measure the world's largest and fastest expansion of housing, commercial buildings, industries, and transportation infrastructure - has engendered unprecedented demand for all construction materials. A rough estimate is that the domestic extraction of construction minerals sextupled between 1980 and the year 2000, and that it had nearly tripled during the first two decades of the twenty-first century (Material Flows Net 2018). And that high overall output multiple (nearly 18-fold between 1980 and 2020) was greatly surpassed by the country's production of cement (30-fold increase between 1980 and 2020) and even more so by the production of plate glass (37-fold increase between 1980 and 2020).

The frenzied pace of China's concretization and its overall scale has been stunning. In 1980, the country produced just short of 80 Mt of cement; a decade later it had more than doubled the total to about 210 Mt; by the year 2000, it rose to 595 Mt; by 2010, that total had tripled and reached 1.88 Gt (nearly 24 times the 1980 total and 57% of the global production for less than 20% of the world's population); by 2020, it rose to 2.4 Gt and the next year to 2.5 Gt or 56% of the global output (USGS 2022a). This mean that China has been recently producing every year about 1,700 kg of cement per capita, more than double the record rate reached during Japan's post-WWII period of concrete use (Steel 2017).

And China's recent per capita cement production rates have been almost exactly 6 times higher than in the United States and 10 times higher than in the EU, with only Qatar (nearly 3 t/capita) and Saudi Arabia (nearly 2 t/capita) consuming more. But perhaps no other comparison illustrates the scale of China's concretization than this one: consumption of cement in the United States totaled about 4.56 Gt during the entire twentieth century - while China emplaced more cement (4.8 Gt) in new construction in just two years in 2014 and 2015, and this pace slowed down only during the pandemic, with some 2.6 Gt of cement emplaced during 2020 and 2021!

Inevitably, such a pace of construction increases the probability that some of the newly poured concrete will be of sub-standard quality, a conclusion confirmed by the obvious dilapidation of China's concrete structures built during the late 1980s and the early 1990s, the first period of China's construction boom. Quality of concrete used to construct many of China's new dams (the country has more than 20,000 structures taller than 15 m, including the world's largest dam, Sanxia) is of particular concern, even more so as thousands of them are located in areas of repeated, vigorous seismic activity. Even if the initial quality of concrete posed no problems, the US experience with the country's extensive concrete-based infrastructure offers a sobering view of what lies ahead for China.

Poor quality and enormous overcapacity have been also the problems in plate glass industry whose output increased 25 times between 1980 and 2010, from about 25 million to 663 million weight cases, or from just 1.25 to 33.1 Mt. China thus produced 60% of the world's flat glass demand of about 55 Mt in 2010 when its plate glass-making capacity rose by 6.5 Mt to 44.5 Mt; that lowered the industry's average capacity factor to just 70%. Moreover, a major share of that excess capacity has been in energy-intensive and highly polluting enterprises producing roughly 20 Mt, or nearly two thirds of the total, making low-quality float glass, but as the construction boom continued, production rose by another 40% during the 2010s to 927 million cases (46.3 Mt) in 2020 (NBSC 2022).

Between 1980 and 2010, the expansion of China's steelmaking has matched almost exactly the pace of GDP growth: raw steel output rose 17.2 times from 37.1 Mt in 1980 to 637.4 Mt in 2010 when it accounted for nearly 45% of the global output. A significant slow-down could be expected, but by 2020, the output of crude steel rose by 67% to 1.064 Gt, a nearly 29-fold rise since 1980. But as the domestic extraction of iron ores increased to only 270 Mt by 2020, a rising share of the input into China's blast furnaces has been coming from imported materials. In 2010, China imported 618 Mt of iron ore, in 2020 the total rose to 1.12 Gt worth $99 billion, supplied mostly by Australia (about 65%) and Brazil (nearly 20%). Chinese iron ore imports now account for nearly 68% of global shipment and for 82% of domestic consumption (WSA 2022a). China has also become the world's largest, and growing, importer of bauxite (44 Mt in 2010, 111 Mt in 2020) as well as copper ores and concentrates. In contrast, the country has been the largest exporter of rare earths and also the largest producer and exporter of molybdenum, magnesium and graphite.

The only major construction material whose production has not seen a rapid increase has been timber. Its output in the year 2000 was actually lower than in 1980 (47 vs. 54 Mm³) but more than doubled by 2020 as the post-1990 reforestation campaigns dominated by plantings of rapidly growing species (pines, poplar, eucalyptus) began to mature (NBSC 2022). Even so, domestic production of industrial roundwood remains inadequate and China remains the world's largest importer of logs and sawn timber: the total for 2020 was nearly 94 Mm^3 or more than half of total consumption. About 80% of this volume was split between sawn and rough industrial wood (sawn wood mostly from Russia, Thailand, and United States; rough wood mostly from New Zealand, United States, and Russia) and fuelwood was nearly 10% of imports (OEC 2022). Rising share of timber imports has been coming from Africa (Gabon, Congo) contributing to the destruction of the continent's tropical rain forests.

Material category that has seen the greatest production increase has been the synthesis of plastics, with a nearly 90-fold rise between 1980 and 2020. Of course, that large multiple is due to a rapid development from a very low base (only 1.23 Mt in 1980), but the output of 105.42 Mt in 2020 was about 29% of the global total. Finally, the need to secure more food and better nutrition for a still-growing population led to substantial gains in the production, imports, and applications of fertilizers. New Haber-Bosch plants were added to raise the output of ammonia converted mostly to urea and to nitrates. When in measured in terms of pure nitrogen, domestic production rose from 10.3 Mt N in 1980 to 34.3 Mt in 2010 and 32 Mt in 2020, with applications rising from 12.1 Mt in 1980 to nearly 30 Mt in 2010 before falling to 25.9 Mt in 2020 (FAO 2022).

Recently falling use of nitrogen fertilizers has resulted from deliberate efforts to reduce China's excessive applications of this macronutrient that led to highly unbalanced fertilization ratios and hence to low efficiency of use and high nitrogen waste (in rice cultivation only a quarter of the applied nutrient may end up in harvested grain) due to leaching and volatilization. Here are the key numbers required to appreciate this imbalance and attempts at its correction. The US N:P:K application ratios (calculated as N:P$_2$O$_5$:K$_2$O) were 1 : 0.46 : 0.53 in 1980 compared to China's 1 : 0.22 : 0.04; by 2020, the Chinese ratios has become more balanced, 1 : 0.38 : 0.11 compared to 1 : 0.34 : 0.37 for the United States.

China has also become prominent importer of waste materials and the United States has been their leading supplier (Magnaghi 2010; OEC 2022). American exports of waste paper to China rose rapidly from just $38 million in the year 2000 to more than $2 billion by 2011. In that year, China also imported the record amount of US scrap iron worth nearly $2 billion, more than one billion dollars of scrap aluminum and nearly $3.5 billion of scrap copper. In total, in 2011 the United States exported more than $11 billion of waste and scrap (materials belonging to the 910 category of North American Industry Classification System) to China.

Perhaps no trade stream illustrates better the stunning reversal of economic fortunes: a decade after China's 2001 accession to the World Trade Organization, the world's largest affluent economy became the primary supplier of waste materials to the second largest, one-party-controlled, economy experiencing rapid rates of growth. This began to change in 2017 when the State Council enacted the ban on the importation of 24 types of recyclable waste beginning at the end of the year (Yoshida 2022). Imports of household waste plastics were banned by the end of 2017, and this was followed by the ban on metal and electrical appliance scrap by the end of 2018. This sudden policy change led to major restructuring of global trade in plastic waste, but in 2020, China recategorized high-grade copper scrap as a resource rather than waste and hence not subject to any import restrictions (Metalshub 2020), and in 2021, it once again expanded its imports of ferrous scrap.

Finally, here are a few comparisons showing how China's enormous expansion of material consumption translated into private possessions, with all rates per 100 urban households and with the ownership statistics starting in 1985 (NBSC 2022 and previous editions). In 1985, nearly half of all urban households had a small washing machine, 17% had a color TV, and less than 7% had a small refrigerator. By the year 2000, these three rates jumped to, respectively, 90%, 112%, and 78%; and nearly 25% also had one window air conditioner. By 2020, ownership of washing machines, refrigerators, and TVs became saturated, and urban households owned two AC units per family, more than a third of them had cars (55% in Beijing) and an average family owned 2.3 mobile phones.

And finally, a personal recollection. These lines were written in October 2022, exactly four decades after my first extended visit to Beijing. I stayed at the Chinese Academy of Sciences dormitory reserved for visiting scholars, located near Beijing Normal University in the city's northwestern Haidian district. One day I walked from there (via Xinwai and Xidan Avenues) to Tian'anmen at the city's center, distance of about 17 km, and during that long morning walk all I saw were four or five black limos reserved for high-party officials or generals: there were taxis (very few) but not a single private car. Four decades later, China is by far the world's largest maker of passenger cars, with annual sales above 20 million units (2017 record was 24.8 million) and the capital has nearly 5.5 million registered vehicles (CEIC 2021). In this case, as with so many other consumer items, it is impossible to calculate the intervening multiplier as the original 1980 values were zero.

I will return to China's material production and trade in the book's last chapter when I will appraise near- and long-term prospects for material supply in general and the material needs of the unfolding energy transition in particular. As the world's largest producer of steel, aluminum, and copper, the three metals dominating energy infrastructures ranging from wind turbine towers to shipping and from transmission to electric motors, and as the dominant producer (with nearly 80% of the global market share in 2021) of photovoltaic panels and lithium batteries for electric vehicles, China already has an outsize role in this transformation, and given its prominence and targeted, government-subsidized development of strategic materials, its prominence is here to stay for decades to come.

Before China's post-1990 rise, Japan was the Asia's largest (and the world's second largest) economy, and thanks to the country excellent statistics, we can trace its rise from a pre-industrial economy (beginning of the Meiji era in 1868) to its post-1960s global prominence. Historical studies of Japan's material flows include an overview by Krausmann et al. (2011) as well as a number of sectoral reviews, including construction materials (Tanikawa et al. 2015), metals (Murakami et al. 2004), copper (Daigo et al. 2009; Terakado et al. 2009; Yokoi et al. 2018), chromium and nickel in stainless steel (Daigo et al. 2010) and platinum group (Kuriki et al. 2010).

Key features of Japan's material trajectory correlate closely with the country's large and abrupt political and economic shifts. The country's total mass of consumed materials increased 40-fold in 127 years between 1878 and 2005, with most of this expansion coming only during the 1950s and 1960s and supplied overwhelmingly by imports (a dependence made even more pronounced by the inclusion of fossil fuels). This great post-WWII surge in material requirements reached its first peak at the time of the first round of OPEC-driven oil price increases (1973–1974), and it rebounded with China's unprecedented post-1990 drive to modernize the country's economy: by 2020, that expansion made China the world's largest producer (and consumer) of virtually all leading materials, from cement and steel, ammonia and plastics, and aluminum to zinc.

Analysis of existing stocks showed that by 2010, 43% of all materials were in buildings (about 9 Gt by 2020), with roads containing 26%, seaports 19%, and dams 8%. Specific composition of these stocks has not changed significantly between 1965 and 2010: construction materials (concrete and asphalt) added up to 95% of the total, timber only 2%, and, as expected, these stocks are heavily concentrated in cities, with 80% of the mass is in 20% of the country's area (Tanikawa et al. 2015). Material flow accounting for metals showed the extreme dependence on imports, with domestic extraction amounting to mere 0.1% of final domestic consumption (Murakami et al. 2004). High dependence on imports (albeit not that extreme) characterizes all other flows of major materials, from fuels to foodstuffs.

Period of generalized decline, stagnation, and low growth began in 1990 when the Japanese stock market suddenly collapsed (and still has not recovered more than three decades later) and when the country's economy entered a protracted period of low economic growth and deindustrialization (with many activities moving to China and Southeast Asia) accompanied by rapid population aging and now also by accelerating population loss (in 2021, it exceeded 700,000 people). Japan remains an affluent society whose high quality of life is based on mass consumption of mostly imported materials. As there is no immediate prospect for any mass immigration that would reverse the country's population decline, it is unlikely that Japan will see any substantial increases of its material consumption in decades ahead.

These national reviews are informative, but they bring only a few surprises as the trajectories of material consumption must broadly correspond to the shifts in economic fortunes. A much more profitable inquiry is to turn to some fundamental attributes of material flows, to their energy costs, and to their environmental impacts. I will do this in two separate sections, the first one quantifying fuels and electricity requirements for many raw and processed materials, the second one introducing results of revealing life-cycle assessments, an analytical tool whose aim is to present costs other than capital and operating expenditures.

4.4   Energy Cost of Materials

Before any materials can start flowing through the economies, energies must flow to power their extraction from natural deposits or their production by industrial processes ranging from simple mechanical procedures to complex chemical reactions. Those energies belong to two distinct streams: direct flows of fuels and electricity used to energize the production processes (producing mechanical energy, heat or pressure, and lighting and electronically controlling a process) and indirect flows (embedded energies) that were needed to produce the requisite materials, machines, equipment, and infrastructures. In modern production systems, the first category of flows is usually dominant.

For example, energy needed to smelt a ton of iron from its ore in a blast furnace (as coke and supplementary coal, gas, or oil) will be vastly greater than energy embedded in the furnace's steel, lining and charging apparatus, and prorated per unit of output. Modern blast furnace can operate without relining for two decades, and during that time, it can produce tens of millions of tons of hot metal, reducing the embedded energy cost per unit of output to a small fraction of overall energy requirements (and in some cases to less than an inevitable errors in counting or estimating direct energy inputs). Similarly, energy needed to create a combination of high temperature and pressure that is required by many chemical syntheses will be far greater that any prorated energy embedded in the initial construction of reaction vessels, pipes, boilers, compressors, and computerized controls (some of which will, again, serve for decades while enabling massive aggregate outputs).

That explains why the second category of flows (indirect inputs) is almost always neglected - and until the early 1970s, most industries had little reason to examine even the direct energy flows of their energy use: for decades preceding OPEC's two rounds of crude oil price increases (1973–1974 and 1979–1980), real fuel and electricity prices in the world's most advanced economies were low, even falling, and hence only the managers in sectors with inherently high electricity and fuel inputs (such as electrochemical processes, high pressure, and high temperature chemical syntheses) were motivated to reduce their overall energy cost in order to improve profits. This common neglect of energy uses ended abruptly in 1973 as OPEC quintupled the world oil price in a matter of months, and concerns about energy's affordability had further increased when the second round of OPEC's price manipulation (driven by the overthrow of Iranian monarchy) pushed the price of an average barrel of the Middle Eastern oil from about $12 in 1977 to nearly $36 in 1980 (Smil 2003; BP 2022).

4.4.1   Energy Analyses

New discipline of energy analysis aimed to trace and quantify energy used not only to extract and process natural commodities and to produce manufactured items but also to grow food and to provide services (IFIAS 1974; Chapman et al. 1974; Verbraeck 1976; Thomas 1979). The chosen label was misleadingly too broad as energy analysis should stand for a much wider range of inquiries. Embedded (or incorporated) energy is a descriptively more correct term but I have always preferred to use a simple label of energy costs. I have been an early practitioner of this method - among other efforts being the principal author of the first comprehensive energy analysis of the US corn, the country's most important crop (Smil et al. 1983) – and hence I have a good appreciation of associated problems and weaknesses.

Most of the appraisals of energy costs have followed one of the two distinct approaches, either a quantification based on input–output tables of economic activities or a process analysis that traces all important energy flows needed to produce a specific commodity or manufactured item. In the first instance, relevant prices are used to convert values of energy flows in a matrix of economic inputs and outputs (for major industrial sectors or, where available, disaggregated to the level of product groups or individual major products) to energy equivalents in order to assemble direct and indirect energy requirements. In contrast to this aggregate approach, the process analysis can focus on a particulate product in specific circumstances as it identifies all direct energy inputs and as many relevant indirect needs as possible, a process that is in itself quite valuable as a management tool.

As with all appraisals that deal with complex inputs and encompass sequential events, the setting of analytical boundaries will affect the outcome of processes analysis. In most instances, the truncation error inherent in counting only direct energy inputs (purchased fuels and electricity) will be small, but in some cases, it could be surprisingly large. For example, Lenzen and Dey (2000) found the energy costs of Australian steel to be 19 GJ/t with process analysis but 40.1 GJ/t with input–output analysis. Similarly, input–output analysis by Lenzen and Treloar (2002) put the total embodied energy in a four-story Swedish apartment building roughly twice as high as a process analysis by Börjesson and Gustavsson (2000).

Another complication is introduced due to increasing shares of globally traded commodities and products: in some cases, additional energies required for imports of raw materials and export of finished products will be a negligible share of the process energy cost; in others their omission will cause a serious undercount. For example, two identically looking steel beams used at two construction sites in New York may have two very different histories: the first one being a domestic product made by the scrap metal-electric arc furnace-continuous casting route in a steel mill in California, and the other one coming from China where Australian iron ore and coke made from Indonesian coal were smelted in a blast furnace in one province and the beams were made from ingots in another one before loaded for a trans-Pacific shipment and then transported by railroads across the continent. Obviously, both the production and transportation energy costs will be very different.

Approximate energy costs of long-distance transportation can be easily calculated by assuming the following averages (all in ton–kilometers for easy comparability, ranked from the highest rates to the lowest): air transport 30 MJ, diesel-powered trucks (depending on their size) mostly between 1 and 2.5 MJ, diesel-powered trains 600–900 kJ, electricity-powered trains 200–400 kJ, smaller cargo ships 100–150 kJ, and large tankers and bulk cargo carriers just 50 kJ/tkm (Smil 2010). Obviously, energy-intensive air shipments will be restricted to high value-added products (foodstuffs, electronics, other small consumer goods), while, as even an extreme example shows, bringing iron ore from one continent to another by a bulk carrier would entail energy expenditure equal to less than 10% of overall requirements for steel production.

Brazilian iron ore from enormous open-cast mines in Carajás is transported 892 km by train to Ponta de Madeira in the country's northeast, loaded on Valemax bulk ore carriers and shipped 2,800 km before reaching steel mills in Tianjin (Brauch et al. 2020). But this shipment will need just 1.4 GJ/t for shipping and 0.5 GJ/t for railway transportation, that is an equivalent of just 10% of typical cost of China's BF-BOF steel that was about 18 GJ/t in 2020 (He et al. 2020). In contrast, energy cost of shipping construction stone (Carrara marble) from Italy to China or California may be equal to 25–50% of energy used to cut and polish it. And another Brazilian example shows that energy cost of long-distance transport can equal the energy cost of production. Energy costs of cultivating soybeans in the country's two leading farming regions, Centre West and South, are, respectively, 4.8 and 3.9 GJ/t, but (in the absence of railways) trucking to the costs consumes 3 GJ/t and shipping to Rotterdam adds another 1.5 GJ/t, a total nearly as large as the cost of cultivation in some areas (da Silva et al. 2010).

These realities should be kept in mind when examining and comparing the values reviewed in this section - or reviewed elsewhere (Gutowski et al. 2013) - as they almost always refer just to the cost of production (growing, processing, assembling) and exclude the costs of transportation, both for raw materials and for finished items. Even with these limitations, energy costs - presented here in a uniform way as gigajoules per ton (GJ/t) of raw material or product - offer a revealing insight into requirements and impacts of diverse materials and they are useful comparative guides. But, as with any analytical tool, they alone cannot be used to guide our choices and preferences of material use without concurrent considerations of affordability, quality, durability, or esthetic preference; if the latter were to be ignored concrete, a material of low energy intensity, would rule the modern world even more than it does. Presentation of energy costs follows the same material order as the preceding chapter, starting with biomaterials and concluding with silicon.

4.4.2   Embodied Energies of Materials

Energy cost of market-ready **lumber** (timber) is low, comparable to the energy cost of many bulk minerals used as basic construction materials. Tree felling, removal of boles from the forest, their squaring and air drying will add up to no more than about 500 MJ/t, and even with relatively energy-intensive kiln-drying (this operation may account for 80–90% of all thermal energy), the total could be as low as 1.5 and no more than 3.5 GJ/t (including cutting and planning) for such common dimensional construction cuts as 2×4 studs used for framing North American houses (Li et al. 2006). These cuts are rectangular prisms whose actual dimensions are smaller (1.5×3.5 in., or 38×89 mm) whose standard length for house construction is 8 or 9 ft. Low-energy cost of wood is also illustrated by the fact that in Canada, energy cost of wood products represents less than 5% of the cost of the goods sold (Meil et al. 2009). Energy costs on the order of 1–3 GJ/t are, of course, only small fractions of wood's energy content that ranges from 15 to 17 GJ/t for air dry material.

Obviously, energy cost of **wood products** rises with the degree of their processing (FAO 1990; Hammond and Jones 2008; Humar 2020). Particle board (with density between 0.66 and 0.70 g/cm^3) may need as little as 3 GJ/t and no more than 8 GJ/t, with some 60% of all energy needed for particle drying and 20% for hot pressing. Energy needs for the production of **plywood** vary by a factor of 2 (depending on the wood and a specific manufacturing procedure), from 7 to 15 GJ/t, with, once again, some 60% of all energy required for drying and 10% for pressing. Increasingly popular **glulam** used in commercial construction requires between 5 and 12 GJ/t.

Paper is a relatively energy-intensive product. Chemical pulping entails pressurized boiling of ground wood in acid solutions followed by even more energy-intensive dewatering of freshly laid fibers making new paper layers: fiber concentrations are just 0.7% at the start of the process (headbox) and water removal raises them to 92% at the end (where paper is reeled). As a result, production of many kinds of paper needs more energy per ton than the smelting and rolling of steel, another fact that most people, entirely unfamiliar with modern material-production processes, found surprising. The process requires about 2.2 t of wood per ton of paper, but it produces about 22 GJ of black liquor per ton of pulp and burning that by-product energy can make a mill a net energy producer.

In contrast, mechanical pulping needs less than 1.1 t of wood per ton of pulp and it is a large consumer of electricity. Integration of the two processes is the best way to maximize energy efficiency. Primary energy requirements for the most efficient pulping are about 10 GJ/t of air-dried kraft **pulp** but as much as 23 GJ/t for the thermo-mechanical process, and the most efficient re-pulping of waste paper con-sumes only 4 GJ/t (Worrell et al. 2008a; Man et al. 2019). Energy cost of papermak-ing varies widely with the final product and with the size and production scale of modern papermaking machines. Their typical length is 150 m, their speed is between 600 and 1,800 m/minute, and their annual output is about 300,000 t of paper (Austin 2010).

Obviously, once installed such large machines are not amenable to drastic changes, and high-energy requirements of the process are an inevitable consequence of removing, so much water by evaporation. Unbleached packaging **paper** made from thermo-mechanical pulp is the least energy-expensive kind (less than 20 GJ/t), fine bleached uncoated paper made from kraft pulp consumes at least 27 GJ/t, news-print no less than 30 GJ/t (Worrell et al. 2008). In contrast, the best printing paper, the coated stock used for arts books may be actually less energy intensive (about 25 GJ/t) because the production and application of the filler (mostly chalk) is cheaper than making fiber from wood. Recycled and de-inked newsprint or tissue can be made with less than 18 GJ/t, but the material is often down-cycled into lower quality packaging materials. A comprehensive study of Chinese papermaking showed average energy cost ranging from maxima of about 40 GJ/t for tissue paper to minima of about 15 GJ/t for packaging paperboard (Man et al. 2019).

Construction aggregates whose production requires only extraction and some physical treatment (sorting, sizing, crushing, milling, drying) have generally very low to low energy costs, and higher fuel and electricity use comes only with pyro processing required to make bricks, tiles, glass, and, above all, cement (Hammond and Jones 2008). Energy cost of natural **stone** products is low, usually just around 500 MJ/t for quarried blocks, somewhat less for crushed stone, but twice as much for roughly cut or split stones, and three to four times as much (up to 2 GJ/t) for accurately cut and polished ornamental stones (marble, granite). Energy costs of **gravel** and **sand** extraction and processing can easily vary by a factor of 2, but even the higher costs leave them in the category of the least energy-intensive materials (far less than 1 GJ/t) when compared in mass terms.

The simplest mining and preparation sequence to produce a fairly clean sand and uniformly-sized sand may require no more than 100 MJ/t and even more costly gravel sorting (or crushing as needed) should have energy cost well below 500 MJ/t. The highest energy input is required for the preparation of industrial sand that is used in glassmaking, ceramics and refractory materials, metal smelting and casting, paints, and now also increasingly in hydraulic fractioning of gas- and oil-bearing shales: its moisture must be reduced (in heavy-duty rotary or fluidized-bed dryers) to less than 0.5%, and this may consume close to 1 GJ/t. **Bricks** fired in inefficient rural furnaces in Asia may require as much as 2 GJ/t and just 1.1–1.2 GJ/t is typical for Chinese enterprises (Global Environmental Facility 2012), while the US production of high-quality bricks requires 2.3 GJ/t (USEPA 2003).

Cement production is fairly energy-intensive because of high temperatures required for the thermochemical processing of the mineral charge. Limestone supplies Ca and other oxides, and clay, shale, or waste materials provide silicon, aluminum, and iron; in order to produce a ton of cement about 1.8 t of raw minerals are ground and their mixture is heated to at least 1,450 °C. This sintering process combines the constituent molecules and the resulting clinker is ground again with the addition of other materials to produce 1 t of clinker that is then ground to produce fine Portland cement. Fly ash (captured in coal-fired power plants) or blast furnace slag can be used to lower the amount of clinker. Additional energy is to needed to rotate large kilns. These inclined (3.5–4°) metal cylinders are commonly around 100 m, and up to 230 m, long, with diameters of 6–8 m and they turn typically at 1–3 rpm, with the charged raw material moving down the tube against the rising hot gases (Peray 1986; FLSmidth 2011; Chatterjee 2020).

Disaggregation of all energy inputs shows that the extraction of minerals (limestone, clay, shale) and their delivery to cement kilns is a minimal burden. Kiln feed preparation is electricity-intensive as crushing and grinding of the charge consumes about 25–35 kWh/t and the grinding and transportation of the finished product (clinker) claims at least 32–37 kWh/t (Worrell and Galitsky 2008). International Energy Agency data show the total use of electricity in cement production ranging from 70 kWh/t in India to 130 kWh/t in Canada, with the mean for G20 nations at about 85 kWh/t (IEA 2021a). This leaves the bulk of energy consumption for the pyro processing, a sequence of water evaporation, decomposition of clays to yield SiO_2, decomposition of limestone or dolomite (calcination) that releases $CaCO_3$, formation of belite (Ca_2SiO_4, making up about 15% of clinker by mass), and finally sintering, production of alite ($Ca_3O \cdot SiO_4$) that makes up some 65% of the clinker mass.

Total energy use in cement production varies with the principal fuel used, origin of the electric supply, and the method of production. Average specific energy consumption in cement industry has declined as a more efficient dry process replaced the old wet method. The highest electricity consumption in the dry process is for the grinding of raw materials and clinker and for the kiln and the cooler, in aggregate more than 80% of the total that averages mostly between 90 and 120 kWh/t of cement (Madlool et al. 2011). Heating of dry kilns (mostly with coal, petroleum coke, and waste materials in the United States, with coal in China) consumes between mostly 3–4 GJ/t; the range is 3.0–3.5 GJ/t for kilns with four or five stages of preheating, while a six-stage process could work with as little as 2.9–3 GJ/t.

World's best practice can now produce Portland cement with the total primary energy inputs of 3.3–3.5 GJ/t, while the rates for fly-ash cement and blast furnace slag cement can be as low as, respectively, 2.4 and 2.1 GJ/t (Worrell et al. 2008; IEA 2021a). In contrast, many plants in low-income countries still need around 4.5 GJ/t of Portland cement. As expected, energy consumption for cement production has the greatest impact and produces most CO_2 emissions in China. Average nationwide rate was still close to 5 GJ/t in 2000 and just over 4 GJ/t in 2007 when the industry claimed 7% of the nation's total energy use. Introduction of more efficient dry kilns reduced the nationwide mean to about 3.3 GJ/t by 2020 and many efforts have been underway to limit the industry's emissions (Liu et al. 2018). Using IEA's mean energy intensity of 3.5 GJ/t, 2020 global production of 4.1 Gt of cement would have consumed more than 14 EJ or 2.5% of the world's primary energy supply.

Energy requirements for **glass** production range mostly between 5 and 12 GJ/t, with little difference between container and flat glass and with about 7.5 GJ/t being a typical value for a float glass line (IEA 2007; Strauven 2016; Revitasari and Susanto 2018). The theoretical minimum for glass-making (chemical reactions among the constituent minerals and the melting) is about 2.4 for borosilicate and crystal glass and 2.8 GJ/t for common soda-lime glass (Zier et al. 2021). As expected, energy cost of ceramic products rises with the degree of pyro processing and the quality of items: unglazed tiles need only 6 GJ/t, glazed tiles up to 10 GJ/t, fine ceramics as much as 70 GJ/t, while sanitary stoneware rates at about 30 GJ/t.

The usual approach for quantifying energy costs of **iron and steel** industry is to include energy costs of coke, pelletizing and sintering of ore, iron and steel making, cold and hot rolling, and galvanizing and coating; that leaves out energy cost of coal and ore mining and transportation of such energy-intensive inputs as electrodes and refractories, as well as the embodied energy cost of scrap metal. Analyses performed (more or less) within these boundaries show that average energy consumption in the global steel industry was about 60 GJ/t in 1950, just over 30 GJ/t by 1975, and 19.5 GJ/t by 2020 (Yellishetty et al. 2010; IEA 2021d).

Reduced charging of coke to blast furnace has been the main reason why the lowest energy requirements in iron smelting declined from about 30 GJ/t in 1950 to as little as 12.5 GJ/t by the late 1990s, the rate only twice as high as the thermodynamic minimum of 6.6 GJ/t needed to reduce iron from hematite (De Beer et al. 1998).

Typical ranges are 10–13 GJ/t for the blast furnace operation, 2–3 GJ/t for sintering, 0.75–2 GJ/t for coking, and 1.5–3 GJ/t for rolling (Smil 2016). WSA's latest analysis of steel's primary energy demand showed 19.2 GJ/t for sections, 25.4 GJ/t for hot-rolled metal, and 30.6 GJ/t for hot-dipped galvanized steel (WSA 2021). Reviews of best industry practices for the entire iron-steel sequence ended up with 16.3–18.2 GJ/t for the blast furnace-BOF-continuous casting route, 18.6 GJ/t for direct iron reduction followed by EAF steelmaking and thin-slab casting, and 6 GJ/t for melting scrap metal in EAF and thin-slab casting (Worrell et al. 2008; Smil 2016). Overall national energy intensity of steelmaking will be thus considerably lower in countries relying more on the most efficient EAF route.

Aluminum production is much more energy intensive. Fuel and electricity consumption in the Bayer process, between 10 and 13 GJ/t of alumina, is a small share of the overall cost dominated by the electrolysis that is done preferably with the cheapest kind of electricity produced in large hydro stations (it supplies about 60% of the industry's needs worldwide). A version using electrodes that are baked *in situ* is slightly more electricity-intensive than the dominant method using pre-baked anodes made from tar pitch or coke in oil. At least 400 kg of anode are required to produce a ton of the metal and anode production needs about 2.5 GJ/t of fuel and 140 kWh/t of electricity.

Efficiency of electrolysis has been steadily improving: between 1900 and 2000, the size of average cell used in the Hall-Héroult process had doubled every 18 years, while electricity use declined from about 50 MWh/t in 1900 to 25 MWh/t by 1950 (Beck 2001). International Aluminum Institute data trace the progress since 1980 when the global mean about 17 MWh/t: by the year 2000, the mean declined to 15.4 MWh, and in 2021 it was 14.1 MWh/t, with China, now by far the largest producer, being below the global average at 13.5 MWh/t (IAI 2022b). Compared to the theoretical minimum of 6.3 MWh/t, this means that the process is now, at best, about 47% efficient.

The average 2021 requirement (13.5 MWh) is equivalent to 48.6 GJ/t when supplied by hydroelectricity (or, in the future, by other forms of renewable generation) but to nearly 150 GJ/t of primary energy when relying on combustion of coal to produce electricity. Additional energy is needed for casting and rolling of ingots and other smelter activities add 7–8 GJ/t, raising the overall rate to as much as 160 GJ/t. That is nearly twice the energy intensity of copper and almost 10 times as much as the least energy-intensive production of steel using the blast furnace-BOF-continuous casting route. Global 2021 production of about 67 Mt of Al would have thus required about 7.9 EJ, or about 1.3% of the world's total primary commercial energy supply. Metal's high electricity requirements steer the location of primary aluminum production to countries with abundant hydro resources, and the four such largest producers (China, Russia, Canada, and the United States) account for half of the world output. In contrast, secondary aluminum made from scrap needs as little as 4.3 GJ/t, a tremendous incentive to recycle the metal.

As already noted, titanium has the highest energy cost among the other relatively commonly used metals (400 GJ/t), followed by nickel at about 160 GJ/t; copper comes second (global average of 93 GJ/t), while chromium, manganese, tin, and zinc have a very similar energy cost of about 50 GJ/t (IEA 2007). Not surprisingly, very low metal concentrations of even the best exploited deposits raise the energy intensities of silver and gold orders of magnitude above common metals: the average for silver is about 2.9 TJ/t (30 times higher than for copper) and for gold it is 53 TJ/t, roughly 300 times the energy cost of aluminum.

Plastics are, without exception, energy-intensive materials, but the published rates have been surprisingly divergent. During the mid-1970s, the first comparative studies of energy used in material production showed final process energy requirements of PE to be as low as 50 GJ/t in the Netherlands and as high as 116 GJ/t in the United States (Berry et al. 1975). For PP the range was much narrower, between 111 and 125 GJ/t, and for PVC the Dutch, British, and US rates were very close at, respectively, 69.8, 66.0, and 60.8–70.8 GJ/t. The most detailed industry-wide analysis of energy needs in the US chemical industry found that in 1997 PE production averaged 77.4 GJ/t (with feedstock energy being 20.1 GJ/t), that PP embodied, on the average, 54.7 GJ/t, and that PVC's total average energy cost was 44.7 GJ/t, with feedstock accounting for 17 GJ/t (Energetics 2000).

A World Bank review of PE energy costs found ranges of 87.4–107.8 GJ/t for HDPE and 74.4–116.3 for LDPE, with the processing energy being as low as 25–28 GJ/t and as high as 45 GJ/t (Vlachopoulos 2009). Berge (2009), in his survey of construction materials, cited European rates of 110–115 GJ/t for PE and PP and 80–90 GJ/t for PVC. And in their review of best current practices, Worrell et al. (2008) found that energy costs of ethylene made from naphtha or ethane were very similar, respectively, 14–22 and 12.5–21 GJ/t. Heavy dependence of modern production of plastics on hydrocarbon feedstocks has not been, so far, a major burden as the industry still claims less than 5% of the world's natural gas and crude oil output.

Rising demand for plastic materials and higher costs, particularly of crude oil, will change this, and in the long run plant-based bioplastics appear to be the only practical answer both to the eventually less abundant petrochemical feedstocks and to the presence of nonbiodegradable materials in the environment. But higher demand for composite materials increasingly used in aerospace and automotive industries will result in higher energy costs even after adjustments are made for the lower density and superior strength of these new materials. For example, lignin carbon fiber used in reducing the weight of passenger cars and other vehicles costs 670 GJ/t and carbon-fiber reinforced polymer (polyacrylonitrite) fiber requires just over 700 GJ/t (Das 2011), more than three times than aluminum.

Because of high pressures and high temperatures required for the catalytic synthesis of **ammonia** from its elements, the Haber-Bosch process used to be among the most energy-intensive sectors in chemical industry: the first commercial Haber-Bosch plants (using coke as their feedstock) required more than 100 GJ/t of NH_3, and pre-WWII coal-based plants still needed about 85 GJ/t. Post-1950 shift to natural gas and low-pressure reforming using reciprocating compressors lowered the rate to 50–55 GJ/t. In modern plants, nitrogen is obtained by fractional separation from air and hydrogen comes from the steam reforming natural gas; natural gas is also used to drive compressors, acting as both feedstock and energy source and the specific energy cost of the best practices has been reduced to rates remarkably close to the stoichiometric minimum of 20.9 GJ/t (Smil 2001; Worrell and Galitsky 2008).

The most important innovation came after 1963 with the introduction of centrifugal compressors: while electricity-powered reciprocating compressors consumed 520–700 kWh/t NH_3, steam turbine-driven centrifugal compressors need as little as 20–35 kWh/t NH_3, a 95% reduction. High-pressure reforming (with pressures above 3 MPa) was another innovation of the 1960s, and by the 1970s, this combination lowered the energy use to just 35 GJ/t NH_3. Combination of new plant designs, higher efficiency of individual plant processes and better catalysts cut the best performances to as little as 27 GJ/t NH_3 in the year 2000. When Worrell et al. (2008) reviewed the best commercial practices, they rated natural gas-based synthesis at 28 GJ/t (roughly a third higher than the stoichiometric minimum) and coal-based process at 34.8 GJ/t.

Naturally, typical performances are higher, around 30 GJ/t NH_3 for new gas-based plants, 36 GJ/t for heavy fuel oil feedstock, and more than 45 GJ/t NH_3 for coal-based synthesis (Rafiqul et al. 2005). IEA (2007) used the actual regional means ranging from 48.4 GJ/t in China to 35 GJ/t in Western Europe, resulting in the global weighted mean of 41.6 GJ/t for the year 2005. At 41 GJ/t that global estimate remained almost unchanged by 2020 (IEA 2021b). Conversion of NH_3 to $CO(NH_2)_2$ (urea) requires 10–12 GJ of energy per ton of nitrogen, and the prilling (production of small uniform spheres from molten solids commonly used to make granular fertilizers), bagging and distribution of the compound raise the total energy cost of this solid fertilizer is typically 55–58 GJ/t N.

Surface-mining of **phosphates** takes only 4–5 GJ/t, but treating the insoluble rocks with sulfuric and nitric acids in order to produce water-soluble phosphorus compounds is much more energy-intensive. Overall energy costs range from 18 to 20 GJ/t for superphosphates (single superphosphate with just 8.8% P, triple superphosphate with 18–27% P) to 28–33 GJ/t for diammonium phosphate containing 20% of soluble P (Smil 2008). Energy cost of potash (sylvinite) extraction is low: in Saskatchewan conventional underground mining followed by milling needs only 1–1.5 GJ/t, and surface mining and milling averages only about 300 MJ/t (NRC 2009).

Because electricity accounts for most of the energy used to produce **electronic-grade silicon,** the energy intensity of the production chain is usually expressed in kWh rather than in MJ/kg, and the conversion to primary energy will depend on the source of electricity. Crystals of electronic-grade silicon are one billion times purer than the metallurgical grade of the element that is used for deoxidation and alloying (Föll 2000). Making metallurgical-grade Si from quartz requires 12–13 kWh/kg, Siemens process consumes about 250 (200–300) kWh/kg, Czochralski process has roughly the same energy need, and processing wafers from ingots adds 240 kWh/kg for a total of about 750 kWh. But the actual rate is much higher because of significant material losses along the processing chain.

Conversion from quartz to metallurgical Si has a 90% yield, but the Si yield for the Siemens process is less than 50%. Monocrystalline Si produced by the Czochralski process from polycrystalline Si (whose making uses at least 400 MJ/kg) that consumes on the order of 1 GJ/kg of electricity for the total of nearly 1.5 GJ/kg (Kato et al. 1997). Conversion of Czochralski crystals to wafers is fairly wasteful, with Si lost from the ingot's top and tail, due to kerf and pot scrap and test wafer cuts. As a result, Williams et al. (2002) found that 9.4 kg of raw Si were needed to make a kilogram of final microchip wafer.

This means that their entire production chain – starting with Si made from quartz and carbon through trichlorosilane, polysilicon, single crystal ingot, Si wafers and actual fabrication, and assembly of a microchip - consumed about 41 MJ for a 2 g chip. This implies total electricity cost of shipped wafers of at least 2,100 kWh/kg per kilogram. Even if using only hydroelectricity, this would prorate to about 7.6 GJ/kg, and energizing the entire process by electricity generated from fossil fuels would push the total primary energy to more than 20 GJ/kg of finished Si wafers. Schmidt et al. (2012) offered a detailed assessment of life-cycle electronic-grade silicon wafers but expressed their results per m^2 of wafers. Their data, converted per kilogram of microchips (assuming 30-mm wafers weighing 125 g), translate to between 2.5 and 7.7 GJ of primary energy with all electricity from renewables, and 7.5–23 GJ/kg with electricity from the combustion of fossil fuels. Total of 20 GJ/kg would be three orders of magnitude higher than for steel made from iron ore and two orders of magnitude more than aluminum made from bauxite.

Typical rates presented in this section can be used (after rounding, to avoid impressions of unwarranted accuracy) to assess global energy needs of major material sectors and to calculate their fractions of total primary energy supply (TPES) of 587 EJ in 2019 (BP 2022). Not surprisingly, steel's relatively high energy intensity (global mean of 19.5 GJ/t in 2019) and its massive output (1.88 Gt in 2019) make it the material with the highest total energy demand, about 37 EJ or 6.5% of the global primary energy supply. Plastics are next (assuming 80 GJ/t and output of 367 Mt in 2020) with roughly 29 EJ (about 5% of TPES), well ahead of construction materials (cement, bricks, glass) with about 20 EJ, or 3.5% of TPES (cement alone was about 14 EJ or 2.5%). Paper production required about 10 EJ (1.7%), aluminum came next (about 8 EJ and 1.4%), synthesis of ammonia (180 Mt at 41 GJ/t) required about 7.4 EJ (1.3%), and phosphate and potassium fertilizers added less than 1 EJ.

A surprising fact that paper is slightly ahead of aluminum is as a result of paper's relatively higher energy intensity and much large mass of annual production. Perhaps the most interesting result concerns the energy cost of inorganic fertilizers: given their truly existential importance, it is reassuring to realize that energy needed to produce them adds up to a surprisingly small share of global supply. Assuming averages of 50, 25, and 10 GJ/t for, respectively, N, P, and K (all including the cost of final formulation, packaging and distribution) would result in total demand of a bit more than 5 EJ in the year 2020 (with nitrogenous fertilizers accounting for about 90% of the total) - or slightly less than 1% of the TPES.

That must be one of the most rewarding energy investments in history because without the application of these compounds, we could not support nearly half of the existing population of eight billion people; I hasten to add that a still high level of global malnutrition (just below 10% in 2020) is due overwhelmingly to unequal access to food, not to inadequate food supply (the latter is the case only in such countries as Ethiopia and South Sudan, beset by chronic violent conflicts). Another case of a highly rewarding energy pay-off is the cost of silicon wafers. As I explained, their embodied energy is orders of magnitude higher than for any of the commonly used materials, on the order of 20 TJ/t compared to less than 20 GJ/t for crude steel.

But the aggregate energy demand remains very small because the steadily increasing crowding of transistors has limited the annual mass of wafers needed to produce all of the world's microchips. In 2020, producers shipped about 8 Mm^2 of semiconductor silicon and that is the equivalent of about 113 million 300-mm wafers (each one has the area of 706.95 cm^2) or (with weight of 125 g/wafer) to only about 14,000 tons of wafers. When multiplied by 20 TJ/t, this implies energy requirement of about 300 PJ, or a mere 0.05% of the world's primary energy demand in 2020. These calculations also make it clear that the modern civilization can afford to consume large and increasing mass of steel and fertilizers and deploy microchips in myriads of finished products because scientific discoveries and technical advances have greatly reduced their energy intensities.

For example, had the steelmaking process remained as it was in 1900 (all pig iron smelted in blast furnaces, all steel made in open-hearth furnaces), then the global production of 1.88 Gt of the metal in the year 2020 would have claimed almost 20% of the world's total primary energy supply - rather than just over 6%. Another perspective is even more impressive: in 2020, the aggregate energy cost of what I have called the four material pillars if the modern world - steel, cement, plastics, and ammonia - was about 88 EJ, nearly 16% of all primary energy supply and slightly less than would have been the cost of steel alone if made with energy inputs typical of 1900. And even when using liberal rates for average energy intensities of all biomaterials other than paper and all metals other than steel and aluminum, I end up with a grand total of no more than 120 EJ, or just over 21% of the world's total primary energy supply in 2020: we create the modern world's material wealth with about a fifth of all energy we use!

4.5 Life-Cycle Assessments

Quantifications of energies embedded in materials are useful and revealing companions to standard analyses of material flows using mass or value. But a brief reflection shows that even the combination of mass, value, and energy does not provide a sufficient basis for informed and balanced assessment because it leaves out the broad category of environmental impacts. Moreover, such assessments should not be limited to costs and impacts at the production stage because many materials are parts of products and structures that will serve for years and decades and full consequences of their use can be appreciated only by appraising their life-long trajectories. As the OECD put it, "all the phases of organized matter and energy that are in some way related to the making and use of a product can also be linked to an impact on the environment" (OECD 1995).

The discipline that is doing precisely that - quantifying objective indicators of the environmental impacts of materials during their entire life span - is life-cycle assessment (or life-cycle analysis, with the identical acronym, LCA), now a widely accepted procedure that has been codified as a part of international environmental management standards that are routinely used to evaluate various burdens imposed by production, use, and disposal (or re-use) of products and by the performance of services (ISO 2006a, 2006b). LCA is now a mature analytical discipline that has its own periodical, *International Journal of Life Cycle Assessment* published since 1995 by Springer, and a growing number of how-to and field-review books (Horn et al. 2009; Jolliet et al. 2013; Klöpffer and Grahl 2014; Hauschild et al. 2018), as well as extensive life-cycle inventory databases maintained by the US National Renewable Energy Laboratory (NREL 2022b), by the European Commission (2022), and by ecoinvent in Zurich (ecoinvent 2022).

Varieties of LCA include complete cradle-to-grave sequence (from raw materials to final disposal) and partial cradle-to-gate assessments (that is for finished products before their distribution for sale). Inventory databases contain information on input and output flows for many materials but a closer look at the recently published LCAs quickly reveals that the choice of assessed items (products or processes) has not been guided either by their importance to modern civilization (in that case studies of steel, other leading metals, concrete, glass, ammonia, and various plastics would dominate) or by the quest to quantify the potentially most worrisome cases (in that case there would be many assessments of heavy metals).

Instead, recently published life-cycle analyses offer a diverse choice: they are now available for a number of common crops, for foodstuffs including bread, milk and tomatoes, for renewable energy production systems (biomass, wind), and for products ranging from tiny microchips to granite sidewalks and entire residential and office buildings. Size, obscurity, or novelty of products has been no obstacle: we have LCAs of cotton shirts (Kazan et al. 2020) and houses (Praseeda et al. 2017), of zirconium sand and nanomaterials (Hischier 2014; Gediga et al. 2019), and thousands of analyses are accessible to those able to pay major database fees. Obviously, some of these findings will be fairly durable, other LCAs will require relatively frequent reassessments: making steel via the blast furnace-BOF-continuous casting route is not subject to the same rate of change as is the production of consumer electronics.

4.5.1 LCA Benefits and Problems

LCA can offer many useful insights but it also has a number of inherent problems, and it is often performed in a truncated or superficial manner. Its most obvious contribution is to provide a revealing reminder of post-purchase (or post-installation) costs of a product, as well as of the fact that in many instances, costs associated with the operation and use of products and infrastructures are far higher than the costs and impacts of its creation; that, obviously, is a key for devising rational response strategies. These realities are well illustrated by examples ranging from LCA of buildings and cars to production and washing of garments.

Given the differences in house sizes and in energy intensity of lumber, steel, or concrete (the three principal structural materials in modern housing), it is not surprising that the life-time energy cost (construction, maintenance, operation, demolition) of residential space has a roughly threefold range ($3–9\,GJ/m^2$). Total energy embedded in single family house can be as low as 200 GJ for a small wooden structure, it averages around 500 GJ for a three-bedroom, wood-framed bungalow and it surpasses 2 TJ for large two-story buildings (Canadian Wood Council 2004; Smil 2008). The quest for energy savings in buildings (regulation of temperature, air flow, and moisture) has led to more rigorous construction codes and to a deployment of a wider range of insulating materials. Fiberglass, foamed plastics, and loose fill (made of plant fibers and cellulose) are commonly used to insulate walls and ceilings, plastic sheets and foils are applied as moisture barriers, and aerated concrete and gypsum boards are used to reduce noise.

LCA of new, highly energy-efficient buildings shows that the shares of embodied energy as high as 57% and, obviously, for a near-zero energy construction (superinsulation, solar heating, and cooling), the ratio is above 90%. In contrast, and as must be expected, LCAs of housing in cold climates show embodied energies of ordinary houses to be a small fraction of life-time total. LCA of three types of residential buildings (concrete, steel, and wood framework) in Beijing (hot summers but often very cold winters) showed that embodied energy was no more than 13–27% of lifelong energy consumption (Gong et al. 2012). And even for a larger (200 m²) Canadian house that cost 1.5 TJ to build, heating and lighting (averaging about 25 W/m²) will claim about 9.5 TJ in 60 years, reducing the construction share to just 14% of the overall cost. By coincidence that share is nearly identical to construction share of a medium-sized American car: it takes about 100 GJ to produce and (at about 8 l/100 km and 20,000 km/year) it will need about 550 GJ of fuel and oil in 10 years and the initial construction will claim only about 15% of the overall cost, and this share would decline further if repairs and garaging were included (Smil 2008). Embodied energies make up even lower shares in life cycles of machines that are nearly in constant operation: only 6–7% for jetliners, freight trains, and cargo ships (Allwood and Cullen 2012).

But the shares of initial and operating costs are not always clear. While several LCAs show the energy costs of operating a personal computer or a laptop substantially higher than those of manufacturing it – the split as low as 74 : 26 and as high as 93 : 7 (Andrae and Andersen 2010) - Williams (2004) ended up with a reverse ratio. According to his analysis energy used to make a desktop with Pentium III processor, 30 GB hard drive, and 42.5 cm monitor added up to 6.4 GJ while during its relatively life of 3 years, the desktop would consume about 420 kWh or electricity or roughly 1.5 GJ of primary energy, yielding the manufacturing-to-usage energy split of 81 : 19.

LCA also makes it clear that over their lifetime many infrastructures will cost nearly as much, as much or more, to maintain than their initial construction. Canadian LCA for a high-volume two-lane concrete highway shows the initial construction costs 6.7 TJ/km and rehabilitation cost of 4.1 TJ/km or 38% of the total 50-year cost of 10.8 TJ/km, and the burdens are actually reversed for a roadway made of flexible asphalt concrete that needs 15 TJ/km to build and 16% more (17.4 TJ/km) to rehabilitate over a 50-year life cycle.

LCA of repeatedly washed garments is yet another excellent illustration of the boundary problem noted at the beginning of the energy cost section, as products made from different materials have different durabilities and maintenance requirements and a complete account of these realities may shift the overall advantage from a material that requires less energy to produce to one that is more energy-intensive to make but whose life-long energy cost may be lower. LCA of America's ubiquitous short-sleeved industrial work shirt (65% PE, 35% cotton) laundered 52 times and then landfilled showed that the shirt's production accounted for 36% of all energy use, 26% of CO_2 equivalents, and 24% of water (Cartwright et al. 2011).

Comparison of pure cotton, cotton/polyester blends, and pure polyester fabric by van Winkle et al. (1978) was one of the pioneering energy analysis studies embracing the LCA. The study found that polyester production required twice as much energy as producing cotton lint (clear advantage cotton) but because of a higher cost of cloth manufacturing the total energy cost of cotton shirt was about 20% higher than that of its pure polyester counterpart (slight advantage polyester), and after including the energy costs of maintenance (washing, drying, ironing) the cotton shirt was about 3.6 times more energy-intensive (clear advantage polyester). Similarly, life-cycle analysis by Kalliala and Nousiainen (1999) found that sheets made from pure cotton and laundered 100 times require 72% more energy than 50/50 cotton/polyester blend because two 100% cotton sheets are needed to cover the life-time of one blended sheet.

Cotton appears even more disadvantaged once non-energy impacts are compared: water requirements of the blend are less than a third of those for pure cotton, and global warming and acidification potentials are about 40% lower than those for producing and laundering two pure cotton sheets. Going even further we can consider the long-term cost of excessive soil erosion in cotton fields, soil quality declines due to salinization in irrigated cotton fields in arid regions and the presence of pesticide residues in soil and water (Smil 2008). All of those are avoided by using a synthetic fiber - but its production depends on a nonrenewable feedstock. But so does the cultivation of cotton: PE synthesis consumes about 1.5 kg of hydrocarbons per kilogram of fiber but growing a kilogram of cotton requires nearly 500 g of fertilizers and 15 g of pesticides made from hydrocarbon feedstocks, as well as liquid fossil fuels for farm machinery.

And one element still missing in all of these comparisons is consumer preference: should the superior wearing qualities of cotton shirt count for nothing? Moreover, a closer look at LCA of textiles makes clear that such analyses cannot be accurate unless the yarn thickness and fabric density are specified, and that from the environmental point of view knitting is better than weaving (van der Velden et al. 2014). And when using LCA in order to compare environmental impacts of different products, their weight may be a misleading common denominator. That is why Shadman and McManus (2004, 1915) objected to an analysis of microchip energy and material cost by Williams et al. (2002), calling the use of weight as the basis for comparison "arbitrary, nonscientific and inaccurate" because "it does not account for performance, utility, and benefits" of compared products. Of course, this is fundamentally just another instance of a missing qualitative perspective, an inherent weakness of LCA comparisons based on mass.

But what LCAs have done is to allow comparisons within the same category of environmental consequences (when choice of materials is possible which one will have the lowest effect on water use or water pollution?) as well as more comprehensive rankings of materials according to several categories of environmental impacts. This is important because material use is one of the three dominant ways in which the humans have been changing the biosphere: food production and energy supply (dominated by extraction and combustion of fossil fuels) are the other two great interventions. And when considered in its entirety the intricate system of extraction, processing, transportation, use, reuse, and disposal of materials encompasses every major environmental interference, from land use changes (ranging from deforestation due to lumber and pulp production to destruction of plant cover and disruption of water cycle due to massive surface ore mines) to atmospheric emissions (ranging from acidifying gases to being a major contributor to anthropogenic warming).

I should stress that because of their different action modes and different qualitative consequences (including many physical changes, chemical transformations and interferences in the structure and function of ecosystems), different spatial effects (ranging from local to global), and different durations (from ephemeral to consequences felt for millennia), there can be no ranking of anthropogenic environmental impacts. There is no unifying metric that would allow us to conclude that soil erosion should be a greater concern than photochemical smog or that tropical deforestation is more worrisome than enormous water demand by modern irrigated agriculture. And, obviously, it would be impractical to perform LCAs that would attempt to quantify every known environmental impact: that is why most analyses settle on a small number of leading indicators of environmental change or degradation.

4.5.2   LCA and GWP

Calculations of overall global warming potential - GWP, a relative indicator of the intensity with which specific greenhouse gases (GHGs) absorb the outgoing terrestrial infrared radiation and of the lifetime of those gases in the atmosphere - have been the most common component of recent LCAs. All GHGs are scored against CO_2 whose GWP value is 1. Values for the other leading greenhouse gases are 28 for CH_4 and 265 for N_2O and the highest equivalents are 23,500 for SF_6 (sulfur hexafluoride, used as insulating medium in electric industry) and 11,100 for C_2F_6, perfluoroethane, used in electronic manufacturing (Myhre et al. 2013). Obviously omitting or undercounting such emissions would have a huge impact on the final value. GWPs of materials are usually expressed in equivalents of CO_2 per unit mass of a material (kg CO_2-eq/kg), but for buildings the values can be prorated per unit area (kg CO_2-eq/m^2).

Calculations begin with specific emission rates expressed in t CO_2/TJ of energy; IPCC default rates are 56.1 for natural gas, 73.3 for crude oil, 94.6 for bituminous coal, 101.02 for lignite, but actual nationwide or regional values can differ by a few percent for hydrocarbons and by more than 10% for solid fuels, with European lignite ranging from 90.7 to 124.7 t CO_2/TJ. With CO_2 being usually the leading contributor, this indicator will have a very high correlation with the mass of fossil fuel used (directly or indirectly to generate electricity) to make a product. For example, making cement will release about 0.5 kg CO_2 per kilogram but making a kilogram of hot-rolled sheet steel will release about 2.25 kg of CO_2.

Given the rising interest in carbon emissions and their intended reduction, it is not surprising that we now know CO_2 emissions for thousands of individual items, and that this data collection keeps on increasing. To make searches easier, anybody can now access available electronic databases or buy paperbacks that summarize our expanding carbon footprint knowledge (Meinrenken et al. 2022; Berners-Lee 2022). These numbers provide many interesting contrasts, no matter if the comparisons are based on a product basis (comparing different masses) or per unit mass (usually per kilogram for lighter items, per ton for bulky materials).

Beverages (mostly water) have low-carbon intensity (<1 kg CO_2/l), common foodstuffs come next (pasta and bread at about 1), basic building and industrial materials have relatively low carbon intensities, but their large annual output makes them leaders in absolute terms (all tons per ton): cement is around 1, plastics 1.5–2.5, steel 2–2.5, aluminum 4–5. Production of complex (and still predominantly mechanical) machines generates an order of magnitude more of the gas, cars mostly 20–30, but everything that is now largely or predominantly electronic is at least as much or an order of magnitude more, carbon-intensive: cordless mouse comes at about 30, laser printers 50–150, laptops at around 200, a smartphone at 400.

The other two commonly assessed variables are acidification and eutrophication impacts. Emissions of sulfur and nitrogen oxides from combustion of fossil fuels, smelting of ores, and other industrial processes are the precursors of atmospheric sulfates and nitrates whose wet and dry deposition acidifies waters and soils. In LCAs, this acidifying potential of products or processes is calculated in terms of grams of hydrogen ions per square meter (g H^+/m^2) or per volume of water (g H^+/L). Eutrophication is the process of nutrient enrichment of fresh and coastal waters (most often by releases and leaching of nitrates and phosphates) that leads to excessive growth of algae whose decay deprives waters of dissolved oxygen and creates anoxic zones that either kill or impoverish heterotrophic life in affected waters (Harper 1992; Ansari and Gill 2013). In LCAs, this impact is usually expressed in kilogram of

PO_4^{3-} equivalents). Perhaps the least frequent variable included in LCAs is the smog-forming potential of products or processes, expressed as $g\ NO_x/m^2$.

As explained in the previous section, production of materials claims roughly a quarter of the world's primary energy supply. Because the combustion of fossil fuels provides most of this energy - in global terms about 82% in 2020, with the rest coming from primary, that is mostly hydro and nuclear electricity (BP 2022) - production of materials is a leading source of emissions of particulate matter (including black carbon), SO_x and NO_x (whose conversion to sulfates and nitrates is the primary cause of acidifying precipitation) and greenhouse gases. Water needed for processing, reaction, and cooling ends up contaminated or is released with elevated temperature by many industries, and most of them also share the necessity of disposing of relatively large volumes of solid waste and small but potentially worrisome volumes of hazardous waste. Fortunately, these common environmental challenges have many common (generic would be a correct adjective here) solutions. At the same time, there are many specific challenges arising from unique physical or chemical properties or from exceptionally demanding industrial processes.

Many studies offer either fairly comprehensive environmental profiles of materials - including, as applicable, assessments of their GWP, acidification potential, water consumption, and water pollution burden – or the comparisons of specific consequences (with GWP being, not surprisingly, a recent favorite). Here are brief environmental impact summaries of some common materials. In 2020, efficient production of a kilogram of construction steel sections generated about 1.6 kg of CO_2 equivalent, nearly 4 g of SO_2 equivalent of acidification potential, negligible amount of eutrophication potential (0.32 g), and just 0.67 g of the photochemical smog-inducing ethene (WSA 2020).

4.5.3   LCA Guidance

What I like best about LCA is that it repeatedly offers some interesting and sometimes also counterintuitive results. Which paving has the lower impact: natural granite-slab or a concrete sidewalk? Mendoza et al. (2012) concluded that - largely because of energy needed to cut and move the stone - granite sidewalk has 25–140% higher impact than a concrete one. Which transport packaging of fruits and vegetables carries the lowest CO_2 equivalent: traditional light (less than 1 kg) one-way wooden boxes made from obviously renewable material whose energy content can be recovered by incineration; even lighter one-way cardboard boxes that could be either incinerated or recycled; or much heavier (2 kg) multi-way plastic crates that must be washed between uses? When the impact is scaled to 10 annual rotations, then the per person GWP of plastic crates is about 10% lower than that of wooden boxes and less than half of cardboard packaging; when scaled to one million annual rotations, plastic crates generate about 332 kg of CO_2 equivalent, wooden boxes about 367 kg, and cardboard boxes 708 kg (University of Stuttgart 2007).

Another counterintuitive result comes from a Thai LCA of single-use thermoform boxes that are widely used for food packaging (Suwanmanee et al. 2013). The study compared boxes made from hydrocarbon-derived polystyrene (PS) with those made from polylactic acid (PLA) derived from corn, cassava, or sugarcane by using country-specific material and energy inputs to calculate three impact variables, GWP (including direct GHG emission and indirect emissions from land-use changes), acidification potential, and photochemical ozone formation. PS boxes have considerably lower overall environmental impact, above all due to considerable indirect impacts caused by corn and cassava cultivation.

Natural materials can be a better choice, and in the case of biomaterials resoundingly so. For example, a comprehensive LCA by Bolin and Smith (2011) showed that a wood/plastic composite decking results in 14 times higher fossil fuel use, three times more GHG, almost three times more water use, four times higher acidification potential, and about two times more smog potential and ecotoxicity than a wooden deck built with the lumber treated with alkaline copper. Advantages of biomaterials matter most in long-lasting structures, and many studies have shown environmental advantages of wooden buildings and wooden house components.

For example, glulam beams embody about 500 MJ/m² of roof compared to at least 1 GJ/m³ for steel beams made from the recycled metal and 1.6 GJ/m³ for steel made from blast-furnace iron (Petersen and Solberg 2002). Similarly, using cross-laminated timber instead of reinforced concrete floors in steel-skeleton buildings could bring substantial energy savings (D'Amico et al. 2021). Wooden floors are also much less energy-intensive than the common alternatives: total energy per m² of flooring per year of service was put at 1.6 MJ for wood (usually oak or maple) compared to 2.3 MJ for linoleum and 2.8 MJ for vinyl (Jönsson et al. 1997). Another study found solid oak boards superior not only to linoleum and vinyl but also to polyamide carpet and wool carpet (Petersen and Solberg 2004).

LCAs indicate significantly lower (as much as 65%) greenhouse gas emissions when compared to reinforced concrete buildings (pre-cast or cast-in-place), and similarly large savings for flooring or furniture. Eriksson's (2004) comparisons of Swedish house show major energy and GWP savings. A four-story wood-frame residential houses required 960 MJ/m² but after subtracting recyclable energy of 1,460 MJ/m², there was an actual credit of 530 MJ/m² compared to net energy requirement of 1,770 MJ/m² for a concrete house, and the GWPs (both excluding usage phase) were, respectively, just 30 and 400 kg CO_2-equivalent/m². And increasing timber content of typical residential buildings in Scotland could save as much as 75% of life-cycle GHG emissions for a small semi-detached house and up to 86%

for a larger detached structure (Edinburgh Centre for Carbon Management 2006). Moreover, wood recovered from demolished houses or discarded furniture could be burned, releasing more, or much more, energy than went into its production.

Additional advantage of managed forests and tree plantations are their contributions to oxygen generation and carbon sequestration, and if trees are harvested in rotation and promptly replanted, such plantations can be long-lasting stores of carbon. Even more importantly, proper forestry management can increase phytomass storage and annual productivity of natural forests as shown by a nearly century-long perspective from Finland. In a periodically studied area in the southern part of the country, growing stock in essentially identical area increased 2.15 times between 1912 and 2005 and annual carbon sequestration averaged about $30 t/km^2$ (Kauppi et al. 2010). In addition, unless the wood cut in a natural forest is used directly as fuel or to make charcoal its carbon sequestration effect can be prolonged by many decades once it is converted into structural lumber or furniture (Berge 2009). Not surprisingly, typical life span of wood products remaining in place vary widely, with the greatest range for construction lumber (Lauk et al. 2012).

Sawn wood remains in place for less than two or three decades in many mobile homes, or it can last easily 80–100 years in well-built wooden houses in the United States and Canada. Life spans of modern wooden (veneered plywood) furniture are generally much shorter, 10–30 years and two decades may be a good average. Wood used in packaging and shipping (boxes, crates, pallets) may be discarded immediately or reused for a few years, and packaging paper and paperboard have similar durability (one to six years, average of about two to three years). Other persistent uses include finishing components of houses (stairs, doors, door and window frames, floors), barrels and casks for wine and spirits, specialty woods for musical instruments and for sport implements, and (a small but rewarding niche) wooden toys and puzzles.

But, as with any complex analysis, users should beware: LCA has plenty of problems and weaknesses and too often it still falls short of really comprehensive cradle-to-grave accounting. An obvious problem involves comparisons of materials made of feedstocks with high energy content with those whose embedded energy comes only from fuels and electricity used in their production. For example, LCA shows that energy embodied in flexible asphalt concrete roadway is roughly twice as high as that in rigid Portland cement concrete – but the difference is almost entirely due to the energy content of the feedstock (Cement Association of Canada 2006). Asphalt (bitumen) is a mixture of heavy hydrocarbons with a high energy value (40 GJ/t), but it is an unsuitable energy source and is obtained as a by-product of oil refining with the expenditure of only 0.5–6 GJ/t of process energy, in most cases no more than the production of Portland cement. Consequently, a more logical procedure would be to compare only the energies needed to construct the two kinds of pavements, and that would yield a very similar value.

Remarkably, many life-cycle assessments take a cavalier approach to life spans and just assume what seems to be a reasonable length or use (often unexplained) spans from previous studies. Perhaps most notably, this has been the case with buildings whose longevity means that arbitrary assumptions may end up with particularly inaccurate appraisals of life-cycle impacts. Modern houses have life spans mostly between 50 and 100 years, modern commercial buildings last between just 12 and about 50 years, but many of them (particularly stores) undergo relatively frequent and extensive remodeling. An inquiry by Aktas and Bilec (2012) put the average lifetime of American residential buildings at 61 with a standard deviation of 25 years. That study established mean lifetimes of major house components: paint at about 7 years, carpets 10 years, linoleum and vinyl floors 22 years, hardwood floors and finishes 42 years, and ceramic 48 years.

Another category of errors arises from assuming generic rates rather than ascertaining nation-specific values. This problem is perhaps best illustrated by examples of highly inefficient industrial processes that abound in most modernizing countries. But perhaps the most important shortcoming of LCA is the fact that in many cases it still presents only truncated and incomplete appraisals, and perhaps no other category of materials illustrates this reality as well as the plastics. While NREL's (2022b) output flows for HDPE quantify about 220 chemicals including organics (from acetone to phthalates), heavy metals, radionuclides, and GHGs, and while some published LCAs of specific plastics, such as the cited work by Suwanmanee et al. (2013), account even for indirect effects resulting from land-use changes caused by cultivation of plants producing renewable feedstocks, there are no LCAs taking into account great longevity of some plastics (on the order of hundreds of years) and their now ubiquitous, and clearly highly disruptive, distribution in aquatic environments.

Their buoyancy, their breakdown into progressively smaller particles, and their eventual sinking through the water column (some remaining suspended at different depths, others settling at the sea bottom) combine to make them a truly global and omnipresent environmental risk to marine biota: they are now found on the beaches of the remotest islands as well as in the abyss. The extent of this new kind of environmental degradation became first clear with the discovery of plastic refuse trapped within the North Pacific Gyre: this was encountered in 1997 by Charles Moore and his crew returning from the trans-Pacific sailing race and soon it became known as the Great Pacific Garbage Patch (Moore and Phillips 2011).

While the highest visible concentrations of plastics (ranging from large pieces to tiny particles) are in surface water and on the beaches, new research makes it clear that the ocean interior conceals enormous amounts of small-sized plastic debris, enough not only to balance but even to exceed the estimated plastic inputs into the ocean since 1950 (Pabortsava and Lampitt 2020). This also means, unfortunately, that both inputs and stocks of ocean plastics are much higher than previously determined. In 2022, UNEP began the process to enact a global treaty to address this enormous, and still worsening, form of contamination (UNEP 2022b).

Dangers to ocean life are posed by every size of discarded plastics: among the largest items are abandoned or damaged fishing nets that can ensnare fishes, dolphins, and often even whales; aquatic birds often mistake small-size pieces of plastic for small fish or invertebrates and regurgitate them to their fledglings: stomachs of many species show a distressing collection of such objects. And microplastics - the smallest pieces (sizes less than 5, 2 or 1 mm) that are manufactured for cosmetics, drugs, and industrial uses and that arise from abrasion and photodegradation of larger pieces - have been found everywhere: in surface waters, in deeper layers, and even at the bottom of the world's deepest ocean trenches (Choy et al. 2019; Jamieson et al. 2019). They can be ingested both by marine biota (and can have serious metabolic and toxic effects) and also by people (in the US annual microplastic consumption and inhalation ranges from 74,000 to 121,000 particles per capita), but more evidence is needed to assess the risks of these exposures (Cole et al. 2011; Ricciardi et al. 2021).

This important example shows how incomplete and uncertain are even our best analytical procedures tracing the requirements and consequences of material production, use, and abandonment; it also provides a strong argument for a much better management of materials and, obviously, recycling should be a key component of these efforts. A closer look at the practice reveals enormous, and inexplicably ignored, opportunities as well as expected complexities (including some counterintuitive realities): recycling is both a practical and a highly effective tool that should be deployed to the greatest extent possible, but that extent must be carefully determined and vigorously enforced in order to avoid an even greater overall waste of materials.

4.6   Recycling

Recycling, together with thoughtful design, is a key factor in using resources more efficiently (Worrell and Reuter 2014). Recycling is the flow that closes a circle and that turns those overwhelmingly unidirectional sequences of modern material extraction-production-use-abandonment into cycles. This is, of course, how the biosphere works: provision of water, carbon, and nitrogen to the Earth's biota could not be sustained without biogeochemical cycles. In a civilization using billions of tons of materials every year, recycling is thus an exceedingly desirable flow that can greatly extend the life span of known mineral reserves, lessen the need for new biomaterials, and prolong the utility of materials turned into manufactured products.

No less beneficial is the associated reduction in the use of embedded energy (differences in overall energy demand per unit of product can be as high as an order of magnitude) and, consequently, often substantial alleviation of many negative environmental impacts that result from production and thoughtless abandonment of materials.

The term covers several outcomes. Material recycling may be a matter of reusing the entire product or component; recovering the original constituents by separating, crushing, or melting the discarded products; or burning the recovered materials to generate heat or electricity. Materials are reused either as salvaged (bricks are perhaps the most common example of this practice) or after minimal refurbishment (massive timbers salvaged from old buildings are often resawn). Most of the recycling is done in order to recover the original ingredients that can be used by recyclers or sold to other industries as raw materials. Metals could be recycled with a minimum degree of waste to produce materials that are qualitatively indistinguishable, or very close, to the original (primary) inputs, but with most materials recycling is more accurately described as down-cycling: high-quality paper becomes packaging stock or cardboard, expensive plastics are turned into cheap items.

4.6.1   Benefits and Rates

Energy savings and reduced environmental impacts are among the most appealing, and the best studied, advantages of recycling. Production of recycled steel will require roughly 70–75% less energy; it will save mining, transportation, and smelting of 1.4 t of iron ore and 740 kg of coking coal; and LCA of common steel products (sections, hot-rolled coil, and hot-dip galvanized steel) showed recycling benefits of up to about 50% for both GWP and acidification potential (WSA 2022c). Benefits of aluminum recycling, although not that high as commonly claimed, are even greater than for steel. A standard claim is that aluminum recycling saves of up to 96% of energy compared to production from bauxite. But that claim would be true only for melting the old metal. Its delaquering (removal of any coatings) will need about 7 GJ/t, melting consumes 7 GJ/t, addition of pure aluminum to adjust the alloy composition takes about 8 GJ/t, and the production of containers (casting, rolling, blanketing, forming) adds 30 GJ/t for the total of about 52 GJ/t, saving 74% rather than 96% (Luo and Soria 2008).

Energy cost of recycled paper varies with the waste product: recycling office paper can be done with just 1.8–2.5 GJ/t (the higher rate included deinking), by recycling boxboard takes up to 6 GJ/6: the latter rate is only 20%, the former more than 75% lower than the cost of new paper (7.5–8 GJ/t) made by kraft or thermomechanical process (Jacobs and IPST 2006; Rogers 2018). Energy savings for recycled high-density polyethylene are 45–50% and 40–45% for PVC. As for the emissions of GHGs, producing a kilogram of hot-rolled steel coil will generate 2.2 kg of CO_2 equivalent while recycling will claim only 1.8 kg, preventing emissions f 0.4 kg CO_2; savings with the recycling of plastics are much higher, close to 2 kg CO_2/kg of both PP and PVC, and aluminum recycling offers the greatest GWP reduction with at least 10 kg CO_2/kg of Al scrap (GHK 2006).

Obviously, lower energy intensity of recycled materials results in lower greenhouse gas emissions, claims less water, and generates less water and air pollution, and Bureau of International Recycling (2016) reviews many environmental benefits of the practice. But recycling is also a quest that is often very difficult to pursue (be it because of logistic challenges, excessive costs, or negligible energy savings) and one that, unlike the flows of carbon or nitrogen atoms in grand biogeochemical cycles, often amounts to a fairly rapid down-cycling as the reused materials appear in less valuable guises. Some challenges are fairly universal, others are material-specific. Recycling of waste predictably generated during large-scale industrial processing (be it smelting; casting; or machining of metals or production of plastics, glass, or paper) is obviously easiest to do, and it is the cheapest and the most rewarding way to reuse valuable waste. In some industries (above all in steelmaking, aluminum smelting and manufacturing of plastic products), this is a major source of recycled material.

In contrast, expensive collection costs burden recycling of all wastes that are produced with low spatial densities, with fluctuating quality and at varying rates: paper and plastic waste generated by urban households are the prime examples in this challenging category. But the greatest challenge is to recycle increasing amount of electronic waste that is, in mass terms, dominated by a few plastics, a few metals, and screen glass but that also contains small, but in aggregate quite substantial, amounts of more than a dozen elements. An overall verdict about the modern civilization's recycling performance is easy: so far, its recycling efforts have been quite inadequate even when compared to what might be labeled as a barely satisfactory effort.

For example, of the 60 metals and metalloids examined by a UN reported more than half (including all rare earths, as well as germanium, selenium, indium and tellurium) have recovery rates of less than 1%, for only five elements the rate is between 1% and 25%, and 18 common (or relatively common) metals have end-of-life recycling rates above 50%, but rarely above 60% (Graedel et al. 2011). America's metal recycling rates (latest data are for 2018) range widely, from the highs of 75% for lead and 60% for titanium to the lows of 17% for zinc and 25% for chromium, with iron and steel averaging 48%, copper 34%, and aluminum 53% (USGS 2022b). Another way to look at the current extent of recycling is to express the recovery of materials as a share of latest annual outputs. For metals these rates cluster in the 30s: 30% for zinc, 33% for aluminum, 35% for lead, 37% for steel, and 40% for copper.

Calculating the amounts of metals that will become available for recycling depends on estimates of the average lifetime of products. For example, for aluminum these range from discarding the items in production year (beverage cans, other containers, and packaging foils) to a decade in consumer durables and to more than 30 years in structural elements, machinery, and some equipment (Hatayama et al. 2009). Unfortunately, some toxic heavy metals, whose release into the environment is particularly undesirable, have very low recycling rates. The only relatively common way of recycling cadmium is by returned Ni-Cd batteries, but their collection rate remains low. Recycling rate of old fluorescent lights containing mercury has been also too low.

I will devote the rest of this chapter to brief appraisals of recycling efforts of four materials, two key metals (steel and aluminum) and plastics and paper, and of electronic waste, a category of discarded materials that would most benefit from much enhanced rates of recycling. Re-smelting of steel is by far the largest recycling effort in overall mass terms, an achievement made easier by the ubiquitous availability of the material, by the ease of its separation (magnetic sorting), and by its perfect recyclability: re-smelted material retains all physical properties of the original metal. In 2019, worldwide steel scrap use reached nearly 500 Mt.

About 38% of it originated in the category labeled by the industry as "own arisings," that is circulating scrap produced by mills and foundries during smelting, casting, rolling, and further processing of the metal (BIR 2021). The rest (62%) of steel scrap was purchased by steelworks, and it originated not only from many domestic sources but from imports, including those from different continents, with the United States (about 18 Mt) and European countries (about 50 Mt) being the largest exporters. National steel scrap/crude steel ratios cover a considerable range, from just 21% in China to 100% in Luxembourg. United States, Japanese, and Russian ratios in 2020 were, respectively, nearly 70%, 33%, and 42%, reflecting domestic availability of old and new steel scrap whose supply has been boosted in the United States by generations of high rates of car and major appliance ownership made largely of the metal. Recycling has become particularly important for making stainless steel (alloyed with Mo, Ti, V, and W) that is in increasingly higher demand.

High-energy pay-off (energy needed to recycle the metal is only about 5% of the total required for its production from bauxite) has made aluminum the world's most widely recycled material: nearly 75% of the metal that were produced worldwide since the 1880s (the global in-use stock of 1.09 Gt) are still embedded in buildings (about 36%), in electrical cables and machinery (about 25%), and about 30% of the total is in transportation equipment (IAI 2021). Global recycled ingot production reached 33.8 Mt in 2019, compared to 64.9 Mt for primary ingots. But there are clear limits to the metal's recycling. Hatayama et al. (2009) calculated that in the coming decades Japan, the United States, EU, and China could reduce their primary aluminum consumption to, respectively, 60%, 65%, 30%, and 85% of the levels prevailing between 2003 and 2005 - but also that by 2050 there will be 12.4 Mt of obsolete aluminum scrap that could not be recycled due to high concentrations of alloying elements.

Global aluminum recycling rates are high (about 90%) for transportation equipment (particularly airplanes) and construction materials, but six decades of American data illustrate how limited the recycling remains (USEPA 2022a). Aluminum now accounts for nearly 2% of all municipal solid waste, with the mass roughly split between goods (durable and nondurable) and packaging (dominated by beverage containers). The latter category has a higher recycling rate (about 35%), but the recycling of beer and soft drinks cans has recently fallen below 50%. In 1960, all aluminum municipal waste was landfilled; by 1980 18% of it were recycled, the ratio peaked at 27% in the year 2000 and since 2010, it has been just below 20%. In absolute terms, this means that in 2018 American landfills received almost twice as much aluminum as in 1960, nearly 2.4 Mt.

4.6.2   Plastics, Paper, E-Waste

But even the declining recycling rates for steel and aluminum are far higher than the returns for plastics. USEPA data on plastics in municipal solid waste show the rising burden of this synthetic waste: there was no recycling in 1970; by the year 2000 less than 6% of all plastics were recycled, by 2018 the rate was up to just 8.7% but the landfilled mass rose more than nine times even as about 16% of plastic was incinerated with energy recovery: plastics have high energy content, mostly between 41 and 45 GJ/t (USEPA 2021a). EU, with its profusion of national rules and EU directives, has seen slowly rising rates of recycling. In 2020, when the consumption reached about 49 Mt, about 30 Mt of plastics were actually collected, 42% of that total were in incinerated (official term is always "energy recovery"), nearly a quarter landfilled (down from about half a decade ago) and about 10 Mt (a fifth of all consumption) were sent for recycling, within EU or abroad (Plastics Europe 2021).

Recycling of omnipresent PET bottles illustrates the challenges of dealing with enormous numbers of lightweight items. In the United States, PET bottle recycling rate peaked in 1995 at about 37%, fell to about 22% by the year 2000 and even slightly lower afterward, and after brief period of rise, it began to decline again, from 29.1% in 2018 to 26.6% by 2020 (NAPCOR 2022), a poor performance that results in billions of bottles landfilled or discarded in cities, along roads, in forests, and on shorelines. Lightness and dispersion of these containers will always make their near-perfect recycling challenging: it takes 20,000 PET bottles to make 1 t of recyclable waste.

Recycling of plastics brings expected benefits of lower energy use - typically about 55–60 MJ/kg of PE, or savings of 60–80% (Vlachopoulos 2009) – and commensurate reduction of environmental pollution, but collection and processing of plastics is much more challenging than the reuse of metals, or paper, because household (or institutional) wastes contain mixtures of five major polymers (sometimes with two compounds forming parts of a single container, often in highly fluctuating ratios) that must be, after time-intensive collection, manually separated. Correct separation is difficult but imperative: a single PVC bottle in a load of 10,000 PET bottles can ruin the entire melt (ImpEE Project 2013). Sorting is followed by cleaning (but removing print and labels cannot be 100% successful) and reduction to uniform pellet sizes ready for reuse, still mostly only in lower-grade applications such as cheap carpets, garbage cans, or park benches.

But the challenge of properly managing plastic waste has been beyond our capabilities, and even with intensified recycling and incineration (that would require coordinated global action), it is likely that by 2050 we would still have up to 100 million tons of additional plastic waste entering the ocean every year and that the aggregate mass of plastics in landfills and in the environment (from soils containing bits of torn plastic coverings now widely used in field farming to deep ocean, from visible accumulations on shorelines to invisible microparticles in food) will be on the order of 12 Gt (UNEP 2018; Lau et al. 2020). That is an astounding total and I do not see how the suggested measures - replacing conventional plastics with alternative materials including natural fibers, biomass-based compostable synthetic biopolymers, and reusable durable nonplastic materials (UNEP 2018) - can reverse the tide.

All affluent countries, as well as China, have done much, and increasingly, better with recycling paper. Global consumption of recycled fibers rose from 152 Mt in the year 2000 to nearly 224 Mt in 2010 and to 250 Mt in 2018, before a slight decline to 244 Mt in 2019 (BIR 2020). With the global 2019 production of 412.5 Mt of paper and board this means that the new materials contained nearly 60% of recycled fibers, with the rate surpassing 70% in Asia. China's traditional shortages of wood led to a common use of non-wood fibers (mainly straw) whose pulping is less energy expensive and to high reliance on recycling. In 2019, it remained the world's largest importer of waste paper, but the total declined by a third while India's imports have been rising.

Collection of household waste paper is expensive and a thorough processing of the material is needed to produce clean fibers for reuse. This includes defibering of paper, cleaning, and removal of all non-fiber ingredients (most often adhesive tapes, plastics, and staples), and de-inking is needed if the fibers are to be reprocessed into white paper. Recovered fibers can be used to make 100% recycled product but more often they are blended with virgin fibers. Reprocessing shortens the cellulose fibers, and this means that paper can be recycled no more than four to seven times. Japan has been the most assiduous collector and reuser of waste paper.

Recovery rate went from 34% in1960 to 50% by 1985, 78% in 2010, and 81.1% in 2021 (JPA 2021). Given the fact that the denominator used to calculate this rate includes a great deal of non-recoverable or nonrecyclable paper (waterproof paper, heavily inked paper, sanitary paper, tissues), the rate of just above 80% is a remarkable achievement. For comparison, the US recovery rate has been also fairly high: in 2018, it reached 68.2% (46 Mt), the highest rate compared to other materials in municipal waste (USEPA 2022b). The recycling rate was highest for corrugated boxes (96.5%), lower for other paper containers and packaging (just 20.8%). Paper recycling is also a good example of limited benefits of the practice as additional recycling may yield only small decarbonization gains or it can actually add to the problem when it relies on electricity generated from fossil fuels or when it uses pulp from old-growth forests (Van Ewijk and Stegemann 2021).

Volumes of e-waste have been increasing faster than those of any other recyclables: they grew by 21% during the five years before the pandemic and reaching 53.6 Mt in 2019 and will see further rapid growth, with mobile Internet users to reach five billion and Internet-of-things-connected devices expecting to double to 25 million by 2025. (GESP 2022) - but in 2019, only 17.4% of e-waste were properly managed and recycled. Recycling of e-waste is particularly challenging because the devices commingle many compounds and elements that must be separated by a sequence of mechanical and chemical operations. Silicon is the core of modern computers but the transistors and microchips could not function without the presence of a multitude of materials, including many heavy metals.

As already noted, elements used to dope silicon include arsenic and phosphorus and also boron and gallium. Printed wire boards, disk drives, expansion cards, electrical supply and connections add up to a hoard of materials, small (and for many elements even miniature) per unit but highly consequential in the aggregate. Most of the mass in electronic devices is steel, glass, plastics, copper, and aluminum (in covers, cases, screens, connections) but more than dozen other metals are also present in tiny amounts, eight of them (As, Cd, Cr, Co, Hg, Pb, Sb, and Se) classified as hazardous (USGS 2001; Williams 2011). But the rewards of e-recycling are obvious: one million of discarded mobile phones contains 14 kg of palladium, 24 kg of gold, 350 kg of silver, and 16,000 kg of copper; or, to use another comparison, the same amount of gold that could be found in 60 t of auriferous ore can be extracted from 5 t of e-waste (Nicolai and Lana 2019).

What happens to electronic devices and to electronic components of many other products is only poorly known. Surprisingly, many people never throw the smaller devices (especially mobiles, also computers) away and keep them at home, too many units are simply thrown into garbage and end up in landfills, some are stripped of some materials, and a substantial share of e-waste is exported to Asia (above all to China and India) and West Africa for dismantling and recycling, a practice associated with some truly horrid labor conditions (working without any protection why handling toxic substances), and severe environmental problems (acids and heavy metals seeping into soils and entering streams and ponds).

Lead is of the greatest concern. Li et al. (2006) used toxicity characteristic leaching procedure (TCLP) to show that the element's concentration in TCLP extracts made from computer wire boards had 30–100 times the level of Pb (5 mg/l) that is used to classify waste as hazardous, while other heavy metals had essentially no potential to cause toxicity through leaching. Current recycling practices make the exposures worse: large shares of collected e-waste are sent from affluent countries to low- and middle-income nations where the devices and components are dismantled, recycled, and refurbished often in absence of any safety regulations. Children are particularly affected (Pascale et al. 2016; Wang et al. 2021; WHO 2021). Analyses document higher incidence of stillbirth, premature births and low birth weight, and exposure to lead has been associated with neurological impairment, behavioral problems, and reduced cognitive and language scores. Other adverse effects range from changes in lung and thyroid function, DNA damage, and increased risk of cancer and cardiovascular diseases later in life.

The most infamous of these recycling center is Guiyu, in China's Guangdong province. This city (area of about 50 km² with some 200,000 people) now accommodates millions of tons of e-waste and decades of informal recycling made it perhaps the world's most toxic place highly contaminated with heavy metals (Huang et al. 2021). Consolidation of recycling operations into a new industrial park opened in late 2015 made the situation more manageable, but it did not eliminate exposures and contamination. But as the US trend shows, it is possible to reduce lead in electronic devices below the levels of acceptable exposure (Intrakamhaeng et al. 2019).

But even if they were Pb-free, computers should not be discarded as their concentrations of some precious metals are equal, or higher, than those present in metallic ores. This is, most notably, the case with gold: there is as much as 1,500 g AU/t of circuit boards, 40 times the metal's concentration in ores mined in the United States (USGS 2001). Concentrations of copper in discarded computers are also high, and the challenges of heavy metal contamination, recycling of rare elements, plastics, and glass extend to the still increasing range of electronic devices and to electronic component that are now part of widely used products ranging from cars to stoves.

Even in the United States as recently as 2012, Kahhat and Williams (2012) could do no better than to offer rather wide ranges concerning the fate of 40 million used and scrap devices, concluding that 6–29% of them were exported, 17–21% were landfilled, and 20–47% were collected for domestic recycling (but the actual recycling rate was much lower). A year later, a new study of domestic and transboundary e-waste flows found that 258.2 million units of used electronics (computers, monitors, TVs, and mobile phones) were generated in 2010 and that 66.4% of that total (171.4 million) and that 14.4 million (8.5% of the total) of used electronic products were exported (Duan et al. 2013).

In mass terms, the annual total was about 1.6 Mt of generated, 900,000 t of collected and 27,000 t of exported e-waste. As expected, TV sets and monitors were the heaviest items, but mobile phones accounted for most of the waste generation and collection on unit base: means were about 176 million units sold, and 119 million (68%) recycled, with the rate as high as 87% for business mobiles. For computers (desktops and laptops), the recycling rate was 74%, for monitors 72%. In 2016, a study of North American e-waste flows estimated that about 30 million computers were acquired in the United States and 22 million (about 73%) were collected in 2010 (CEC 2016). The latter study did not attempt to quantify mobile phone flows and the dominance of this e-waste still keeps growing.

Their commercial introduction during the 1990s (I am beginning the count only with rise of digital cellular networks, leaving out pre-1991 heavy and inconvenient devices with limited connectivity) was followed by a rapid conquest of the global market. More than 100 million mobile phones were sold worldwide for the first time in 1997, one billion total was reached in 2009, 1.42 billion were sold in 2015 and, so far, the peak in 2018 was 1.556 billion with 1.433 billion sold in 2021, with the total number of users rising to 3.6 billion in 2020 and up to 4.3 billion (more than half of the global population) in 2023 (Fleming 2021).

Most of the mobile phone contracts are for two years and many people replace their models every 24–30 months, some go with the latest models (every 12–15 months), others keep the devices for 3–4 years. Even a conservative average of 2.5 years would mean that some 600 million units are now discarded every year. The best estimate for the total waste electronic and electrical equipment (WEEE) was 53.6 Mt in 2019, 21% increase in the five years since 2014 and pointing to as much as 74 Mt by 2030 (WEEE Forum 2019). In Europe an average household has 72 electronic items of which 11 are no longer in use or are broken and every year another 4–5 kg of unused electrical and electronic products per capita are hoarded before being eventually discarded.

National accumulations of hoarded obsolete e-waste keep rising: in France more than 100 million mobiles sleep in drawers. Prompt discarding has also reached stunning levels: in the United States more than 400,000 mobiles are dumped every day and end up incinerated or landfilled, contributing some 40% of all heavy metals in municipal waste. Obviously, aggregate amounts of metals in annually discarded mobiles are considerable. As already noted, every kilogram of printed circuit boards (they account for about 20% of the total device mass) contains 250–450 g of copper, 15–60 g of nickel, 1.5–4 g of silver, and about 1.5 g of gold (Sahan et al. 2019).

And, finally, an important reminder that (all call for circular economy notwithstanding) the maximum practicable recycling might not be the best option: as every action there is an optimum that, when properly assessed, might be far below the possible maximum. This is particularly relevant for managing municipal solid waste with its considerable social cost. Its total consists of costs incurred by households to prepare material for recycling collection, of municipal budgets devoted to managing waste and recycling collection (after subtracting any revenue earned from the sale of collected recyclables), of costs associated with landfilled waste disposal or incineration (after adjusting for benefits associated with manufacturing final goods with recycled materials). This kind of assessment was done for Japan with a surprising result: after considering all economic and environmental costs and benefits of the practice, the estimated optimal recycling rate in Japan was put at just 10% (Kinnaman et al. 2014). Of course, this rate will differ according to specific conditions, but it suggests that it is easy to strive for targets that might be far above the optimum.

5

Are We Dematerializing?

Tomomarusan/Wikimedia Commons

Materials and Dematerialization: Making the Modern World, Second Edition. Vaclav Smil.
© 2023 John Wiley & Sons Ltd. Published 2023 by John Wiley & Sons Ltd.

Oxford English Dictionary (OED) defines the verb *dematerialize* as becoming free of physical substance. This definition has obvious supernatural and spiritual connotations: it has been used in science fiction (in the United States most famously in the Star Trek) where objects and persons disappear by means of an unexplained process; dematerialization of Christ's body is to explain the apparent imprint of torso, limbs, and wounds on the fabric of the shroud of Turin (Berard 1991); and a famous Yoghi asserted that a man will be free only when he "is able to dematerialize . . . his human body . . . and then materialize it again" (Yogananda 1946).

Fortunately, my concern is with more mundane matters and OED's example of the real-world dematerialization refers to replacing physical records or certificates "with a paperless computerized system." This is an example of dematerialization that appears to be factually correct and that has been, in many instances, already accomplished in most high-income economies. Printed stock certificates (including many older large and ornately designed versions) became just strings of ones and zero in electronic data banks. Recent decades have seen many other cases of such dematerializations: medical records and library catalogues have migrated from handwritten paper files and stiff paper cards to electronic files; music recordings went from records to tapes to CDs to streaming; books can be listened to or read on-line; photographs do not need films and development; some people have not touched paper monies for years; banking has done away with long paper trails; museums can be visited virtually sitting at home; instead of printed paper tickets, be they for trains, airplanes or events, are just electrons on screen.

All of these changes resulted in visible and tangible reduction of materials that we have to handle, but careful evaluations would be needed to quantify the actual aggregate net material savings. There is no need for CDs made of polycarbonate plastic and for CD players, but streaming songs or entire operas requires requisite electronic devices (desktops, laptops, tablets, and mobiles), reliable Web access and enormous (and redundant) data storages, all depending on uninterrupted electricity supply. Dematerialized music recordings are a perfect example of relative (and apparent) dematerialization, just one of many immaterial e-replacements whose final net effects can range from appreciable material savings to no substantial over-all reduction of material uses (because of the need for less obvious deployment of other materials in new infrastructures needed to provide the substitutes) or even to increased consumption of specific materials (most notably, consumption of office paper has actually gone up with large-scale computerization).

But a much more common category of dematerialization is not a complete elimination of a particular material but a substantial reduction of material used, be it per a finished item, per a service rendered (per unit of power delivered by a prime mover, per unit of computer performance), or per a unit of national eco-nomic product. Studies of dematerialization began during the 1990s but the topic has never received as much attention as the more recent studies of decarboniza-tion. Papers by Herman et al. (1990), Wernick et al. (1996), Cleveland and Ruth (1999), Bartelmus (2003), van der Voet et al. (2005), Ausubel and Waggoner (2008), Shao et al. (2017), and Petrides et al. (2018) review the field's progress. My principal concern is to deconstruct and assess commonly used dematerializa-tion measures and to quantify some longer-term trends that showed both appreci-able reduction and gains of material use.

Relative dematerialization has been one of the key trends in modern manufactur-ing whose quest for higher productivity and lower prices has brought reduced use of materials (be they traditional and inexpensive, or modern and costly) for products ranging from beverage cans to jetliners, and from simple structural elements in buildings to computers. Herman et al. (1990, 334) argued that from an environmen-tal point of view it is not the reduced unit mass of products that matters (as smaller and lighter items could be of inferior quality and their more frequent replacement could generate more waste) but that dematerialization "should perhaps be defined as the change in the amount of waste generated per unit of industrial products."

This might be a desirable and revealing choice but collecting the necessary information about all possible waste streams and coming up with relative valua-tions of different emissions or effluents (as mass reduction of one waste stream can be replaced by a much smaller mass release of a compound that is more harmful or more durable) is obviously challenging. LCA's goal is to provide precisely that kind of information but, as shown in the fourth chapter, such assessments should be used with caution. Similarly, two other measures of dematerialization – as declining consumption of goods or lower use of energy per unit of economic product – face a number of intractable data challenges relating to the accounting for overall economic output.

Adjustments for inflation can be done fairly easily and with acceptable uncertainty, conversions to purchasing power parities are far more questionable. Moreover, energy use per unit of GDP may be a common measure of economy's overall energy intensity but (even when setting aside the uncertainties inherent in converting various energies to a common denominator) a closer look shows that it is fundamentally flawed, that its narrow interpretation gives very limited insights, and that it is only a poor proxy for tracing both historical and recent changes of material consumption in growing economies.

5.1   Apparent Dematerializations

When explaining dematerialization, the OED should have used the conversion from blueprints to computer-assisted design (CAD) as a far more consequential example of disappearing uses of paper than the replacement of printed stock certificates by electronic versions. That apparent dematerialization had eliminated roomfuls of workers at their slanted drafting boards and it replaced large numbers of paper blueprints filed in heavy storage steel cabinets by electronic graphics displayed on screens and saved initially on tapes and then on magnetic devices, hard drives, and various portable storage media. CAD applications, beginning with two-dimensional designs and progressing to complex three-dimensional capabilities, were used first in military projects; GM began its CAD development during the late 1950s and commercial aircraft makers joined during the 1960s (Weisberg 2008).

Total number of blueprints needed to build more complicated machines was steadily increasing, first between the 1890s and 1920s as internal combustion engines and steam turbines began to replace steam engines, and then (between 1900s and 1930s) as gasoline and diesel engines were energizing more advanced car and truck designs. And the complexity of airplane design had increased the number of blueprints by another order of magnitude. A WWII bomber required about 8,000 drawings – and the adoption of jet propulsion (starting in the 1940s for military designs and a decade later for civilian aircraft) and the introduction of wide-bodied jetliners during the 1960s brought another order of magnitude increase.

Boeing 747, conceived in 1965 and launched in 1969, required about 75,000 drawings adding up to nearly 8 t (Lacitis 2019). That is why during the 1970s, Boeing developed a mainframe-based CAD system that eventually contained more than one million drawings and that was replaced, after more than 25 years of service, by a server-based Sun Solaris operating system (Nielsen 2003). During the 1990s, such airplanes as Boeing 767 and Airbus 340 became the most impressive embodiment of pure CAD. Even the earliest CAD systems often halved the time spent on a particular task and later advances reduced labor requirements to between 10% and 20% of the original need.

No less important have been the paper savings brought by automated electronic redesign as previously laborious adjustment and changes needed for redrawing entire blueprints were turned into just handfuls of mouse clicks. For example, during the design of B-29 (Stratofortress, America's largest and most complex long-distance bomber) more than 4,000 test drafts were thrown away for every final draft making up the plane's final production version (Herman 2012). In contrast, CAD makes it easy to try out a large number of variants and adjustments as appropriate software follows any ripple changes through the entire design and amends everything.

As CAD became the norm, it replaced experimental and trial drafts and proposal and final blueprints in virtually all categories of construction, manufacturing, and network design by disembodied electronic versions. In less than two generations advanced countries had experienced a nearly universal dematerialization of design. Of course, the dematerialization claim is correct only when looking at what has been eliminated, and that is why I have used the qualifier "apparent." Thanks to CAD, large and small sheets of paper are gone, as are drafting tables and utensils (protractors, rulers, compasses, and curve rulers) and large steel storage cabinets.

But creating and preserving dematerialized blueprints requires extensive infrastructures of modern electronic computing, redundant mass data storage and communication, ranging from specialized software (written by using other computers) to large flat screens to the roomfuls of servers required to process Internet traffic (globalization of design and the reliance on construction subcontractors on different continents, be it for making airplanes or cellphones, call for routine international sharing of proposals, specifications, and drafts) – to say nothing about the increased demand for electricity and hence for the requisite infrastructure of its reliable supply (generating by combustion of fossil fuels or harnessing of renewable flows, transforming the current and transmitting it along HV and distribution lines to final consumers).

Consequently, even in the case that appears to be a perfect example of dematerialization, the reality is nothing but a complex form of material substitution. Diffusion of CAD has reduced wood harvesting, production of pulp, paper, and wooden and plastic drafting implements and steel storages – but it has generated additional demands for complex (and energy-intensive) silicon-based material infrastructures of modern electronic world and, secondarily, for materials needed to energize those systems by reliable generation of electricity. Given the intricacy of modern electrical and electronic infrastructures, their multiple uses, and their very different longevities (software or screens may be replaced in a matter of months, thermal power stations will stay put for decades), it would be very difficult to come up with a net assessment of material balances involved in such complex substitutions. Even if such exercises were much easier to conduct, there is little doubt that their results would have a wide range of outcomes.

Similarly, microprocessors have helped to dematerialize an increasing assortment of products and services, from books (e-books accessible on computers and on numerous e-readers) and educational materials (on-line language lessons and tutorials with teachers in a different continent) to banking statements and supplier invoices, and from correspondence (e-mail) and transportation tickets (e-tickets on cellphone screens) to phone directories (even though fat yellow page books are still available) and maps. And new forms of e-art do not require canvass, paint, stone or clay, e-palettes offer millions of hues and while the world's great museums will not go the way of stock certificates visiting many of them can be now done entirely (room by room) without any travel and any physical presence via virtual e-tours now offered by many leading museums and collections of artifacts.

Obviously, electronic infrastructures (including material and labor back-ups) will be very different for creating a virtual tour of a small museum and for designing and maintaining a system for reservation and e-ticketing by a large international airline. And it would be even trickier to account for the consequences of these dematerializations. Access to on-line reservations for every airline flying a particular route has created more competition, reduced prices, and contributed to increased frequency of flying: paper tickets have vanished but more materials are needed to support complex infrastructures of international air travel. Similarly, an e-tour of a museum may actually prompt a viewer to take a trip in order to see a real thing, but we will never know the net impact of virtual activities on overall material requirements of the museum itself: the only certain conclusion is that they will not go away.

5.2   Relative Dematerializations: Specific Weight Reductions

Reduction of material inputs in production can be accomplished in four principal ways: by gradual improvements that do not involve new materials; by substitutions (often relatively rapid) of constituent materials by lighter or more durable alternatives; by intensified recycling, particularly effective in those cases where re-using brings major energy savings; and by introduction of entirely new devices that perform desired functions with only a fraction of mass needed for their predecessors. The first approach usually requires nothing more than a smarter design that reduces specific material input without compromising desired function: lighter steel beams or engineered timber used in modern construction are good examples of these design changes but the most ubiquitous instance of this development are lighter beverage cans.

Beverage cans are also a good example to illustrate the second and the third category of changes, relatively rapid substitution of materials and intensive recycling. Replacement of steel by aluminum reduced the mass of soft-drink cans but given a high energy intensity of primary aluminum production that substitution would have led to higher energy costs – but more common use of recycled aluminum has also lowered the need for new materials required to produce primary inputs and resulted in substantial overall energy savings. In mass terms, steel recycling offers the best global example of this welcome consequence as the practice obviates large-scale deployment of large amounts of materials needed to produce the metal from its ores.

As for the last category, perhaps the best example of adopting entirely new solutions has been the displacement of heavy and bulky vacuum tubes first by tiny transistors and then by transistors crowded onto silicon chips to make microprocessors. In reality, these processes of relative dematerialization are not mutually exclusive and, as in the just described case of beverage containers, strategies that combine two or more approaches have been common. But this does not mean that the quest for dematerialization has been the paramount factor of product evolution. History of virtually any kind of manufactured product could be presented as a story of declining costs and improving performances, be they measured by convenience of use, greater power, longer durability, or higher reliability (with their course following expected logistic curves as additional gains bring diminishing returns).

5.2.1 Beverage Cans

In some cases, reduced material use has been spearheading these trends or it has been a critical factor driving the improvements: mass reduction, together with safer handling, has been an explicit design goal in replacing glass by metals in distributing beer and soft drinks. First came steel beverage cans, then aluminum cans and both kinds of containers have been progressively redesigned in order to use thinner body sheets and smaller tops. In other cases, relative dematerialization has been a welcome consequence of innovations motivated by other goals. Substitution of tiles by plastic flooring simulating tile design was not driven by mass considerations by more affordable cost of materials and installation, and heavy glass, cast iron, and ceramic objects (bowls, pots) present in traditional kitchens were not replaced by plastics and thinner metals primarily because of their weight but because they, too, were more expensive and often less safe to handle.

Evolution of soft-drink cans also illustrates both a protracted process of substitution and a surprisingly large magnitude of eventual improvements. Steel cans for storing many kinds of food have a long history that goes back to the beginning of the nineteenth century, to the first commercial steps to produce canned food by Nicolas Appert in France and Peter Durand in the United Kingdom (Shephard 2000). But steel containers began to be used for beer and non-alcoholic beverages only during the 1930s and reached their highest popularity during the 1950s. The first aluminum cans were introduced by Coors in 1959; Royal Crown Cola was the first soft-drink maker to use them in 1964, Coca Cola and Pepsi Cola followed in 1967.

A decade later, steel cans were on the way out and none of them have been used for beer since 1994 and for soft drinks since 1996, but they have retained their dominant position in preserving fruits, vegetables, and meats. Aluminum cans are made from alloys (containing about 5% Mg, less than 2% Mn and traces of Fe, Si, and Cu), their insides are often protectively coated, and they have crimped-on tops opened by pull tabs. At 85 g, the first aluminum cans were surprisingly heavy; by 1972, the weight of a two-piece can was just below 21 g; by 1988, it was less than 16 g; by 1998, it averaged 13.6 g; by 2011, the lightest cans weighed just 12.75 g, but in 2020, the can that dominates the North American marker (12 oz or 355 ml) weigh 14.9 g.

Lighter cans are being made. Already in 2012, a European maker introduced a 330 ml can (11.03 oz, a European standard) that weighed just 9.5 g (CanTech International 2012). And after years of effort, in 2021, Crown achieved a 4% average global reduction in the weight of standard 12 oz can (their announcement makes it equal to 330 ml but it should be 355 or 354.88 ml to be exact) and their commemorative Libra can is just under 10 g (Crown 2021). Only marginal reductions remain possible: after all, today's cans are made with wall thickness of mere 0.097 mm, as thin as human hair (IAI 2018).

As already noted in the preceding chapter, aluminum can is the most frequently recycled container (indeed the most recycled small object) with more than half of the annual US output (no matter how the rate is measured) now coming from recycled metal. Combined with reduced unit weight, this is certainly one of the more impressive examples of relative dematerialization. But this often-cited example of relative dematerialization is dwarfed by a much more consequential and yet (curiously) overlooked evolution of prime movers, the shift from animate to inanimate energies and the subsequent advances of engines and motors. This neglected but fundamental development deserves a closer look.

5.2.2   Engines and Turbines

Muscles, converting chemical energy in food and feed into kinetic energy of human and animal labor, were the dominant, animate, prime movers in all pre-industrial civilizations. Mechanical (inanimate) prime movers had limited roles in traditional societies but became the enablers of modern economic development characterized by industrialization, mechanization of agriculture and urbanization, and resulting in unprecedented gains in quality of life. The first mechanical prime movers were simple wooden machines, water wheels (first introduced during the antiquity), and windmills (used since the early Middle Ages) converting energies of flowing water and wind into reciprocating or circular motion that was used to mill grains, press oil from seeds, saw wood, pump water, or blow air (Smil 2017).

Combustion engines convert chemical energy of fossil or biomass fuels into mechanical energy of reciprocating or rotary motion. External combustion in steam engines (heating water in boilers to supply steam, the working medium, to the machines moving parts) came first, very inefficiently during the first half of the eighteenth century, and then with increasing efficiencies after 1760 until the machines reached their performance plateau by the end of the nineteenth century. Internal combustion engines were commercialized after 1860, starting with inefficient and heavy stationary engines burning coal gas, proceeding to much lighter gasoline-fuelled automotive engines (pioneered during the late1880s) and to relatively heavy but inherently more efficient engines designed first by Rudolf Diesel during the 1890s (Smil 2010).

The twentieth century brought two other categories of internal combustion engines, gas turbines powered either by liquid fuels (kerosene for jet planes) or by gases (by natural gas in large stationary gas turbines), and rocket engines using a variety of liquid or solid propellants. The third class of modern prime movers has been used to convert kinetic energy of water, wind, and steam into electricity: steam turbines date back to the 1880s, their modern high-capacity designs receive superheated steam from large coal or natural gas-fired boilers and they have been the dominant means of generating the world's electricity for more than a century. Large hydro turbines have supplied substantial shares of electricity in many countries with rich water resources and during the past decade, modern triple-bladed giant wind turbines have become important electricity providers, particularly in some European nations (Smil 2017).

Declining specific mass/power (g/W) ratio has been, besides the growth of their unit capacities, perhaps the most remarkable long-term trend affecting all categories of modern prime movers. The rate's inverse (W/g), commonly known as power-to-weight ratio (or specific power), is one of the most revealing characteristics used when assessing and comparing the performance of engines or machines, and its variant, thrust-to-weight ratio, is used for gas turbines and rocket engines deployed in commercial and military jet propulsion and in launching payloads to space. The baseline for comparisons is the performance of the first fuel-powered mechanical prime movers, steam engines of the eighteenth century. By 1750, after decades of marginal improvements, standard-sized Newcomen's inefficient steam engine developed about 15 kW, weighed nearly 9.6 t, and had mass/power ratio of about 640 g/W (Smil 2008).

Three decades later, even Watt's famous improved design (patented in 1765) remained a very heavy machine (9.2 t, 15 kW) rated at just over 600 g/W. That was the same order of magnitude as the specific mass for the two most important animate prime movers, hard-working men and large draft horses. Although the power of brief (anaerobic) human exertions can be as high as 400 W (prorating to less than 200 g/W), men can work steadily at rates between 60 and 80 W, averaging about 1,000 g/W of useful labor. Heaviest working horses (English Clydesdales, French Percherons) weigh nearly 1 t and develop more than 1 horsepower (745 W), that is, again, specific power close to 1,000 g/W during prolonged pulls (Smil 2017).

By the mid-nineteenth century, the specific mass of many steam engines was below 500 g/W and by the late-nineteenth century, the best design (triple-expansion engines) rated less than 100 g/W, still too heavy to energize affordable road transport. Internal combustion engines changed that. In 1874, Otto's first internal combustion engine (non-compression, stationary) had a nearly four-meter tall cylinder and it was very heavy: at about 900 g/W, its mass/power ratio was several times that of the best contemporary steam engines. Compression lowered the engine's size and weight and reduced the ratio to about 270 g/W by 1890, less than that of small stationary steam engines used in workshops and factories (Smil 2005).

The first order-of-magnitude improvement came with the introduction of gasoline-powered engines designed during the 1880s by Karl Benz, Gottlieb Daimler, and Wilhelm Maybach. By 1890, Daimler and Maybach introduced their first four-cylinder engine whose mass/power ratio was less than 30 g/W (Beaumont 1902). In 1901, Maybach revealed Mercedes 35, the prototype of all modern vehicles whose four-cylinder engine had power of 26 kW but whose aluminum block and honeycomb radiator lowered its weight to just 230 kg, yielding mass/power ratio of 8.8 g/W. The second order-of-magnitude improvement came during the twentieth century with the transformation of cars from expensive oddities to affordable machines.

The world's first mass-produced car, Henry Ford's Model T launched in 1908, was initially powered by a 2.9-liter four-cylinder engine with mass/power ratio of about 15 g/W, but by the time the car's production ended (in 1927, although the production of replacement engines continued until August 1941), the ratio declined to less than 5 g/W. By the early 1930s, even such powerful engines as Pierce Arrow's powerful V-12 rated less than 4 g/W and gradual changes before and after WWII had eventually brought the mass/power ratio of commonly used automotive engines to less than 2 g/W during the 1950s and to less than 1.5 g/W by the mid-1960s: even Ford's 298 V8 for 1965 Mustang rated just 1.4 g/W. During the first decade of the twenty-first century, most passenger cars had engines with mass/power ratio between 1 and 1.5 g/W and the most powerful turbocharged engines rated well below 1 g/W.

Similarly, low ratings were necessary to transform early airplanes carrying just one or two people to progressively heavier passenger and military aircraft. Wilbur and Orville Wright designed the engine that powered their first successful flight on December 17 1903 (it was built by skilled bicycle mechanic Charles Taylor): its weight was 91 kg and it eventually produced 12 kW, resulting in mass/power ratio of 7.6 g/W. The two world wars brought new record ratings, the first one with Liberty, America's first mass-produced aero engine with 1.34 g/W in 1917, the second one with Wright R-3350 radial engines powering B-29 (Superfortress) bombers with just 0.74 g/W (Gunston 1986). Its successors, while Wright R-1820-82A, deployed just before the conversion of large-scale commercial flight to jetliners during the 1960s weighed only 0.59 g/W.

Inherently heavier diesel engines – that have conquered heavy road, rail, and ocean transport (and have also made important inroads into the passenger car market, especially in Europe) – have seen similar weight reductions (Smil 2010). Rudolf Diesel's third prototype, a massive (nearly 4.5 t) upright stationary machine with piston stroke of four meters that was used for specification tests in 1897, had maximum power of 13.5 kW and mass/power ratio of 333 g/W, three times higher than the best contemporary large steam engines and an order of magnitude above the improving gasoline-fueled designs (Diesel 1913). The first stationary commercial unit sold in 1898 was still very heavy (296 g/W) but weights began to decline with mobile applications. By 1910, marine diesels were around 120 g/W and during the 1930s, German engineer brought the mass/power ratio of Jumo Junkers 205 aircraft engine (595 kg, 647 kW) to a remarkably low level of 0.92 (Wilkinson 1936).

Today's diesel engines in passenger cars are only slightly heavier than their gasoline counterparts, rating between 1.3 and 1.5 g/W, and in heavy road service (buses, trucks) the ratios are as low as 3 g/W (Smil 2010). Much heavier machines that are the prime movers on non-electrified railways weigh between 3 and 10 g/W, and the world's most powerful marine diesel engines that propel massive container ships rate nearly 30 g/W: 80.1 MW 14-cylinder turbocharged Wärtsilä's RT-flex96C, whose stroke length is 2.5 m, has specific mass of 28.72 g/W (Wärtsilä 2006).

Starting in the 1940s, jet engines (gas turbines) reduced the specific mass of the lightest prime movers by yet another order of magnitude, to less than 0.1 g/W. Whittle's pioneering W.1 turbojet of the early 1940s needed 0.38 g/W, early commercial turbojets of the mid-1950s rated about 0.25 g/W, and the latest turbofan engines weigh less than 0.1 g/W (Smil 2010). Standard way to relate the mass of aero engines to their performance is by calculating thrust-to-weight ratio: it had increased from 1.6 for Whittle's prototypes of the early 1940s to nearly 4.0 for turbojet engines of the late 1950s to 5.2 for GE9X, the largest turbofan introduced in 2016 (Sampson 2019). This means that the world's most powerful turbofan engine needs less than a third of mass of Whittle's prototype to deliver a unit of propulsive power.

But even such low ratios would not be enough to escape the Earth's gravity and only rocket engines have mass/power ratios lower than that. In terms of thrust/weight ratio, rocket engines have advanced from less than 20 for PGM-11 that propelled Redstone, America's first ballistic missile in 1958, to nearly 140 by the early 1970s when the Kuznetsov group designed NK-33, the world's most powerful liquid oxygen/kerosene engine that never carried out its intended mission to the Moon. To compare these specific masses by using the same units as in the case of car and airplane engines, when Saturn C 5 rocket propelled Apollo 11 on its path to the Moon (in July 1969), its mass/power ratio was (even after counting all the fuel) merely 0.001 g/W.

Reduction of specific mass has been also impressive in the case of the largest stationary prime movers: mass/power ratios of steam turbogenerators installed in thermal power plants are two orders of magnitude lower than the rates for the best late-nineteenth century steam engines. Those machines, rating nearly 100 g/W, were used in all early coal-fired electricity-generating stations but by 1891, just seven years after its patenting, Charles Parson's early design of steam turbine rated no more than 40 g/W, by 1914, that ratio was reduced to 10 g/W, and modern large turbogenerators have specific only a bit above 1 g/W while the largest stationary gas turbines are only a bit heavier at around 2 g/W (Smil 2005).

5.2.3   Electronics

Finally, a few paragraphs about relative dematerialization of electronic devices in general and computers in particular. These improvements are among the most appreciated instances of dematerialization process, and in the case of computers, this trend has been even more significant because their declining mass has been inversely related to their performance: the machines are not only smaller, their power has increased as they became lighter, a result of the remarkably long duration of Gordon Moore's law formulated in 1965, at a time when the most complex integrated circuit had 50 transistors. Moore's original forecast was for the doubling transistor density per integrated circuit every year but a decade later, he extended the doubling period to two years (Moore 1965, 1975).

Comparisons of declining computer weight often start with the Apollo 11 guidance system that was used to steer and then to land the capsule on the Moon in August 1969. That computer had just 2 kb RAM and with mass of 32 kg that prorated to 16 g/b (Hall 1996). In 1981, IBM's first personal computer had 16 kb RAM and mass of

11.3 kg or 0.7 g/b (IBM 2022). In 2011, my new Dell Studio laptop, used to write the first version of this book, had 4 Gb RAM and it weighed 3.6 kg, resulting in a specific mass of mere 0.0000009 g/b. This means that between 1969 and 2011, in 42 years, the mass/RAM ratio was reduced by seven orders of magnitude, a truly stunning speed of relative dematerialization. By 2022, when I began to rewrite and update this book's original version, there was more progress but it stayed within the same order of magnitude. The new version was done on Dell Inspiron 7706 bought in 2021: it has 8 Gb RAM, it weighs 2.43 kg, and hence its specific mass is 0.0000003 g/b.

In quantifying the aggregate effect of this relative dematerialization, we should start in 1981 with IBM PC because the Apollo computer was not a commercial system and it was not really portable. In 1981, the aggregate RAM of some 2 million computers sold in that year was just over 30 Gb and their mass added to about 22,000 t; 40 years later, the aggregate RAM of about 340 million computers sold in 2021 (Gartner 2022) was on the order of 2.7 Eb (10^{18} b, assuming average 8 Gb per machine) but their mass was about 750,000 t: consequently, an aggregate memory gain of seven orders of magnitude (90 million times) was accompanied by only a 34-fold increase of total mass.

This is an exceptional example of relative dematerialization associated with an enormous operational improvement and performance/mass gains amounting to at least one (10-fold) or several orders of magnitude are limited to devices that are functionally dominated by microprocessors. Modern digital cameras, whose design was transformed by the introduction of microchips and that became popular before the rise of high-quality cellphone performance (Casio and Cannon were two leading makes), were an order of magnitude lighter than their predecessors of just 30 years ago (my comparison is in terms of image quality and hence it ignores small Instamatic-type devices that have been available since the 1960s). These compact cameras weighed just around 100 g while standard SLR cameras of the 1970s and 1980s weighed about 1 kg and considerably more with telephoto or macro lenses and with bags needed to protect the lenses and to carry rolls of film.

Mobile phones offer an even better example as they have combined an impressive rate of relative dematerialization with no less impressive gains in communication quality and overall functionality. In 1973, Motorola's first mobile phone weighed 1,135 g; in 1984, the first commercial device, Motorola's DynaTAC 8000X, weighed 900 g; in 1994, Nokia's units still had mass around 600 g; by 1998, the company had devices weighing only 170 g; and in 2000, the average phone weighed 120 g and the lightest design just 96.35 (GSM Arena 2022). Since 2005, the average weighted mass has remained mostly within 110–150 g range: new phones are thinner but their displays are larger. Apple's popular iPhones reflect that change: the original design released in 2007 weighed 135 g with a 6.2 cm-wide display, 2022 iPhone SE released in 2022 has mass of 144 g and display width of 6.73 cm (Social Compare 2022).

Tupy (2012) made a list of 16 devices or function that are replaced by applications installed on an iPhone: camera, PC used to receive e-mail, radio, fixed phone, alarm clock, newspaper, photo album, video recorder, stereo, map, white-noise generator, DVD player for movies, rolodex, TV, voice recorder, and compass. Even when leaving out all those functions that are far from being true equivalents of what they have replaced (truncated versions of news do not replicate a real newspaper and tiny screen cannot match viewing large-screen HD TVs), we are left with an array of equivalent, or even superior, substitutions: a mobile phone can be a perfect watch and alarm clock, it works fine as a Sirius radio, its portable list of contacts neatly replaces old-fashioned rolodex, GPS is a superior locator and direction finder compared to a compass and a sheaf of maps, it will do as voice recorder, and the latest designs take better-quality pictures than did most standard SLR cameras, to say nothing about eliminating purchases and development of film and printing pictures.

Even when adding just the weight of modern, that is mostly electronic, versions of replaced devices – a watch, a portable alarm clock, a small portable radio, a pocket voice recorder, and a digital camera, each of them weighing on the order of 100 g – we end up with total mass of 500 g compared to a bit more than 100 g for a cellphone; more liberal replacement criteria could double the mass of replaced devices and adding up the weight of their mechanical predecessors from the 1950s would raise the aggregate to several kilograms and bring the relative dematerialization gain to (or very close to) two orders of magnitude. Keep this impressive comparison in mind: I will return to it in the penultimate section of the last chapter!

But where microchips are not the dominant component of the total design, there has been no even remotely similar mass decline, and in some cases, microprocessor-driven dematerialization has been actually accompanied by substantial increases of the overall mass. Passenger cars (and other two-axle four-wheel vehicles) are perhaps the best example of the countervailing trend. Although they remain complex assemblages of mechanical components (containing on the order of 30,000 parts), all of their key functions (above all, timing of engine operation) and many subsidiary tasks (with the deployment of airbags being perhaps the most critical one) are controlled by microprocessors using software that is more complex than that onboard fighter jets or jetliners (Charette 2009). Modern cars are mechatronic hybrids with electronics accounting for up to 40% of the cost of the most expensive vehicles. But I will show in the following section that the total mass of average American vehicles used for passenger transport has been steadily increasing, driven not by engineering necessities but by personal preferences.

And among the enormous number of quotidian consumer products whose mass-produced varieties contain no microprocessors (from basic tools to sofas, from toothpaste to glass, from socks to concrete needed for a building's foundation), there has not been a single example where mass per unit of product or per an indicator of performance has improved by several orders of magnitude. Moreover, even reductions on the order of one magnitude (resulting in identical product performance while reducing its mass to around 10% of its original value) are extremely rare, and in most cases relative dematerialization has amounted to less than 25–30% compared to the same types of products available one or two generations ago.

5.3 Consequences of Dematerialization

Before taking a closer look at the material consequences of dematerialization, I should stress that this important indicator gives a necessarily limited and imperfect insight. A more realistic measure that would also consider environmental impacts of weight reduction (changing land use, carbon emissions, and aquatic eco-toxicity) might show that weight loss may not be the most important consideration. For example, small amounts of heavy metals in mass-produced electronic devices may have far greater aggregate impacts than larger mass of plastics (van der Voet et al. 2005). In order to appraise these impacts, we would need specific comparative LCAs that, in some cases, would have to be repeated frequently as new materials and devices become parts of dematerialization processes.

Secular declines of specific mass/power ratios of products and prime movers could be used to make many revealing comparisons similar to that made in the last section when I equated mass of aluminum saved because of lighter beverage cans with a number of new Boeing airplanes. None of these comparisons is more stunning than a calculation of total computer mass when assuming that the average mass/RAM ratio of 340 million personal computers sold worldwide in 2021 would have remained at the early 1980s (the first IBM PC) level of 0.7 g/b: the total weight of those personal computers would have been 1.9 Tt, the mass two orders of magnitude larger than the annual anthropogenic mobilization of all materials.

Dematerialization gains have been far lower in all instances where the modern electronics has played just a supportive (control and optimization) role and where the designs could not avoid large-scale reliance on high-performance metals. Prime movers (engines and turbines) offer perhaps the best examples in this category (Smil 2010). Everything else being equal, gas turbines powering today's intercontinental jetliners (large turbofans) would be at least 50% heavier if their thrust/power ratio remained at the level of turbojets of the early 1960s level. Internal combustion gasoline engines in today's cars would have to be at least three times as heavy if they were built as the best designs of the early 1930s. And if massive container ships were powered by the best steam engines (whose efficiency rarely surpassed 15%), the weight of those old prime movers needed to provide the same power would be three to four times heavier than the weight of the latest, highly efficient (on the order of 50%), marine diesels (Smil 2010).

These contrasts also make it clear that relative dematerialization has been a key reason for falling costs, greater affordability, and mass diffusion of successive generations of products, as well as for preventing some truly unimaginable levels of environmental pollution and degradation. Obviously, it would have been physically impossible for the global sales of personal computers to reach more than 300 million units in 2021, indeed not even 30 million machines, if their mass/RAM ratio had remained at the early 1990s level: even the most affluent and the most innovative society would not be able to acquire and to handle the material throughput created by such a demand.

Similarly, today's passenger cars would not be so powerful if their mass/power ratio got stuck at the level of the 1930s. We could build and drive large numbers of very heavy cars but their high capital and operating costs would have prevented any truly mass ownership, with the most affluent countries now having more than 600 cars for every 1,000 people. And electricity generation powered by steam engines could never match low cost of supply from massive turbogenerators with very low mass/power ratio. Continued reliance on steam engine-powered plants would have greatly restricted average consumption of electricity, would have prevented widespread ownership of electric and electronic appliances, and could have supported only limited household and public lighting and a relatively modest deployment of industrial electrochemical processes.

Of course, lower specific mass of products and lower mass per unit of performance of prime movers have not been the only major reasons behind falling costs and more widespread ownership of products and a greater affordability of service. Technical innovation usually works in complex, interactive (albeit often unanticipated) ways and it becomes difficult, if not impossible, to assess relative contributions of individual factors. Lower costs of energy needed to produce, process, and transport materials have been a key contributor and the other principal factors included declining cost of progressively more mechanized manufacturing, frequent (and in some cases almost continuous) re-design of commercially successful product, and introduction of new techniques and processes, some of incremental nature, others representing radical departures.

But lower energy costs would have been impossible without new products and processes: just think of the machinery needed for the cheapest surface coal mining or of the impact of modern rotary rigs as they replaced old percussion tools used in hydrocarbon extraction, or of the possibilities for economic long-distance electricity transmission brought by direct current high-voltage lines. In turn, new materials needed for these products required often much higher energy inputs than did the production of their predecessors: aluminum vs. steel and plastics vs. glass are the two most obvious examples. And while many innovations that lowered operation costs were deliberately developed to fill a needed niche, others had originally nothing to do with eventually widespread commercial applications: Czochralski's invention of growing pure silicon crystals, the material foundation of modern solid-state electronics, is a perfect example (Scheel and Fukuda 2004).

There can be no doubt that relative dematerialization has been a key (and not infrequently the dominant) factor promoting often massive expansion of total material consumption. Less has thus been an enabling agent of more. In an overwhelming majority of cases these complex, dynamic interactions of cheaper energy, less expensive raw materials, and cheaper manufactures have resulted in ubiquitous ownership of an increasing range of products and in more frequent use of a widening array of services. As a result, even the most impressive relative weight reductions that have accompanied these consumption increases have been only rarely translated into any absolute cuts in the overall use of materials.

Consequently, it is not at all surprising that a study of 150 economies during the four decades between 1970 and 2010 showed that temporary absolute dematerialization coincided with economic recession or with periods of very low growth, with industrial and construction minerals and metal ores most likely to decrease (Shao et al. 2017). Nor it is unexpected to conclude, as Giorgios Kallis did, that

> decarbonization and dematerialization are incompatible with economic growth; and that strict policies for decarbonization and dematerialization will have a negative effect on growth (Kallis 2017, 10).

5.3.1 Rebounds

This counterintuitive phenomenon was first described in relation to energy consumption by William Stanley Jevons, an English economist, during the mid-1860s:

> **It is wholly a confusion of ideas to suppose that the economical use of fuels is equivalent to a diminished consumption. The very contrary is the truth.** As a rule, new modes of economy will lead to an increase of consumption according to a principle recognised in many parallel instances (Jevons 1865, 140; emphasis is in the original).

Savings arising from increasing efficiency of energy conversions have been a major factor (together with rising disposable incomes and with technical advances resulting in higher efficiencies, simpler operation, and longer durability) in promoting their more frequent use and driving up the overall use of fuels and electricity.

Examples abound: just compare the number of lights and electronic gadgets in an average American household in 2020 with the total of their much less efficient predecessors in 1930. In 1930, even in the richest countries a typical household had just a small number (6–8) weak (40–60 W) lights, a radio, and perhaps a small electric stove; 90 years later it had more than two dozen lights (many being 100 W or more), an array of electrical appliances (including such major energy consumers as a refrigerator, stove, toaster, washing machine, dishwasher, clothes dryer, and air conditioner), and a still growing range of electronic gadgets ranging from TVs (with increasingly larger screens) to game boxes, personal computers, and mobile phones. European ownership of household energy converters is not nearly as extensive as in the United States and Canada (air conditioners and clothes dryers are largely absent), and their typical sizes remain smaller and the same is true (with the exception of near-universal air-conditioning) in Japan, while Chinese gains have shown unprecedented rate of diffusion, especially as far as refrigerators, TVs, air conditioners, and mobile phones are concerned (NBSC 2022).

Evolution of major household electricity converters is just one of "many parallel instances" noted by Jevons, where relative dematerialization means that less enables more. Progressively lower mass (a major contributor to decreased cost) of specific products, be they common consumer items or powerful prime movers, has contributed to their increased use as well as to their deployment in heavier (more powerful, larger, more comfortable) machines. Inevitably, this has resulted in a very steep rise in demand for the constituent materials – even after making an adjustment for increased populations or for a greater numbers of businesses. Perhaps the two best examples of how impressive relative dematerialization has led to rising demand for materials (and for more energy-intensive and more costly materials!) are offered by the items at the opposite side of size spectrum – by mobile phones and by passenger cars.

Mobile phones offer the most obvious recent example of this increase. Their truly explosive diffusion has more than made up for their impressive evolution from heavy brick-like units (the first Motorola prototype weighed 1.1 kg) to still substantial early commercial design (averaging about 600 g in 1990) to slim designs of the early 2020s, some now slightly thinner than the standard pencil diameter of 7 mm and weighing mostly just between 150 and 200 g. The net outcome of this impressive relative dematerialization (typical mass down by 85% in 50 years and 75% since 1990) has been a much expanded demand for energy-intensive materials.

With average mass of 600 g in 1990 (when only about 11 million units were in use worldwide) and 165 g in 2020 (when there were about 7.25 billion phones), the aggregate mass of cellphones rose from less than 7,000 t in 1990 to about 1.2 million tons, a gain of two orders of magnitude. As has been pointed out many times, the tiny devices contain such rare earth elements as lanthanum, praseodymium, europium, gadolinium, terbium, and dysprosium (to produce colors on the liquid crystal display and to give screens their glow) and metals ranging from cobalt to tantalum and from copper to nickel (in their electronics and batteries) as well as gold and silver (in their printed circuit boards). Every kilogram of printed circuit boards of mobile phones contains 250–450 g of copper, 15–60 g of nickel, 1.5–4 g of silver and about 1.5 g of gold (Sahan et al. 2019).

This translates to typical concentrations of 33% of Cu, 2.5% Ni, 0.4% Ag, 0.14% Au, and 0.03% Pd: no metallic ore comes even close to being so rich in valuable metals as is the printed circuit waste – and yet (unlike with old bulky but simple landline devices that lasted for decades!) mobile phones are replaced every 15–30 months and (as already noted) only a small fraction of discarded mobiles is recycled: all in all, a perfect example of relative dematerialization leading to mass-scale mobilization, unproductive accumulations, and waste of precious minerals.

5.3.2 Materials for American Cars

Evolution of American passenger cars provides perhaps the best-documented example of an impressive relative dematerialization, in this case that of the machine's very prime mover, **that has not been accompanied by any aggregate decline in consumed materials**. Impressive improvements of power/mass ratio of American gasoline engines that have been achieved thanks to a combination of higher compression ratios, better fuels (preventing engine knocking), and lighter and yet more durable materials (above all by the use of aluminum in engine blocks) slowed down the growth of demand for materials needed to mass-produce those prime movers – but even if we set aside the increased rate of ownership, this desirable effect has been more than negated by the combination of higher average power of those relatively dematerialized engines and a greater mass of dominant machines.

A century of American data makes it possible to make accurate comparisons of realities that prevailed in 1920 and in 2020. In 1920, the US car market was dominated by Ford's Model T whose 15 kW (20 hp) engine had low compression ratio of 4.5 : 1, the mass of 230 kg (depending on the engine parts included this rate may be off by ±10%), and hence mass/power ratio of 15.3 g/W (Ford 2010). By 2020, the company's bestselling passenger car, Ford Fiesta, was powered by engine rated at 93 kW (with high compression ratio of 10 : 1) and weighing only about 136 kg resulting in mass/power ratio of just 1.46 g/W. Even when using 1.5 g/W as a highly conservative average for light-duty vehicles made in 2020, it would mean that the typical mass/power ratio of gasoline engines commonly deployed in passenger cars was reduced by about 90% in 100 years.

But (as already seen from the comparison of the two bestselling Fords) these notable improvements of typical mass/power ratio of gasoline car engines were accompanied by substantial increases of average engine power. Ford Fiesta has 6.25 times more powerful engine than did Model T, but that multiple underestimates the overall rise of US automotive power as pick-up trucks and SUVs gained a larger share of the market. Average power of new cars sold in 1975 was 102 kW and then (as a result of rapid crude oil price increases) it declined to around 80 kW during the early 1980s, but soon after crude oil price fell, it began rising again. Average power of new vehicles has been above 150 kW since 2003 and it reached a new record at 184.2 kW in 2020 (USEPA 2021b). Typical power of new US cars thus rose more than 12-fold between 1920 and 2020 (from 15 to nearly 185 kW).

Not surprisingly, there has been a high degree of correlation between average power and average car mass: during the 1920s and 1930s, cars got heavier as they acquired fully enclosed metal bodies and a growing array of amenities and that necessitated the use of more powerful, and heavier, engines. In 1908, Ford's Model T weighed just 540 kg, 30 years later Ford's popular Model 74 sedan weighed 1,090 kg. Since 1950, light duty vehicles got heavier because they got both larger and equipped with more features. This trend was reinforced by the rapid adoption of SUVs and by a growing habit of using vans and pick-ups as passenger cars carrying usually a single person. Automatic transmissions (now in about 95% of all new American vehicles), air conditioning, servomotors operating windows and mirrors, audio systems, and better insulation added to average car's mass. As long as crude oil prices remained low, America's car designers were quite unconcerned about vehicle weight, only about (often rather dubious) appearance.

As a result, in 1975 (the year when the USEPA began to monitor the mass of newly sold cars), when the typical mass of European cars was between 750 and 900 kg, the mean weight of US cars and light trucks was 1,840 kg. The two round of rapid oil prices increase during the 1970s favored smaller vehicles and the average mass of new US cars declined to 1,450 kg by 1981, but once the world oil price collapsed in 1985, larger cars and light trucks became again more popular and introduction of SUVs has worsened the trend. By 2004, new passenger vehicles were heavier than in 1975 as they reached a record average weight of 1,860 kg and subsequent slow growth raised the weight to a new record of 1,885 kg by 2020.

This means that average vehicle weight increased 3.5-fold between 1920 and 2020 (USEPA 2021b), and because America's automotive market is splintered among scores of brands and hundreds of models, the best-selling vehicle in 2020 (and for 40 years before that) was not actually a passenger car but a heavy (2,130 kg) pick-up, Ford F-150. These ratings, summarized in Table 5.1, show that while engine's typical mass/power ratio declined by 94% between 1920 and 2020, average engine power increased 12.5-fold: this means that the increased engine power alone had erased about 75% of the material savings resulting from vastly improved

Table 5.1 *Weight, engine power, and mass/power ratios of US cars, 1920–2020.*

Year	Curb	Engine		
	Weight (kg)	Weight (kg)	Power (kW)	Mass/Power (g/W)
1920	540	230	15	15.3
2020	1,885	170	188	0.9
Multiples	3.5	0.74	12.5	0.06

mass/power ratio of these prime movers, and that this small material gain was completely negated due to more than trebled average curb mass!

This failure to reduce car mass has been even worse when assessed in per capita terms because the rate of vehicle ownership did not remain constant between 1920 and 2020: passenger car registrations increased from fewer than 8 cars/100 people in 1920 to 27/100 people cars in 1950. Afterward, it became common for a family to own two or more vehicles: in 1950, only about 3% of US households had a second car; by 1965, that share reached 24%; and by 2020, it was nearly 60%, with about 24% of all families having three or more cars and there were 88 vehicles per 100 people. Between 1920 and 2020, relative registrations thus increased almost exactly 11-fold and even after completely eliminating the effects of the intervening population growth (from 106 to 330 million people) which would have, everything else being equal, resulted in 3.1 times as many vehicles on the road, the combined effect of higher weights and higher rate of ownership **had increased average per capita mass of materials deployed in passenger cars registered in 2020 nearly 40-fold when compared to their mass in 1920.**

And similar weight and ownership trends have been evident even for small Japanese and European vehicles. In 1973, the first Honda Civic imported to North America weighed just 697 kg while 2023 Civic (LX) with automatic transmission (and with air conditioning standard) is, at 1,313 kg, more than 88% heavier, and its engine is 3.2 times more powerful (118 vs. 37 kW). And while some small archetypal post-WWII European cars weighed less than 600 kg (Citroen 2 CV 510 kg, Fiat Topolino 550 kg), the average mass of European compacts rose to about 800 kg by 1970 and to just over 1,400 kg by 2020 (ICCT 2021).

Car ownership in Japan rose from less than one car per 100 people in the early 1960s to 63 cars/100 people by 2020, and Germany 2020 rate of ownership was the same as in Japan. Higher rate of ownership and greater vehicle weight thus combined to increase the mass of European or Japanese car fleets by an order of magnitude, much as in the United States. So much for any dematerialization of that most ubiquitous of all large modern machines! And new designs – hybrid drives and electric cars – will not result in much lighter vehicles as they have to accommodate more complicated power trains or heavy batteries. For example, GM's 2012 Volt weighs more than 1.7 t and electric Ford Focus weighs 1.67 t. And there is yet another variable that must be considered in assessing material consumption related to passenger vehicles, the average fuel efficiency: obviously, its steady improvement, that is declining fuel demand, would have significant effects on materials needed for extraction, transportation, and refining of crude oil and for the distribution of refined products.

Just the opposite had happened in the United States where the average performance of new vehicles was getting worse for more than four decades, and then, after a decade of improvements, it stagnated for another two decades. In 1936 (the first year for which the mean fuel consumption can be reliably calculated), the performance of new vehicles average was more than 15 mpg and by 1973, it declined to only 13.4 mpg (Sivak and Tsimhoni 2009), a remarkable case of technical retrogression during the era of rapid technical advances. The trend had changed only thanks to the OPEC-driven rise of crude oil prices: in 1975, the US Congress adopted the Corporate Average Fuel Economy (CAFE) regulations and the first mandated efficiency increase took place in 1978.

By 1985, the pre-1975 performance had more than doubled, to 27.5 mpg. But then, as crude oil prices fell (from nearly $40 per barrel in 1980 to only about $15 by 1986), the CAFE standard was left unchanged for more than two decades and with the influx of SUVs, exempt from the standard, the adjusted fuel economy of new vehicles (reflecting actual driving experiences rather than laboratory testing) was merely 19.8 mpg in 2000, 22.6 mpg in 2010, and by 2020, the rate rose to 25.7 mpg, still below the 1985 mean (USEPA 2021b)! This has been an astonishingly poor performance illustrating inexcusably slow rate of progress since the mid-1980s, the decades that witnessed so many important and impressive technical advances.

Finally, there has not been any progressive automotive dematerialization in the United States because the average distance driven annually per capita has been increasing, making higher demands on fuel production and distribution, on car depreciation and maintenance and on road repairs. While the average distances driven annually by Europeans and by Japanese had leveled off at relatively low levels, in the United States that per capita mean had more than tripled between 1950 and 2000 (from about 4,800 km to more than 15,500 km) and by 2019, before it was reduced by the pandemic, it reached about 18,560 km (USEPA 2021b).

The conclusion is clear: relative dematerialization of internal combustion engine and the use of lighter elements and compounds in vehicle construction had only slowed down the rate of overall material consumption claimed by making and operating vehicles. By the year 2020, the combination of expanded vehicle ownership (nearly 11 times after adjusting for population growth) and of greater average mass of heavier sedans, SUVs, and pick-ups (3.5 times) had increased the average per capita mass of American vehicles nearly 40-fold compared to 1920! And the real material burden has been even larger because of a substantial increase in annually driven average distance, higher average speed and first actually declining and then only slowly improving typical efficiency: these realities have increased wear-and-tear on the highways and necessitated higher production, refining, and distribution of fuel. As a result, it would not be an exaggeration to offer a multiple of 50 as an approximate per capita increase of material demand due to passenger driving in the United States between 1920 and 2020.

5.3.3 Materials in Flight

Analogical calculation involving relative dematerialization of prime movers (thanks to their impressively higher efficiency and the use of new materials) and massively rising aggregate consumption of materials could be made for modern airline industry. Thrust-to-weight ratio of the largest turbofan engines (such as those used to power Boeing 787 and Airbus 340) is now roughly 60% higher than that of the turbojets used by the first jetliners of the late 1950s – but this significant dematerialization of performance has been accompanied by introduction of more powerful engines needed to propel many larger airplanes.

Boeing's pioneering 707 was powered by Pratt & Whitney's JT3D engine whose rated thrust was about 80 kN. In contrast, the most powerful engines widely used during the second decade of the twenty-first century had thrust between 311 kN (Rolls Royce and Pratt & Whitney engines uses on Airbus 380) and 512 kN (GE's GE90-115 AD 9), that is nearly four to more than six times as powerful (Pratt and Whitney 2022; GE Aerospace 2022). Engines used on mid-size jetliners that dominate intracontinental flights (Boeing 737 and Airbus 320 models) had thrusts ranging mostly between 100 and 150 kN, that is typically 25–50% more powerful than the first commercial turbojets (Smil 2010).

Aircraft industry has been the foremost pioneer of material substitutions using light-weight metals and compounds, the trend that has now culminated in increasing reliance on carbon composites. But this relative dematerialization has not resulted in lighter commercial fleets as their composition shifted toward larger numbers of small regional jets, and toward larger aircraft. Canada's Bombardier introduced its first regional jet (Canadair CRJ-100/200) in 1992; a decade later, US airlines had more than 900 regional jets and by 2019, their total approached 2,400. Inevitably, smaller airplanes have greater mass per passenger than the larger ones operating on comparable routes: CRJ-100/200 rated 480 kg/passenger when using maximum take-off weight (MTOW) of 24,000 kg and seating for 50 people

(MHI RJ Aviation 2022). In contrast, the same ratio for Boeing 737–100 was just 391 kg/passenger (44,225 kg and 113 passengers).

Long-term trend toward larger aircraft is best illustrated by the evolution of Boeing 737, the most successful jetliner in history introduced in 1967: by mid-2022, the company delivered more than 11,000 planes with each successive model weighing more than its predecessor. In 1967, the MTOW of Boeing 737–100 was 44.2 t; by 1981, the 300 series had MTOW of 56.5 t; planes of the 800 series, in service since 1998, weigh 79 t; 900ER flying since 2006 weighs 85.13 t; MAX 10, launched in 2017, weighs 89.8 t, almost exactly doubling in weight in 50 years of the plane's (seemingly never-ending, controversial, and disaster- and performance-beset) development (Modern Airliners 2022). Despite better engines and lighter materials, the first 737 had typical capacity of 113 people and hence its MTOW was (as already noted) 391 kg/passenger, while for the twice heavier MAX 10 seating 204 people, the ratio is 440 kg/passenger, a 12% deterioration despite better engine and lighter materials.

But two developments above all prevented any absolute dematerialization: large increase in the total number of operating aircraft and an enormous increase in the frequency of flying. The US commercial jetliner fleet and the total of passenger-kilometers flown would have grown by no more than 84% if they had just matched the expansion of the country's population; in reality, between 1960 and 2020, the total fleet of commercial jetliners increased nearly 4.1-fold and passenger-miles logged on just domestic flights had multiplied more than 24-fold, from about 31 to about 754 billion (BTS 2022). Obviously, relative dematerialization of jet engines, and of aircraft designs, has had only a mildly retarding impact on the overall growth of material demand of America's aviation industry.

But tracing only the numbers and weights of airplanes and frequency of air travel adds up to an incomplete account of the sector's increasing material intensity. Enormous amounts of additional materials were required indirectly due to growing airplane production capacities (with parts sourced and transported globally), to steady increases of fuel consumption for more frequent flights, to much expanded associated aircraft services (cargo handling, catering, and airplane maintenance), and to the growth of airports and other associated infrastructures. And America's post-1960 rapid aviation expansion is now being replicated in China where new airplanes, new airports, and soaring frequency of flying have totally marginalized any material savings resulting from the reliance on the latest models of US and European aircraft.

The second example of tracing long-term trends in consumption of a product affected by relative dematerialization follows the uses of American aluminum cans and it shows that even the combination of reduced unit mass and relatively high rates of recycling is insufficient to make up for increased per capita consumption. Moreover, this example shows how difficult it is to maintain high rates of recycling even in the case of a product whose reuse is well established and commercially highly profitable.

Per capita consumption of aluminum cans in the United States rose from 149 in 1979 (the first year for which these numbers are available) to 312 cans in 2020, more than a twofold increase (CMI 2020). With average weight of 14 g, the average 2020 per capita use translates to about 4.4 kg of aluminum or about 47% more than in 1979, the total weight of heavier cans (20 g/can) translated to about 3 kg of the metal.

Again, relative dematerialization of a product – a 30% decrease in unit weight in 40 years – has been completely negated by rising per capita consumption (and it might have been a contributing factor to an absolute increase of this specific material consumption). Attentive readers will remember how I stressed the unequalled benefits of aluminum recycling (above all the 95% reduction in specific energy use) and for years the practice had seen encouraging growth. In the United States, aluminum cans began their rise only in 1963 and their fastest growth came after 1967 when Coca Cola and Pepsi converted to their use. Recycling began slowly but by the early 1980s, its rate surpassed 40%, and it reached the peak of 67.9% (with foreign-made cans included) in 1992 before beginning its undulating descent to about 50% in the year 2000. By 2014, it rose to 56.1%, but by 2020, it was down to 45.2% (CRI 2004; AA 2021). Discarding 55% of US aluminum cans means that in 2020, some 56 billion of them (or close to 800,000 t) were simply thrown away or landfilled. Clearly, American consumers and the country's aluminum industry have been unable to come close to the record recycling rate achieved by Brazil (98%) or to Japan's 77% or Italy's 72% (CMI 2017).

Finally, I must note those reductions of specific mass that have been accompanied not only by increased overall consumption of materials but also by increased specific energy cost of substitutes. By far the most common instance in this category has been the substitution of steel by aluminum. Steels have specific density between 7.7 and 8.0 g/cm^3 compared to 2.6–2.8 g/cm^3 for aluminum alloys. Again, the US automobile industry makes it possible to quantify the consequences of this shift that began on a large scale during the late 1970s when aluminum began to replace cast iron in intake manifolds. That substitution was accomplished in a matter of years and the next wave brought cylinder heads, pistons, blocks, transmission cases, and wheels cast from aluminum alloys.

As a result, aluminum weight of an average American car rose from 38 kg in 1975 to 117 kg by the year 2000 (about 7% of total weight) and 208 kg in 2020, with further rise to 258 kg anticipated by 2030 (DuckerFrontier 2020). Assuming that iron and steel components would fill the same volume (in reality the two volumes would not be identical but the difference does not substantially affect the calculation), 208 kg of aluminum in an average 2020 car replaced about 600 kg of steel. But while steel is nearly three times (2.9 to be exact) as heavy as aluminum, its energy cost is much less than a third of energy needed to produce the lighter metal. That is true even when assuming higher energy cost of specialty steel (up to 30 GJ/t) as the average primary energy used in making aluminum whose worldwide average in 2021 was about 105 GJ/t (IAI 2022b). Metal substitution results in desirable relative dematerialization but not in energy savings.

Unless all aluminum is made by using renewable energies, all weight reductions associated with replacing steel by aluminum result in higher overall energy costs. At the same time, it must be pointed out that this difference could be reduced by lower operating costs of lighter automobiles, rail cars, and trains and that this reduction will increase with the degree of substitution. Aluminum is still a minor constituent of cars (in 2020, it was about 11% of an average US vehicle) but not in rail transport: wheels and driving assemblies are still made of steel but both for large freight cars used to carry bulk minerals and other materials and high-speed passenger trains (Japan's *shinkansen*, French TGV, or Spanish AVE that travel at velocities close to, or even above, 300 km/h) have their hoppers and their bodies made of the lighter alloys.

Replacement of cast iron and steel by plastics brings the same combination of advantages (lighter components, reduced energy cost for machines where these components make a substantial part of the total mass) and disadvantages (invariably higher energy requirements for production, with most plastics costing anywhere between four to six times as much per unit weight than common steel alloys). Again, the substitution makes the greatest difference in the case of lifetime operation cost of energy-intensive means of transport. This is particularly true in the case of jetliners whose interiors now include (for larger planes) more than 5 t of advanced thermoplastics and other synthetic and composite materials in floors, ceiling panels, overhead baggage bins, window frames, seats, food and drink galleys and trolleys, lavatory modules, and class dividers (Composites World 2006).

I could cite many other instances where substitute materials, chosen for their lighter specific weight as well as for other superior or unique properties, have lowered the mass of finished products but required substantially higher energy inputs. Common examples range from plastic flooring replacing wooden boards or tiles in buildings to titanium taking place of steel in aircraft components, and from polyethylene bags replacing sturdy canvas to plastic garbage bins (also made from polyethylene or from polypropylene) taking place of previously ubiquitous much heavier galvanized (zinc-coated) steel metal trash containers. An obvious question to ask is: what effects have all of these relative product or sectoral dematerializations had on the national scale?

5.4   Relative Dematerialization in Modern Economies

The earliest stage of intensive modernization (starting during the latter half of the nineteenth century in most countries of Atlantic Europe and in North America, post-1950 Japan, post-1980 China) is marked by rising demand for construction materials – dominated by traditional bulk inputs of timber, stone, sand, and mineral aggregates and by expanding production of cement and crude oil–derived paving materials – due to rapid and extensive development of material-intensive transportation, industrial, and residential infrastructures. This stage is also characterized by rapid increases in consumption of primary metals due to unprecedented rates of urbanization, industrialization, and the construction of railways and (in maritime nations) of new steel-hulled ship fleets.

The next phase of material consumption experiences rising material demand due to the continuing expansion of industrialization and the next phase of urbanization driven by electrification, mass adoption of automobiles; in turn, those industries require materials for increased fossil fuel extraction, electricity generation, car industries, and road construction. Another major factor is a growing use of agricultural inputs (fertilizers and machinery) as higher disposable incomes boost demand for food, in general, and for animal products, in particular and accelerate the process of the sector's mechanization and applications of agrochemicals. United States of the 1920s and then again of the 1950s, leading European economies and Japan of the 1950s and 1960s, and post-1980 China are excellent examples of this demand stage.

Inevitably, the next stage of material consumption arrives. Uses of some traditional materials (most notably lumber, also fiber crops including hemp, flax, and jute) decline (some even in absolute terms), growth of demand for major materials slows down, eventually saturates and starts declining first in relative terms and in some cases even absolutely. These cessations and reversals of long-lasting upward trends result in temporary consumption plateaus typically followed by prolonged declines of specific material intensities (Herman et al. 1990; Bernardini and Galli 1993). Primary energy demand follows the same trajectory.

This secular decline of traditional materials (as well as energy) consumed per unit of economic product that appears during the second stage of material consumption has been perhaps the most often noted, and praised, category of relative dematerialization. Mass reduction would be perhaps the most preferable term and the process can be measured as a declining mass of a specific product (such as the already documented lighter mass of standard aluminum cans) as a lower annual use of a commodity per capita (for example, average US annual per capita consumption of wood declined during the twentieth century by two-thirds, from about 791 to 282 kg) or as decreasing mass of a particular material or of all materials used annually in a national economy per unit of economic product. A notable example in this latter category would be the US consumption of all metals that had declined from about 27 kg/$ of GDP in the year 1900 to less than 15 kg/$ by the year 2000, a 45% reduction in a century (Matos 2009).

5.4.1   Relative Material Consumption

Of course, many of these declines have obvious physical limits and they will prevent further relative dematerialization and that, coupled with rising demand, may lead to overall rise of material consumption: the already quantified case of American aluminum beverage cans is a good example of this common reality. American per capita cement consumption illustrates how fundamental physical necessities prevent any long-term large-scale dematerialization of a major construction material. The rate rose from about 100 kg in 1945 to nearly 400 kg in the early 1970s; for the next two decades it fluctuated between 300 and 350 kg; in 2005, it spiked with the pre-2008 housing construction boom to 430 kg; and it fell to a post-1950 low of 239 kg in 2009, and recovered to 323 kg in 2021 (USGS 2022a). Clearly, there has not been any secular dematerialization as far as this fundamental construction commodity is concerned, only generations-long fluctuations.

And the same is true of the US steel: its apparent annual per capita consumption rose from 125 kg in 1900 to more than 400 kg during the 1930s, it peaked at 530 kg in 1973, declined and rose again to 411 kg in 2005, fell to just 181 kg in 2009, but by 2021, it was once again close to 300 kg. Again, an impressive relative per capita dematerialization (nearly halving the rate between 1973 and 2023) but no evidence and, given the country's enormous neglected infrastructural repairs, no prospects for any continuous decline in years and decades ahead. Consumption of American steel is also an excellent example of the need for more inclusive accounting that would consider not only the trade in specific materials but also in finished products that embody them: in the case of steel, these are primarily vehicles, machinery, and white goods (refrigerators and washing machines). World Steel Association estimates for 2019 show the United States as the net importer of 28 Mt of steel in finished products (importing 49 Mt, exporting 21 Mt), enough to raise that year's apparent per capita consumption by 26%, from 297 to 374 kg (WSA 2022a).

Similar (and not easily made) adjustments should be made for all major materials in order to take into account the unprecedented post-1990 expansion of global trade: it may be now in a (temporary?) retreat but it has reached such levels that in many cases it may be more illuminating to look at the global levels of use rather than at specific national levels that may have reached new lows thanks to deliberate closures of domestic material-intensive industries and growing, even total, dependence on imports. For example, Denmark has become a much praised example of a country that has made great green strides: nearly half of its electricity comes from wind, and despite its small size, the country is EU's fourth largest producer of biomethane (used mostly for electricity generation) from agricultural wastes (13.4 million pigs, more than two for every Dane).

As a result (and even when taking into account Denmark's relatively significant methane releases from about 1.5 million heads of cattle) the country's per capita greenhouse gas emissions (in terms of CO_2 equivalent) were cut by nearly half between 2000 and 2020 while Germany, despite the two decades of its expensive *Energiewende*, achieved only about a 30% cut, the EU as a whole reduced its emissions by about 24%, and the US emissions declined by about 12% (Olivier 2022). But a closer look shows that Denmark imports about 18% of its electricity (which is the EU's most expensive) and, most importantly, that it does not even have several major industrial sectors that are prominent even in the economies of its Nordic neighbors.

Norway, thanks to its inexpensive electricity, is a major producer of aluminum; Sweden and Finland have relatively sizable steel industries; Finland is also a paper-making super-power – but Denmark does not produce any aluminum, does not make any iron and steel, and it also does not synthesize (despite its intensive crop production) any ammonia, does not produce any float glass, having no forests to speak of, it does not make any paper and, unlike Sweden and Finland (or half a dozen other small EU countries), it does not make or assemble any cars. No wonder that its domestic use of major materials would show impressive dematerialization: it is easy to be green and post-industrial by letting others to take care of material- and energy-intensive inputs and finished products!

At the same time, there can be no doubt that modern economic development, characterized by shift of the dominant sector from agriculture to manufacturing and then to services, has resulted in relative dematerialization, as well as in relative decline of energy use, in every affluent nation. These shifts are expected outcomes in the evolution of material and energy uses in modern economies as they move into post-industrial era – and they should be seen as a part of more general relative reordering of consumption that accompanies rising affluence. More than 150 years ago, Engel (1857) observed that the share of disposable income spent on food declines as income rises, an inevitable outcome given a simple fact that average daily per capita food intakes compatible with healthy and long life should remain below 9 MJ (less than 2,200 kcal).

In low-income societies, that share is close to half of family expenditures, but in 2021, it could be purchased (even when including plenty of more expensive animal protein) with 10.3% of average US disposable income (USDA 2022). Similar saturation levels and inversions have become evident in many other instances of specific consumption. Even in the absence of technical advances, average household expenditures on energy for cooking, lighting, and heating would get eventually saturated, and with the continuing efficiency improvements they have been falling but, inevitably, their share of disposable income is much higher for the low-income families (Drehobl et al. 2020).

As for the material consumption, rising incomes obviously do not produce commensurate increases in demand for a wide assortment of manufactured products ranging from appliances to furniture. Once these demands are saturated, the spending shift from goods to experiences brings relative dematerialization of outlays. At the same time, widespread possession of a widening range of consumer goods and deliberately engineered rapid obsolescence of many products are two notable factors that militate against dematerialization even in the most affluent societies already suffused in goods, and the net outcome can be determined only by taking a longer look at aggregate demand in modern economies.

How closely do they follow the described pattern of vigorous rise followed by saturation and relative decline? Or, to rephrase this question in macroeconomic terms, what has been the extent of decoupling between GDP growth and overall consumption of materials? Has it been a weak decoupling when material consumption grew at a slower rate than GDP, or a strong one when material consumption declined in absolute terms even as inflation-adjusted GDP kept on growing? Data limitations do not make it easy to answer this question. GDP series in constant monies are readily available, and affluent countries have historical series of products going commonly back to the last decades of the nineteenth century. But (a few notable exceptions aside) trade data needed to establish actual domestic consumption are much less complete (or are available only in value terms), totals for major material categories (construction aggregates, metals, other minerals) are either unavailable or have to be reconstructed from incomplete series, and until recently very few countries had long-term series of aggregate material consumption.

As already explained, comparative reconstructions by Adriaanse et al. (1997) and Matthews et al. (2000) offered only fairly short-term (1975–1994 and 1975–1996) aggregates for domestic processed output, domestic hidden flows, and total domestic output of all materials (including and excluding oxygen flows) in Germany, Japan, Netherlands, and the United States, with the second report adding Austria. Any general conclusions based on these studies were limited by the brevity of the time span as well as by some obvious national peculiarities. Germany's specific total domestic output (kg/Deutschmark) declined by about 20% between 1975 and 1990 but the reunification with the East (and resulting inclusion of much less efficient East German industries and, even more so, of massive hidden flows associated with former East Germany's lignite mining) brought a temporary reversal of that trend. And Japan's impressive decline of total domestic output per constant yen ended after 1990 as the country's economy entered its first lost decade.

The same two limitations apply to the EU's new material accounts whose systematic publication began in the year 2000. This data series is the most detailed set of material flows quantified according to common rules for a large group of countries but the brevity of available data and many national specificities preclude a simplistic conclusion regarding any possibly long-lasting trends. Moreover, EU's material aggregates also include all fossil fuels – an analytical choice that inevitably boosts the totals of both direct and hidden flows for countries that are highly dependent on coal – but exclude fertilizers, materials whose inputs made a critical difference to the prosperity of Europe's agricultural sector.

Despite these limitations, these statistics, when aggregated for the entire union, show a notable shift in resource productivity, the reverse indicator of relative dematerialization measured in €/kg (Eurostat 2022b). Between 2000 and 2007, the resource productivity remained essentially unchanged as the rising economic product (up by 16%) was closely underpinned by rising domestic material consumption (up by 14.5%). The Great Recession of 2008–2009 ended this near-lockstep rise: by 2014, EU's economic product was back above the 2007 level – but the domestic material consumption was 20% lower and this divergence resulted in resource productivity rising by 26%. This was followed by slow changes as resource productivity gained less than 6% between 2014 and 2021, with domestic material consumption remaining flat or slightly rising (in 2021 it was slightly above the 2012 level).

Calling the 2000–2014 shift an uncoupling is inaccurate, it has been merely a readjustment of the material intensity of the union's economic product: after a temporary decline that resulted in rapidly improved resource productivity, the aggregate material consumption has stabilized and in 2021 it was almost exactly the same as in 2010. The conclusion is clear: post-2000 EU has seen significant relative dematerialization but no significant decline of aggregate domestic material consumption.

5.4.2 American Trends

I have already noted that longer perspectives are available for several countries but all of these series include fossil fuels (whose extraction dominated the mass of domestically produced materials during the latter half of the nineteenth and the first half of the twentieth century) and food crops, materials excluded from this book, and show the flows only in highly aggregate forms. Consequently, the best choice to examine long-term relationship of material use and economic development is to rely on more than a century of disaggregated and reliable American data. As already explained, this series of material inputs into a national economy is maintained by the US Geological Survey: its first version ranged from 1900 through 1995 (Matos and Wagner 1998) and it was subsequently extended and by 2022, updates for most commodities are available until 2018–2019 (Matos 2009; Kelly and Matos 2016 with updates).

The series contains data for agricultural and forestry (wood, paper and paperboard, and recycled paper) products, metals and minerals (subdivided into primary and recycled metals, and industrial and construction materials) that were used to make products in the United States: it leaves out food and fossil fuels (as large shares of these materials are consumed directly) but it includes all phytomass that is used in manufacturing (ranging from cotton to wool, and from timber to recycled paper). In order to quantify the dematerialization of the US economy during the twentieth century, I used inflation-adjusted US GDP in constant (2020) dollars and total material consumption series prepared by Matos (2009). Material intensity of the US economy shows first a rise from 260 g/$ in 1900 to 380 g/$ in 1929 followed by four decades of fluctuations (low of 200 g in 1945, high 410 g in 1960) and then a steady decline from 370 g in 1970 to 240 g in 2000, that is ending the century more or less where it began – but, obviously, this similar aggregate rate hides vastly different composition of material use.

The post-WWII doubling of average material intensity is not surprising given the fact that the total input is dominated by minerals used in construction and that the first two post-war decades saw an unprecedented expansion of their use for new housing and industrial and transportation infrastructures (Interstate highways, new airports, and regulation of waterways). Their mass is dominated by stone and sand and gravel, and their share of the US material input rose from 54% of the total in 1945 to 72% in 1960. After excluding construction materials, the trend becomes much clearer, with maximum material intensities peaking during the second decade of the twentieth century at 170–210 g/$ (2020) and then, except for a brief post-WWII bounce, declining steadily to about 110 g by 1960 and 60 g/$ by the year 2000, resulting in an overall decline of nearly 75% in 90 years. This rate was almost perfectly matched by the decline in metal intensity (primary and recycled inputs) of American GDP that was reduced from the peak level of about 33 g/$(2005) during the 1920 to just 10 g by the year 2000.

The only greater drop was for wood intensity of GDP, from about 110 g/$ in 1910 to less than 25 g by 1950 and to just 6 g in the year 2000. Use of construction materials per unit of GDP rose, in a fluctuating manner, from about 100 g/$ in 1900 to the peak of roughly 300 g in the early 1960s and the subsequent decline brought it to less than 175 g in the first years of the twenty-first century. This overall reduction contained trends ranging from the already noted continuous and large decline of wood to slower and less pronounced reductions of primary metals and to a temporary rise and subsequent fall of demand evident for both paper and for industrial minerals. Interesting, but expected: secular reduction of material intensity of GDP in modern societies reflects the changing nature of modern economic product (with services accounting for a steadily increasing share of the total) and the ways we quantify it (with intellectual inputs valued far above the manual labor).

But a different metric reveals a very different result: in per capita terms, America's usage of all material inputs rose from about 1.9 t in 1900 to 5.7 t in 1950 and to just over 12 t in the year 2000, a 6.4-fold increase in a century. Per capita consumption of cement rose from just 40 kg in 1900 to 270 kg in 1950, it was 392 kg by the year 2000, and 310 kg in 2020, still nearly eight times more than 120 years ago and 15% more than in 1950. Per capita rates for all nonconstruction minerals rose from almost 1.2 t in 1900 to 3 t in 2000, a 2.6-fold rise and the consumption of metals grew from about 135 kg/capita in 1900 to 510 kg/capita in the year 2000, a nearly fourfold increase. And while the apparent per capita consumption of steel (125 kg in 1900) declined by nearly 20% between 1975 and 2000 (to about 425 kg; but recall that accounting for steel embedded in finished products raises this rate significantly), apparent consumption of aluminum in the year 2000 (nearly 25 kg/capita) was twice as high as the 1975 level. The verdict is clear: the twentieth century America did not see any impressive declines of per capita consumption, neither when measured by an all-encompassing aggregate or when assessed on sectoral basis or by looking at major individual flows.

And the real increase in per capita use of materials has been even greater than the USGS compilations of domestic material flows indicate because of the growing imports of manufactured goods made of material-intensive goods. Just think of metals and plastics embedded in machinery and cars (imported cars now account for a third of all sales), of aluminum and plastics in jetliners (Airbus has a strong presence in the United States), of heavy metals, rare earths, and lithium in electronics (the United States does not make any mobile phones, now by far the best-selling electronic items). Another instructive was to look at the twentieth century material demand in the United States is to estimate the needs for houses and their contents.

In 1900, a typical size of a new American home was 90 m² (1,000 ft²); it changed little during the century's first half, but since the end of WWII, average size of a newly built US home has increased substantially, with the mean area rising from 100 m² (1,100 ft²) in 1950 to about 190 m² in 2000 even as the average household size had declined during those years from 3.7 to 2.6 members (USGS 2008). Consequently, in 1950, average built area was about 27 m²/capita compared to about 75 m²/capita in the year 2000, a threefold gain (Wilson and Boehland 2005). And the mean size kept on growing, peaking at 229 m² in 2015 before it declined slightly. And by 2020, the average size for custom-built houses had surpassed 450 m², nearly five times the size of an average Japanese home, while "mansions" in excess of 600 m² became fairly common.

Average increase in the total house mass was even higher than indicated by the growth of living area because the houses have better insulation, heavier double (even triple) windows, larger garages (by 2020, nearly 80% of all new family houses had a double-garage), larger driveways and often extensive associated structures (decks, patios, storage sheds, and swimming pools), and are crammed with more furnishing that now spills outdoors with entire sets of outdoor chairs, sofas, tables, and accessories and with more consumer products and are finished with heavier materials (marble bathrooms, stone floors, and granite kitchen counters).

These disparate (and highly income-dependent) gains are not easy to quantify, but there is no doubt that the post-1950 per capita mass gain of an average family house has been higher than the threefold increase of the living area. Changes in qualitative terms have been no less important, as many lighter but more energy-intensive materials displaced traditional choices: synthetic wall-to-wall carpets or plastic floors become used instead of wood boards or parquets; aluminum siding replaced boards or mortar; steel, aluminum, and plastics displaced wood in window and door frames and in outside and garage doors; and plastics, rather than metals, became the dominant choice in pipes.

As for items of indoor mass consumption, the first wave of their acquisition of household items that began during the 1920s and it was virtually complete by 1960 when some 95% of all US households had an electric or gas stove, a refrigerator, and a radio – but still only 40% had a washing machine, clothes dryers were only in every fifth home, and only about 10% of households had central air conditioning and dishwasher (Felton 2008). By 1980, a large majority of households had those items and they began to acquire electronic goods (the third wave): CD players, computers, video games, flat-screen and HD TVs, and cellphones, all possible only thanks to tiny microprocessors and other highly energy-intensive electronic components, with many houses having more than one of each.

The first two decades of the twenty-first century brought a more pronounced wave of reduced material consumption in the world's affluent economies. To illustrate it with key US data, between 2000 and 2020, absolute (unadjusted for embedded imports) consumption of steel fell by 37%, its per capita consumption declined by about 40%, and steel intensity of GDP dollar was more than halved, and analogical rates for aluminum were about 30%, 40%, and 60% – and in some European nations, these dematerialization rates were as high or even higher. These impressive declines have been caused largely by the deepening deindustrialization of affluent Western economies and by the shift of material-intensive industries to lower-income countries of Asia, and particularly to China: the country's economic modernization began slowly during the 1980s, advanced during the 1990s, and accelerated after its admission to the World Trade Organization in 2001.

This unprecedented period of globalization and displacement of North American and European domestic production by imports means that significant shares of the new material demand in modernizing countries are getting embodied into exports to affluent countries. The best available accounts show that the gross energy use embodied in China's exports had more than tripled between 2000 and 2007 (from 8.6 to 29.6 EJ) and that it has subsequently stabilized at about 30 EJ (Zhang et al. 2021).

That total is nearly as large as Russia's annual primary energy use and it was equal to about 5% of the world's primary energy consumption in 2020!

Consequently, in this new world economy, the global scale is the most appropriate way to demonstrate that the modern civilization is becoming more, not less, massive even in per capita terms. Leaving fossil fuels aside, the world's aggregate material flows (cultivated products, metals, and nonmetallic minerals) had grown 1.6-fold during the first two decades of the twenty-first century, and this also means that the world of 2020 consumed about 30% more materials per capita than in the year 2000. China led the wave, nearly tripling its aggregate material consumption, but the countries where the overall demand doubled, or nearly doubled, include Asia's other three most populous economies, India, Indonesia and Pakistan. And even after leaving Chinese domestic uses out of the comparison, between 2000 and 2020 the global demand for steel rose by 15%, for aluminum by 20%, for copper and ammonia by 30%, and for cement by 90%.

An important contributing factor helping to explain this continuing rise of demand for materials has been the increase in per capita energy consumption, above all in Asia but also in Africa: these gains require steel, aluminum, copper, and cement for new infrastructures (fuel extraction and processing, electricity generation and transmission) and metals and plastics for new mass-produced energy converters (from cars to air conditioners). Per capita energy consumption is still unacceptably low in most of Africa and among the low-income groups of Asia and Latin America, but between 2000 and 2020, it had more than quadrupled in Vietnam, doubled in India, and grew by 55% in Indonesia. And, obviously, expanding food production in all of these modernizing countries is another source of new material demand.

Clearly, there is no recent evidence of any widespread and substantial dematerialization on the global scale – be it in absolute mass terms or when measured as average per capita. The world's richest economies have become relatively less material-intensive, although those declines are lower when properly corrected for materials embedded in imports. Still, there is no doubt about many instances of relative dematerialization and about subdued growth of raw material inputs in affluent societies that have already put in place extensive and highly material-intensive infrastructures. In contrast, there are enormous unmet material needs in much of Latin America and in most of Asia and Africa.

Even China's extraordinary expansion of material needs still has not reached its upper limit and despite the continued availability of better housing (with improved water and sanitation, sufficient living area, and durable construction) in sub-Saharan Africa (from 2000 to 2015 its share doubled from 11% to 23%), most people in cities still live in unimproved housing, and the population of the region's urban areas is expected to more than double by 2050 (Bisanzio et al. 2019).

But even as the world's aggregate material consumption keeps on rising, perhaps no other material transformation has received so much attention during the first two decades of the twenty-first century as the decarbonization of global energy supply. This shift is, of course, the essence of the now slowly unfolding worldwide energy transition from fossil carbon (coal and liquid and gaseous hydrocarbons) to renewable forms of electricity generation (dominated by wind and solar), carbon-neutral liquid biofuels for transportation and green hydrogen produced by electrolysis of water to be used in industrial processes ranging from ammonia synthesis (and that compound can be also used as fuel) to steelmaking (Fan and Friedmann 2021).

In the final chapter, I will review enormous material needs required to accomplish this global transition and here I will explain that, contrary to today's common impressions, there is nothing new about decarbonization of energy supply, that this transformation has been going on for generations, that it has achieved some remarkable successes, and that it has been accompanied by even more successfully pursued desulfurization. In the early 2020s, how many people recall – amidst today's preoccupation with global warming and CO_2 emissions from the combustion of fossil fuels – that four decades ago sulfur emissions from burning coal and refined fuels were considered the world's foremost environmental problem?

5.5   Decarbonization and Desulfurization

Decarbonization of national and global energy supply can be seen as a form of dematerialization that is particularly welcome from the environmental point of view because it reduces specific emissions of CO_2, the most important greenhouse gas, as well as the generation of acidifying sulfur and nitrogen oxides and particulate matter. In the past, the shift had relied primarily on burning fossil fuels with lower carbon content, producing biofuels whose conversion does not generate additional CO_2 and generating more primary electricity. The process was an organic result of gradual energy transitions from solid to liquid to gaseous fossil fuels and of rising shares of hydro and nuclear generation in total energy supply. In turn, these transitions have been driven by the need for increasingly higher power densities of final energy uses that are required to support urbanization, industrialization, higher intensity of transportation, and more affluent life styles.

The energy density gradient for fuels shows roughly a threefold difference between the extremes (all are gross heating values): dry wood has 18–20 GJ/t, most steam coals rate between 24 and 29 GJ/t, refined oil products are within a tight range of 45–46.5 GJ/t, and a ton of natural gas contains 54 GJ/t (Engineering ToolBox 2008). Higher hydrogen content explains this progression of higher energy density. Combustion of pure carbon releases 32.8 GJ/t, combustion of pure hydrogen liberates 142 GJ/t: fuels with higher hydrogen shares will have higher energy densities, and decarbonization can be quantified either as the decline of carbon content or the rise of hydrogen content, or expressed as a ratio of hydrogen and carbon atoms present in specific fuels and in aggregate national or global energy consumption.

Wood, composed mostly of cellulose, hemicellulose, and lignin, is usually about 50% C and its H content is only about 6%. Bituminous coal usually contains 65% C and 5% H by mass, transportation fuels refined from crude oil (gasoline, kerosene, and diesel) have 86% C and 13% H, while methane (CH_4) has, obviously, 75% C and 25% H by mass. When assuming 19 GJ/t as a mean energy density of absolutely dry wood, the fuel's complete oxidation would release about 30 kg C/GJ. Combustion of good bituminous coal will emit roughly 25 kg C/GJ (which means that the substitution of coal for wood will release 10–15% less carbon per unit of energy); typical emission factor for refined liquid fuels is about 20 kg C/GJ (or 20–25% lower than for coals) and complete combustion of methane will liberate only 15.3 kg C/GJ, nearly 20% less than gasoline, 40% less than typical bituminous coal, and 50% less than wood.

When using averages of 50% C and 6% H, wood would have atomic H:C ratio of about 1.4. But that is not the ratio of atoms whose combustion is converted to thermal energy: a large part of wood hydrogen is never oxidized because hydroxyl (OH·) radicals that are part of cellulose and hemicellulose escape to the air during the early stages of combustion (Shafizadeh 1981). Actual H:C ratio of wood is variable but no higher than 0.5 compared to coal's 1.0, 1.8 for common fuels refined from crude oil (gasoline and kerosene), and 4.0 for methane (actual ratio for natural gas is marginally lower due to the presence of higher homologs, C_2H_6 and C_3H_8, besides CH_4). For comparison, when Ausubel (2003) charted the historic shift toward decarbonization, he worked with the following H:C average: 0.1 for wood (too low, I would argue), 1 for coal, 2 for crude oil, and, naturally, 4 for methane.

Decarbonization of modern energy supply began first in the United States (already during the second half of the nineteenth century), but the process became widespread and proceeded at accelerated rates only after WWII. When the comparisons are made on the basis of energy content (not by mass), high-carbon fuels (wood and coal) accounted for 94% of the global total in 1900 and still 73% in 1950, but by the year 2000, their share was down to about 38%. Decarbonization was also helped by rising shares of primary (hydro and nuclear) electricity: its share in the global primary energy supply rose from just 2% in 1950 to about 10% half a century later. But integrations of individual fuel components show that during the twentieth century, coal supplied significantly more energy than crude oil, roughly 5.3 vs. 4.0 YJ (Smil 2017).

And that thanks to China's rapid post-1990 expansion of coal extraction, coal's share in global primary energy supply has been actually slightly higher in 2010 than it was in 1990, and natural gas, the fuel with the least carbon content, supplied still no more than 25% of all fuel energy in the year 2000 and nearly 30% in 2020 despite the fact that its global extraction had nearly tripled between 1980 and 2020, from about 1.4 to nearly $3.9\,G\,m^3$ (BP 2022). As a result, overall carbon intensity of the world's fossil fuel supply declined only moderately during the course of the twentieth century: in 1900, it was about 25 kg C/GJ of fossil fuels; by 1950, it was about 22.5 kg C/GJ; and in the year 2000, it was basically unchanged at about 22.3 kg C/GJ, altogether only a 15% decline during the twentieth century (and 20% when including all primary electricity in the global supply total). Afterward, higher natural gas extraction negated the effect of China's rising coal production and the global rate fell by about 10% in two decades, to 20.2 kg C/GJ by 2020.

What is true about relative and absolute dematerialization of global material consumption is true about relative and absolute decarbonization of the global energy supply. In both cases relative declines have been unmistakable, impressive, and they will continue, but as there have been no absolute global declines in materials consumed so there have been no worldwide reductions in carbon extracted and hence in CO_2 emitted. Just the opposite has been true, as global carbon emissions from fossil fuel combustion rose from less than 550 Mt C in 1900 to about 1.6 Gt C in 1950, to almost 6.8 Gt in the year 2000; then, after a 35% increase in a decade, they rose to a new record of 9.32 Gt C (or 34.2 Gt CO_2) in 2010 and in 2019, before the global economic slowdown caused by the pandemic, it reached a new record of 10.3 Gt C or 37.9 Gt CO_2 (Olivier 2022).

And long-term national trajectories of dematerialization also have their counterparts in decarbonization. All of the world's major economies show large multiples of CO_2 emissions during the course of the twentieth century: 8.6 for the United States, 62.8 for Japan, and (the following figure is correct!) 35,700 for China. But all high-income economies have seen recently a clear moderation in the growth of CO_2 emissions and many of them have experienced substantial absolute declines. Germany's emissions peaked at the time of unification in 1990: subsequent closing of inefficient coal-fired East German industries was followed by *Energiewende* (starting in 2000) promoting renewable energy uses: by 2019, Germany's emissions were 32% below their 1990 peak. UK's emissions peaked in 1991 and by 2019 they were 40% down, Japan's emissions peaked in 2007 (down by 12% in 2019), and the US emissions peaked in the year 2000 and by 2019 they were 16% lower. In contrast, Chinese 2019 CO_2 emissions were triple the 2000 level and are yet to peak, as is the case with India's carbon.

The verdict is clear: although the generations-old process of relative decarbonization of the world's fuel supply had experienced a significant slowdown during the second half of the twentieth century, it has been accelerating in all affluent economies and it is bound to continue at faster rates coming decades, above all thanks to rising shares of natural gas and to accelerating conversion to non-carbon alternatives. Even so, there are no imminent prospects for any rapid substantial reductions in the absolute level of global CO_2 emissions: in the next chapter, I will show that all such claims are unrealistic and that the worldwide transition to zero carbon economies will be a prolonged, and costly, process. In contrast, the story of desulfurization had, thanks to technical advances, a relatively rapid, ending.

5.5.1 Desulfurization

Except for some natural gases that are almost pure methane, or mixtures of CH_4 and higher homologs of the straight-chained alkanes series (mostly C_2H_6 and C_3H_8), all fossil fuels contain some sulfur, with shares ranging from less than 0.5% in the lightest (sweet) crude oils to more than 4% in low-quality lignites: 1.5–2% is the most common range for bituminous coals and globally traded crude oils (Smil 2008). Combustion of fossil fuels generates more than 90% of all anthropogenic sulfur with the rest originating largely in the smelting of copper, zinc, and lead. Emitted SO_2 oxidizes to sulfates and the resulting acid precipitation (and also acidifying dry deposition) affects soils, plants, and both aquatic and terrestrial ecosystems, and damages exposed metals as well as limestone and marble. Its effects became particularly noticeable across parts of Western and Central Europe, Eastern North America, Europe, and China.

Rising combustion of fossil fuels was accompanied by increased emissions of SO_2, from about 20 Mt in 1900 to just over 58 Mt by 1950, and the peak level of about 150 Mt was reached in in 1979–1980 (Smith et al. 2010). But, unlike in the carbon case, this absolute rise of sulfur emissions was accompanied by pronounced relative desulfurization: I have calculated its rates declining from about 465 g S/GJ in 1900 to about 400 g S/GJ in 1950 and to 270 g S/GJ by 1980. Higher shares of hydrocarbons in global energy supply had, on average, much lower sulfur content than coal, and high-S crudes and natural gases rich in H_2S were routinely desulfurized before combustion. After 1980, this desulfurization trend was amplified due to the adoption of flue gas desulfurization (FGD) in large coal-fired power plants.

FGD removes SO_2 by reactions with basic compounds (mostly lime or ground limestone); $CaSO_4$, its main product, is either land-filled or it can be used in wallboard manufacture. FGD was first commercialized during the 1970s and by the year 2000 nearly 40% of the world's coal-generating capacity had some level of FGD (Smil 2008). Emission declines were impressive: in the United States, they were cut nearly by nearly half (48%) between 1970 (the first year the USEPA began to made consistent calculations) and the year 2000, and by 2020, they were reduced by 90% (USEPA 2022c). Globally, relative SO_2 emissions were reduced to about 175 g S/GJ by the year 2005 as the absolute emissions declined to about 107 Mt (Smith et al. 2010).

As emissions were reduced in Europe and North America, they were rising rapidly in China and India: between 1990 and 2005, both China's and India's SO_2 emissions doubled, respectively, to more than 32 Mt and nearly 10 Mt/year; combination of deep cuts in the West and large increases in the East resulted in a slight global lowering of SO_2 releases, from 121 Mt in 1990 to 112 Mt in 2005. But the rise of Asian emissions introduced a greater uncertainty into the global estimates. For example, a comparison of half a dozen estimates of SO_2 emission by China, the world's worst polluter, shows that the difference between the lowest and the highest value was 38% for the year 2000 and 35% for the year 2005 with, predictably, the lowest totals claimed by China's Ministry of Electric Power (Lu et al. 2010).

Whatever the actual total might be, it seems that even the Chinese emissions finally began to decline after 2007 because of FGD installed at many of China's new (and some older) coal-fired power plants, a reduction that has been confirmed by the Ozone Monitoring Instrument aboard NASA's Aura satellite (Li et al. 2010). Another event that helped to reduce the global SO_2 emissions was the worst post-WWII economic crisis in 2008–2009, and Klimont et al. (2013) put the 2010 total as low as 103 Mt, nearly a third lower than the global record in 1980, and a rate that implies specific emissions of only about 118 g S/GJ, a third lower than the 2000 value. The latest published estimates, based on observations (concentrations in the atmosphere and in precipitation) and on six different global aerosol models quantify global and regional trends from 1990 until 2015 and while they confirm substantial declines, they end up with higher totals (Aas et al. 2019).

They show that global SO_2 emissions were reduced by 55 Mt (31%) between 1990 and 2015, from 175 to 129 Mt, with the largest decrease taking place during the 1990s in Europe. Between 2000 and 2015, emissions in Europe and the United States decreased by a similar total amount (13–14 Mt or by 40–50%) and East Asian emissions decreased by about 6 Mt (13%) since 2005 while India's emissions kept on increasing. Monitoring (and modeling) uncertainties cannot change this basic conclusion: unlike in the case of CO_2 emissions, global reduction of sulfur emissions has taken place not only in relative terms (per unit of energy released from combustion) but also in absolute terms and it has done so despite the rising combustion of coal because an effective technical fix provided by FGD and because of rising importance of sulfur-free natural gas.

But despite this welcome trend, anthropogenic SO_2 remains a major pollutant, with large electricity-generating plants and oil refineries responsible for about two-thirds of all emissions tracked by satellites and with most of the rest coming from large metal smelters. Monitoring shows that Norilsk (northern Siberia) is the world's largest SO_2 hotspot: smelting of nickel and palladium released nearly 2 Mt SO_2 in 2018, and each of the four other clusters of coal-fired electricity-generating plants and petrochemical complexes in South Africa, Iran, Saudi Arabia, and India emitted more than 500,000 t SO_2 (Dahiya and Myllyvirta 2019).

6

Material Outlook

Matteo Ceruti/Adobe Stock

Materials and Dematerialization: Making the Modern World, Second Edition. Vaclav Smil.
© 2023 John Wiley & Sons Ltd. Published 2023 by John Wiley & Sons Ltd.

ake1150/Adobe Stock Photos

Those readers who have persevered (and have, along the way, complained about too many numbers) have now reached the point where they should be impressed by the magnitude and complexity of the global material edifice erected by the modern civilization since the middle of the nineteenth century, and no less so by the magnitude of incessant material flows required to operate it and maintain it. Although there was no shortage of admirable extraction, construction, and consumption feats in pre-1850 history, only the creation, transformation, and expansion of modern civilization made human societies dependent on enormous, incessant, and now also truly global flows of materials. After millennia of reliance on a limited range of materials – wood and other biomass, stone, gravel, sand and clays, and a dozen of metals (both common and rare) – that were processed in relatively simple ways, we now use thousands of materials (natural, processed, or anthropogenic) whose qualities are tailored to a still widening array of specific requirements.

This growing diversity and dependence have been accompanied by some remarkable efficiency gains that can be impressively demonstrated by long-term tracing of specific energy and material uses, that is by declining energy intensities and by relative dematerialization, as well as by moderation, or even near-elimination, of undesirable environmental impacts. At the same time, growing populations and improving quality of life translated into steadily rising demand for virtually all traditional materials: the only notable exceptions have been such ancient biomaterials as flax or hemp, but in aggregate every modern nation now needs more wood, stone, sand, metals, and nonmetallic minerals that have been used since the antiquity than it did 100 or 50 years ago.

And to these traditional materials we have to add greatly expanded production of new materials – including synthetic fertilizers, polymers (plastics), and metals and nonmetallic elements not previously exploited by pre-industrial societies. In all of these cases there has been no global dematerialization in absolute terms, but there have been some absolute declines and many more relative dematerialization on the national level. We use many (old and new) materials more efficiently or we found better substitutes, and as a result we have seen many instances of their declining use when their demand is measured per capita or per unit of economic product.

These realities lead to many obvious questions. How long will these now so well-entrenched trends continue? How fast can low-income nations modernize their economies by considerably expanding their use of materials and reach the consumption levels now prevailing in affluent nations where we have seen many cases of relative dematerialization and even some absolute consumption declines? Post-1980 China, and before it Japan and South Korea, and now also Vietnam and (to a lesser extent) India provide examples of rapid material surges – but how many other nations badly in need of economic development and material expansion can achieve similar rates of improvement?

More fundamentally, how long can this global expansion of material uses, for some elements and compounds now counted in billions of tons per year, continue? What are the natural resource limits to further multiplications of mineral extraction and to modern syntheses based on nonrenewable raw materials? Are there any insurmountable economic obstacles that will force us to moderate our consumption of materials? How well can we cope with enormous material demands of the global energy transition? How realistic are the calls, now favored by many governments, to have it accomplished in less than three decades (and conversion to non-carbon electricity generation already by 2035)? Even more fundamentally, are there environmental constraints that would make it impossible to replicate the twentieth-century material expansion during the twenty-first century? And then there are the most important questions of all: should we advocate some rational limits to material consumption? And if so, how should we be going about it?

Some of these questions can be answered on the basis of indisputable physical evidence or by referring to the most likely course of realistic technical advances. Consequently, in the opening sections of this closing chapter, I will review first contrast the best available evidence concerning the availability of principal material resources with the extent of possible future requirements, and then look at the opportunities for material substitution, enhanced recycling and reuse. Answering other questions posed in the preceding paragraph cannot be done without fundamental value judgments or without assumptions regarding the extent and pace of anticipated material demands and substitutions. This applies, above all, to the speed of national and global decarbonization efforts that, according to some recent targets, are to proceed at unprecedented pace.

Complex linkages of many dynamic determinants of future material growth, including the variables subject to relatively sudden shifts, make all long-range forecasts inherently uncertain. Decades ago, I decided to abstain from offering such prognoses and, true to this resolution, will not provide any estimates of the most likely levels of global or national demand for particular materials by the year 2030 or 2050, nor will I present any scenarios of future rates of dematerialization or reductions in environmental impacts attributable to material extraction and use. What I will do instead is to appraise the needs arising from recently promoted targets and take closer looks at probable trends of some fundamental variables that will determine the future rates and consequences of material use.

I will start with concerns about the availability and extraction of natural resources whose incessant flows have become indispensable for the maintenance and advancement of modern civilization. In the next section, I will stress many encouragingly impressive opportunities for wasting less: potentials for both material and energy savings are nowhere near an early exhaustion. These opportunities will be further enhanced by new materials; many of them needed for the complex transition from carbon-dominated global energy supply to carbon-free primary energies, an enormous and costly transformation whose true dimensions have not been sufficiently appreciated. And this transition, impossible without mass-scale deployment of construction materials, metals, nonmetallic elements, and synthetics, is not the only factor that will limit the extent of future dematerialization. That is why I will close the book (without normative prescriptions) by raising questions about some fundamental departures from the twentieth-century model of mass consumption, and about the chances of more rational arrangements.

6.1 Natural Resources

This is not a place to review the evolution of concerns about the availability of natural resources and their adequacy to maintain the expansion of modern civilization. In the U.S. systematic interest in these matters began during the 1950s (The President's Materials Policy Commission 1952), advanced greatly during the 1960s (Landsberg et al. 1963; Landsberg 1964; NAS 1969), and it has been intertwined with energy and environmental concerns since the 1970s. Boulding (1970) called for the end of linear economy that extracts resources, transforms them into commodities, and spews them out into polluted reservoirs, a process he considered inherently suicidal.

Global concerns about exhaustion of mineral resources have been growing during the 1970s and 1980s (Meadows et al. 1972; Barney 1980; McLaren and Skinner 1987; Ehrlich and Holdren 1988), developed into arguments between catastrophist and cornucopian camps (Simon 1981, 1995) and became combined with an even broader concern about the extent and rapidity of global environmental change. The latest phase of these concerns began with prediction of an imminent peak in global oil extraction that was to be followed by steep decline in production (Campbell 1997; Deffeyes 2001). I called these dire predictions of oil's imminent exhaustion as products of a catastrophic cult (Smil 2006b), a label that was fully justified by the following course of global oil extraction: before the pandemic disrupted the demand, the global crude oil production in 2018 and 2019 set new records.

That there has been no global oil production decline – and that the one that may come will be unrelated to any exhaustion of oil but to the policies mandating decarbonization of global fuel supply – has not stopped the transference of such fears to other resources. For some time, these concerns were driven by China's seemingly insatiable demand for all minerals that led to a temporary reversal of steadily declining cost of raw materials for the world's affluent consumers (Sullivan et al. 2000). In 2012, gold was added to the list of about-to-peak extractions (Kerr 2012) – but a decade later, the world was producing annually nearly 30% more of it (USGS 2022a).

In fact, some commentators began to apply the "running out" meme universally: Heinberg (2010) declared that peak of everything is upon us and Klare (2012), more moderately, saw "the end of easy everything." He did not see any imminent scarcity but high prices and ferocious competition as companies and nations will fight among themselves for dwindling resources. Running out meme is here to stay – "These elements could disappear from the world" (World Economic Forum in 2017), "The World Is Running Out of Elements" (Discover magazine in 2020) – but less sensational assessments have turned to concerns about supply disruptions and import dependence. In the United States, the latter has risen recently to new highs: USGS review of 64 mineral commodities showed that in 2021, the United States was completely dependent on import of 17 of them (from arsenic to yttrium), and that the dependence was above 75% for another 21 commodities, from potash to selenium (USGS 2022a).

6.1.1   Reserves and Resources

In contrast to the complexity and contentiousness of long-term concerns about the impacts of running out of minerals and depending too much on imports, near-term outlook for the supply of essential mineral resources is neither highly uncertain nor difficult to write about. The only prerequisite of such a review is to make clear the often misunderstood fundamental difference between reserves and resources (McKelvey 1973). Resources denote the total mass of material (element, compound, mineral, and ore) in the Earth's crust, and the distinction can be made between resources on land and under sea. Obviously, these aggregates are not known with a high degree of certainty, and their estimates tend to increase with more drilling and more extensive targeted exploration. Even if we were to know the exact amount of a particular resource that information would not allow us to calculate the time of its ultimate exhaustion because no mineral could be ever completely exploited: long before reaching such a point, the costs of removing it from excessive depths or isolating it from deposits where it is present in minuscule concentrations would make its recovery quite uneconomical.

That is why the reserve category is of a much greater practical interests as reserves are the shares of resources that could be extracted from known locations with known costs and by using available techniques. This means that resources in place are given and finite (but poorly known) while reserves are moveable totals, accurately known but constantly changing: they are transferred from resources through investments and technical advances, through the continuous process of exploration and through the deployment of better recovery, mining, and processing techniques. Specific reserve totals, for a country, a continent or the world, are commonly divided by relevant annual production totals to calculate reserve/ production (R/P) ratios.

For example, according to the USGS, the global R/P ratio for copper was just 27 years in 1999, but that ratio did not imply that there will be no copper left to mine by the end of 2026. In 2021, the ratio was up to 42 years (reserves rose by 35%) but, again, this does not mean the end of copper mining by 2063. The ratio's ups and downs indicate how nimbly the industry creates reserves relative to annual production. Put in another way, the R/P ratio reflects transformative effort of moving minerals from resource to reserve category in order to assure production for a given number of years, no too short to engender uncertainty, not too long as there is no advantage of verifying reserves whose extraction may not take place for a century. In order to appraise long-term prospects, we have turn to the best estimates of global mineral resources, but those, too, are subject to often significant shifts. Staying with the copper example, between 2000 and 2021, global resource estimates increased by about 30% (USGS 2022a).

In the early 2020s, we were in no danger of running out of any major mineral, not imminently (in years), not in the near term (in one or two decades), most likely not even for the remainder of the twenty-first century. But the latter conclusion does not exclude temporary supply shortages, significantly higher prices, and the need for better resource management and new extraction techniques. Remarkably, sand, the most commonly excavated nonmetal mineral, is among the materials that need better overall management. In many countries sand remains a common-pool resource accessible to overexploitation or illegal extraction and trade (Torres et al. 2017).

Problems arise mostly with the extraction of active sand bodies in rivers or in coastal and near-shore zones where excessive removal of readily available sand can rapidly exhaust available resources while causing erosion with potentially serious on-site and downstream consequences (UNEP 2022c). Such effects were clearly demonstrated by using high-resolution satellite imagery to quantify sand extraction on the Lower Mekong River in Cambodia where the material's removal is nearly twice as large as previously estimated (Hackney et al. 2021). This excess led to negative sand budget (removal exceeding deposition) that has been driving major bed incision.

Local and regional constraints aside, resources of common construction materials – sands, clays, and stones – are immense, eliminating any concerns about their availability on a civilizational (10^3 years) time scale. Resources of silicon, the material of electronic age, are obviously in the same super-abundant category. Survival of billions of people requires adequate fertilizer applications to produce enough food and there is no danger that we are going to run out of the atmospheric nitrogen needed for the synthesis of ammonia. And even if decarbonization will eliminate abundant natural gas as the source of hydrogen, the element can be produced by electrolysis of water and that route is bound to get more affordable once we adopt it on a large scale (IEA 2022a).

As far as fertilizers go, that still leaves the concerns about running out of phosphorus (Beardsley 2011; Li et al. 2019). Perhaps most notably, this renewed fear (general and phosphate-specific) was raised in *Nature*, the world's leading science weekly, where a World View column concluded that

> Prices of global raw materials are now rising fast. This . . . is a genuine paradigm shift, perhaps the most important economic change since the Industrial Revolution. Simply, we are running out . . . . There is now no safety margin . . . there is the impending shortage of two fertilizers: phosphorus (phosphate) and potassium (potash). These two elements cannot be made, cannot be substituted, are necessary to grow all life forms, and are mined and depleted. It's a scary set of statements . . . What happens when these fertilizers run out is a question I can't get satisfactorily answered and, believe me, I have tried. There seems to be only one conclusion: their use must be drastically reduced in the next 20–40 years or we will begin to starve (Grantham 2012, 303).

I commented that Grantham could have tried just a bit harder (Smil 2012). Literature search would have led him to *World Phosphate Rock Reserves and Resources* published by the International Fertilizer Development Center (Van Kauwenbergh 2010). This detailed assessment of the world's phosphate reserves found them to be adequate to meet fertilizer demand for the next 300–400 years. Similarly, the 2022 edition of the USGS's *Mineral Commodity Summaries* (USGS 2022a) puts the world's phosphate rock reserves at 71 Gt (R/P ratio of 322 years as of 2020) and resources at more than 300 Gt and it concludes that "there are no imminent shortages of phosphate rock."

At the same time, the world needs to use phosphatic fertilizers more efficiently not only because such efforts save money but because excessive applications of phosphates are a major cause of aquatic eutrophication. Indeed, call for minimized losses should be a universal admonition not primarily because of any fears of imminent resources exhaustion but because all wasteful uses of materials have both economic and environmental costs. And much larger gains in reducing applications of all fertilizers could be achieved by reducing average per capita meat consumption in all affluent countries where fertilizers are now used primarily not to grow food but animal feed.

Similarly, there are also no looming shortages of the world's most important metallic ores. In 2022, USGS lists the following global R/P and resources ratios: bauxite, 82 years, 55–75 Gt; copper ores, 34 years, 2.1 Gt; iron ores, 53 years, resources of more than 800 Gt. Copper's R/P ratio provides an excellent example of how the process of translating resources into reserves (through discovery and mine development) has kept up with massively increased demand: in 2000, the ratio was lower (29 years) than in 2020 and its wrong interpretation would have the world running out of copper before 2030. R/P ratios are similar for nickel and manganese, and they are considerably higher for titanium, vanadium, and silver. Nor are the rare earths actually rare, their crustal abundance is not particularly low but the sites with profitably extractable concentrations are less common.

China's near-monopolization of their supply and a temporary reduction of their exports created this impression of rarity. The response to these restrictions has been a surge in exploration in North America, Australia, and Africa and opening of new mines. In the year 2000, China produced 85% of the world's output of 76,500 t (and the United States just 6.5%), by 2020, global production had nearly quadrupled (to 280,000 t) but China's share declined to 60%, U.S. share rose to 15% and new entrants with substantial production included Australia, Myanmar, and Vietnam. There are also opportunities for substitution (Holliday et al. 2012; Omodara et al. 2019) and for recycling (Nemoto et al. 2011; McMahan 2019).

A global assessment by Gordon et al. (2006, 1213) concluded that "there is no immediate concern about the capacity of mineral resources to supply requirements for the geochemically scarce metals." Similarly, after reviewing the needs of energy production, obviously a critical sector, Graedel (2011, 332) found that "at least for the foreseeable future, and from a global perspective, there are no significant supply challenges for any of the parent metals related to energy." Four years later, he saw supplies platinum group metals, gold, and mercury possibly limited by environmental considerations, and those of chromium, niobium, tungsten, and molybdenum by supply restriction (Graedel et al. 2015). In any case, future supply will need higher energy requirements and hence higher production costs: richness of many exploited metallic ores has been declining for generations, with relative degradation averaging about 1% a year.

Naturally, for some ores in some countries the decline has been much faster; for example, Malaysian tin mining companies are now removing twice as much waste as two decades ago (Sim and Rusman 2011). As a result, Norgate and Jahanshahi (2011) estimated that the combination of this declining quality (excluding Al and Fe whose ores are expected to remain unchanged) and higher demand might lead to a quadrupling of energy requirements for the worldwide production of principal metals (Al, Cu, Fe, Ni, Pb, and Zn) by 2030, from about 2 EJ in 2010 to 8.7 EJ. On the other hand, such calculations do not factor in continuing technical advances and future shifts in demand. By far the greatest challenge to meet these coming material requirements will be due to the requirements associated with the unfolding energy transition.

6.2 Materials for Energy Transition

As I already demonstrated, there is nothing new about gradual decarbonization of global energy supply, and in the past, the process has not been hampered by any insurmountable material shortages. But that experience does not provide good guidance for the future because the past changes were the combination of expected, gradual shifts: advancing technical innovations (higher efficiencies of prime movers), replacement of obsolete conversions (above all the shift from large coal-fired electricity-generating plants to gas turbines), and the introduction of new modes of electricity and fuel supply (wind turbines, photovoltaics, biofuels). In contrast, the anticipated energy transition, driven by the need for rapid decarbonization as the best way to prevent excessive global warming, should proceed at an unprecedented scale and it should achieve substantial results in just 10–15 years and be completed by 2050.

I have concluded that these are unrealistically ambitious goals because the proponents of rapid transformation do not sufficiently appreciate the scale and the complexity of the global task (Smil 2017, 2022). This transition requires not just further, and much more extensive, electrification of final energy uses (be in transportation, industries, services, or households) but also fundamental shifts in ways we produce and use non-carbon fuels as well as replacing carbon fuels that now dominate many basic industries (steelmaking, production of cement, and synthesis of ammonia and plastics) by hydrogen or by heat generated by nuclear fission of concentrated solar power.

In sum, it is a transformation of an entire system of energy use, not just a replacement of some constituent parts, and here I will focus only on some fundamental material needs required by the transformation in order to show that the speed of requirements implied by optimistic scenarios will be the greatest near-term challenge. Material demands of global decarbonization fall into three basic categories. The first one is much-expanded production and deployment of such traditional materials as copper and aluminum (above all for electrified transportation and for new high-voltage lines) as well as elements and compounds required for mass-scale installations of wind turbines and photovoltaic panels.

No other basic metal will be as decisive as copper, the key element of expanding electrification. The basic numbers illustrate the need for the electrification of road transport: copper requirements average about 23 kg for an internal combustion (IC) engine, 40 kg for a hybrid vehicle, 60 kg for a plug-in hybrid, and 83 kg for battery-powered car, with 220–370 kg needed for an electric bus. In addition, every EV charger will add 0.7 kg, and a fast charger up to 8 kg of the metal (IDTechEx 2017). A recent forecast expects the global demand of 25 Mt in 2020 to double by 2035 and keep on growing slowly afterward, with the United States expected to be importing 57–67% of its needs during the most intensive period of the unfolding energy transitions. While the future demand will depend on many factors, it seems likely that even the combination of expanded mining and intensified recycling may not be enough to prevent supply shortages.

The second demand is to produce new materials on unprecedented scales: green hydrogen is the foremost example. Of course, there is nothing new about large-scale industrial need for hydrogen: in 2021, global hydrogen demand (dominated by cracking heavier oils in refineries and by feedstock for ammonia synthesis) rose to 94 Mt and its production embodied about 2.5% of the world's final energy consumption, with all but a tiny fraction of it came from reforming methane, a process releasing CO_2 (IEA 2022a). That is known as black hydrogen while the green hydrogen has to be made by electrolysis of water using only renewable forms of electricity.

This process works with efficiency of 70–80% and it requires 50–55 MWh to release a ton of hydrogen. Smelting of steel and the synthesis of ammonia are the leading candidates for conversion to green hydrogen. Stoichiometry determines the needs. For steel, it is

$$Fe_2O_3 + 3H_2 \rightarrow 2Fe + 3H_2O \left(160 + 6 = 112 + 54\right)$$

Substituting 1.2 Gt for 2Fe (other 0.6 Gt are made in electric arc furnaces recycling old steel) results in about 64 Mt of hydrogen for green steelmaking. For the Haber-Bosch ammonia synthesis, it is

$$N_2 + 3H_2 \rightarrow 2NH_3 \left(28 + 6 = 34\right)$$

Substituting 150 Mt for 2NH$_3$ results in about 26.5 Mt of hydrogen for green ammonia synthesis.

This means that just the conversion of existing steel and ammonia production would require annual output of about 90 Mt of hydrogen, and with 50–55 MWh required to produce a ton hydrogen that would translate to 5 PWh of uninterrupted renewable electricity, or about 18% of the worldwide generation (28.5 PWh) in 2021. Both of these demands are expected to grow, and much more hydrogen would be needed for other sectors that now rely for refined fuels (all forms of transportation) or natural gas (space heating, process heat in industries ranging from glassmaking to food preservation) as well as for electricity generation.

The extent of possible requirements is illustrated by the EU's hydrogen roadmap that calls for three major deployments of the element (FCH 2019): to decarbonize the gas grid that delivers more than 40% of heating in EU households and 15% of EU electricity generation; as the most promising decarbonization option for trucks, buses, ships, trains, large cars, and commercial vehicles; and to produce high-grade heat for many industrial processes as well as to become a feedstock for some of them (most notably for ammonia synthesis). As we have yet to embark on any large-scale substitutions, eventual extent of hydrogen uses remains uncertain, but even conservative assumptions show the enormity of potential demand.

In 2020, the world's total final energy consumption amounted to about 9.5 Gt of oil equivalent, or about 400 EJ (IEA 2022c). If green hydrogen were to replace all fossil fuels now directly used in industrial production, half of fossil fuels in transportation (leaving the rest to biofuels and electricity) and a third of all other uses (commercial, residential, with, again, the rest coming from renewable electricity and biofuels), then we would need to produce about 1.1 Gt of it every year, or roughly 12 times the current natural gas- and coal-based production. For comparison, IEA concluded that "if all projects currently in the pipeline were realised, by 2030 the production of low-emission hydrogen could reach 16–24 Mt per year" of which 9–14 Mt would be based on electrolysis and 7–10 Mt on fossil fuels with carbon capture and use (IEA 2022c).

Consequently, even according to this optimistic forecast, by 2030, the world's green hydrogen supply would be just a few percent of the eventual need! And the eventual 10-fold expansion would not be without its problems, and by that I do not mean the need to maintain constant and rigorous safety precautions when distributing and using this highly combustible fuel (flammable when mixed even with small volumes of air), but the element's potential role as indirect greenhouse gas. Because hydrogen reacts with tropospheric hydroxyl radicals (OH), we have known that its increased emissions would perturb the atmospheric levels of methane and ozone, the second and third most important greenhouse gases, and would make hydrogen an indirect greenhouse gas with a global warming potential of 5.8 over a 100-year time horizon (Derwent et al. 2006).

Recent investigations showed that increased hydrogen concentrations would increase the levels of tropospheric ozone and water vapor (particularly in the stratosphere) and atmospheric lifetime of methane and its impact on climate. Depending on the level of future hydrogen leakage, benefits of reduced CO_2 emissions would be reduced and the indirect effect of these changes would mean that the global warming potential for a 100-year time horizon would be 11 ± 5, that is double the previously estimated value (Warwick et al. 2022). Obviously, efforts to minimize any leakage during production and transportation would add to the cost of extensive reliance on hydrogen.

And the third demand is to come up with new materials whose deployment would simplify and accelerate the transition and make it more affordable. There are now many studies assessing and estimating material needs of global energy transitions, including those published by the Institute for Sustainable Futures (Dominish et al. 2019), European Union (Carrara et al. 2020), Manhattan Institute (Mills 2020), World Bank (Hund et al. 2020), International Renewable Energy Agency (Gielen 2021), and International Energy Agency (2022b). Overall demand for materials and the rate of its growth will depend on the specific targets (such as keeping global temperature increase below 2 °C or letting it go higher) and on their timing, now with target years always ending in five or zero: zero carbon electricity generation in the EU and in the United States by 2035, net-zero carbon of total energy supply by 2050 (Moore 2022).

In any case, decarbonization will be highly mineral-intensive because the generation of zero-carbon electricity needs vastly more materials than fossil-fuel-based options and so does the replacement of fossil fuels used for industrial heat or as feedstocks, a key part of the transition that is yet to begin on a meaningful scale. So far, decarbonization has made its greatest progress by increasing the share of renewable electricity generation. Of the two dominant choices, wind and solar, the contribution by wind turbines has been more important: in 2021, they generated 33% more electricity than photovoltaic panels, and wind and solar supplied about 10% of the world's electricity. But these renewable alternatives are much more material-intensive than any form of fossil-fueled generation.

The difference is particularly large when material needs of wind turbines are compared with material needs of gas turbines, the most efficient, the most compact, and the most flexible of all options for generating electricity by the combustion of a fossil fuel. Complete installation of relatively small gas turbines (40–50 MW of capacity, the largest units are now above 500 MW) requires 250,000–290,000 kilograms of material (for the turbine itself and for the associated power generation and drive packages) or about 6 t/MW (Siemens 2019). Gas turbines need steel and aluminum, cobalt, magnesium, nickel, and titanium alloys, while steel dominates all auxiliary structures.

In contrast, a large 5 MW wind turbine will weigh in excess of 400 t: rotor 110 t, hub 54 t, and nacelle (a massive streamlined covering that houses generator, gearbox, drive train, and brakes), 240 t. Its materials will be steel, aluminum, copper, and plastics for blades (Engström et al. 2010). Welded steel shell tower (with the hub at 12 m) will add about 560 t of the metal and hence the above-ground components will add up to 964 t and that works to nearly 200 t/MW, roughly 30 times the material needed for a gas turbine. A more recent evaluation comes up with a very similar rate of at least 170 t/MW, dominated by steel and cast iron (about 80%), followed by glass fiber and composites, plexiglass, and copper (Carrara et al. 2020). But that is not all.

While gas turbine assemblies can be placed on a simple flat concrete foundation (or inside an existing industrial structure), erecting tall wind turbine towers (and mounting nacelles and rotors) requires a flat area in order to get access for giant cranes and the towers must be anchored in massive foundations built of reinforced concrete (Engström et al. 2010). These foundations will typically consume about 60% of all materials, raising the total material rate per installed unit of power to at least 400 t/MW. This means that the overall material demand of a wind turbine would be (even after making general allowance for building new gas turbine foundations) at least 50 times larger than for a gas turbine of the same capacity.

And the comparison would need still further adjustments. Gas turbines, with their small space needs, can be installed within many already developed industrial sites or old power plants and readily connected to the existing grid, and they can generate on demand (starting in less than 10 minutes). In contrast, wind turbines will require new connectors, transformers, and often new transmission lines to connect them to grid, and wind generation is inherently intermittent and hence it requires additional reserve capacity (that is more material) to be available as needed. If these considerations were taken into account, the comparison of material intensities would look even more unfavorable for wind generation.

Photovoltaic electricity generation does not require any massive concrete and steel structures, but its material demands also greatly surpass those of fossil-fuel-based generation when compared with per unit of installed capacity. To cover large expanses of land with PV panels requires about 60 t of concrete and 65 t of steel per MW for supports, about 45 t of glass, 8 t of plastics, and 5–7 t of aluminum and copper (Carrara et al. 2020). The mass of silicon required for PV cells has been steadily declining and it is now only about 4 t/MW and it may be halved by 2030. The mass of listed materials adds up to nearly 200 t/MW, and, once again, this excludes all connectors, transformers, and transmission lines: material requirements will be particularly high for building new high-voltage transmission lines in order to create national grids (still absent in the United States, Canada, or Brazil). Moreover, production of PV cells requires considerable amounts of rare elements.

As for the electric vehicles (EVs), a single lithium battery weighing about 450 kg and contains about 180 kg of steel, aluminum, and plastics as well as 50 kg of graphite, 40 kg Cu, 27 kg Ni, 13 kg Co, and 11 kg Li, and when considering typical concentration of these elements in respective ores, brines or deposits – Cu, Co, and Li <0.1%; Ni ~ 1%, C (graphite) ~ 10% – its production requires about 40 t of ores to extract just these four elements (Mills 2020). Even with a conservative estimate of 5 : 1 as typical overburden:ore ratio that would add up to at least 200 t of materials that has to be handled for every battery, or no less than 20 Gt for annual production of electric cars to reach the current level global deliveries of cars with IC engines (about 80 million in 2021). Depending on the speed of replacing combustion engines by electric motors in passenger cars, recent extraction levels of these mineral would have to triple or quadruple within two to three decades.

IRENA's review of critical materials stresses the need for large amounts of neodymium and dysprosium used in permanent magnets found in electric motors (hundreds of millions will be needed for EVs) and generators (hundreds of thousands will be needed for wind turbines), and it highlights the concerns about political and economic risks of relying on relatively small numbers of major producers and processors of these rare element (Gielen 2021). Moreover, rapid adoption of EVs will also result in growing demand for cabling required for charging stations (Edmondson and Holland 2020). And while we can make meaningful illustrative calculations of future material demand by wind and solar electricity generation (based on assumptions about their eventual rate of adoption), it is much more difficult to quantify the future material needs of displacing fossil fuels in transportation and industrial production.

In the former case, competing alternatives make the future of transportation fuels highly uncertain, even in the case of passenger cars. Adoption of EVs has been accelerating, and in 2022, IEA forecast that the global market value of electricity for EV charging will more than 20-fold by 2030, an equivalent of about one-tenth of today's diesel and gasoline market value (IEA 2022d). But arguments have been made in favor of other options, above all fuel cells (Toyota's favorite), direct hydrogen combustion, ammonia, and, an old stand-by, biofuels. If we assume that EVs will dominate the market, then the future material requirements are staggering. In 2021, the global stock of EVs was 16 million, but there were nearly 1.4 billion IC engines on the world's roads, in passenger cars, trucks, and buses. Replacing the recent global output of IC vehicles (more than 90 million a year) by EVs would call (with 6.6 million EVs sold in 2021) for nearly 15-fold increase in the supply of essential materials for batteries, battery packs, and electric motors: those range from aluminum, copper, and nickel to expensive cobalt and rare earths.

Similarly, it remains unclear how we will eventually replace process heat produced by the combustion of fossil fuels in industries ranging from ammonia synthesis (natural gas and coal) to ferrous metallurgy (coke and natural gas). Will it be solely or largely by hydrogen or by a mixture of hydrogen, synthetic fuels (some even made by using captured CO_2), concentrated solar power, and electric resistance heating? One thing is certain: no matter which option will dominate, material requirements to generate a unit of heat will rise compared to today's prevailing uses of fossil fuels, above all the highly efficient combustion of natural gas.

Moreover, we will have to start from scratch: in 2021, the world produced 94 Mt of hydrogen, more than 93 Mt by steam-methane reforming. Green hydrogen will be made mostly by electrolysis of water using wind- or solar-generated electricity and the International Energy Agency estimates that the world would need 100 Mt of it by 2030 in order to meet 2050 decarbonization target (IEA 2022a). Again (leaving aside the attainability of this goal), material implications of going from less than 1 to 100 Mt in eight years are obviously profound.

And we also cannot specify the eventual outcome of changing energy foundations as far as such key economic sectors as air and marine transportation are concerned. In either case the electric option is highly constrained by low energy density of batteries: in 2022, Panasonic's nickel–cobalt–aluminum 2170 battery widely used in electric cars had power density of 730 Wh/l compared to 10,300 Wh/l for kerosene and diesel fuel used, respectively, by jet engines and in shipping. Biofuels may be a partial option for flying but not for supplying the total demand of the global jetliner fleet, and hydrogen and ammonia are seen as the best candidates for long-distance shipping. But, yet again, no matter what combination of options prevails, replacement of processed fossil fuels in these sectors will further increase demand for materials, including lightweight alloys and highly energy-intensive composites.

When adding material demands of the global energy transition to the expected material needs of even moderate economic growth required to support the still expanding global population, it is obvious that the only way to meet these demands with limited environmental impacts is to promote any measures aimed at relative dematerialization by reducing all material inputs per products or per unit of performance. This should start by better design, include better ways of production and distribution, and make maximized reuse and recycling the essential part of the process. In the next section, I will summarize some notable options for wasting less.

6.3 Wasting Less

Despite many recent improvements at various stages of design, production, use, and reuse of materials, opportunities for wasting less, be it raw commodities or finished products, remain both substantial and ubiquitous. A systematic review of these opportunities would easily fill another book: here I will offer just some notable examples illustrating the diversity of realistic improvements and their significant rewards. I will do so by concentrating on four categories of reducing waste: by avoiding the use of a material through better design; by more rational, less wasteful, manufacturing; by more intensive and more efficacious recycling; and by rewarding material substitutions.

But before turning to these specifics, I must emphasize that the most important indirect way to rationalize and to moderate modern material consumption is to keep on pursuing many outstanding opportunities for reducing energy cost all along the material chain, starting with mineral extraction and processing, through transportation and manufacturing to distribution and reusing or recycling of unwanted components and products. Remaining differences between the current best practices and the energies of formation indicate the extent of potential savings: for ammonia, the gap is only about 30%; but for steel, it is almost three times as much. Of course, in practice it is impossible to come very close to theoretical minima required to liberate metals from ores or to synthesize compounds from their constituent elements, but considerable savings are achievable without any stunning innovations and savings on the order of 20–25% with the application of the best available practices would be common for most major materials.

Avoiding the use of materials should be a high priority in all cases where such omissions do not result in products that are less safe, less convenient to use, or less durable. Opportunities for saving by design are ubiquitous and they would have the greatest and longest-lasting effect in new construction: suboptimal design of buildings means that 30% reduction in overall material use is commonly possible. The easiest choices to avoid material use are in packaging. Intercontinental transport requires adequate protection but excessive packaging of consumer items (be it tools, food, or gifts) has become such a ubiquitous attribute of modern affluent societies that people have become inured to this often gratuitous waste, they just remove the packaging and discard it. This is a relatively recent development: even during the 1970s every small tool was not encased in a blister pack glued to a paper backing, making it more difficult to recycle the two very different materials.

Paper bags, easily recyclable, dominated grocery shopping, and vegetables and fruits did not come preloaded in clam-shell containers or in small plastic pails. Japan is an unequaled paragon of this packaging excess, with items wrapped and placed in several layers of paper and plastic: a croissant is put into one mini-bag, a *manju* bun into another, both bags are closed with (often golden) tie and placed in a bigger plastic bag. Numbers involved are not trivial. Single-use plastics, mostly bags, containers, cups, and bottles, but also straws and cutlery, account for more than a third of the world's plastic production, and annual generation of plastic packaging waste alone is now about 50 kg in the United States, and about 35 kg not only in the EU and Japan but also in China (UNEP 2018). The shift to excessive packaging has been driven by the quest for reducing sales job and speeding up checkouts but its material, and environmental, consequences have been detrimental.

One of the best systematic reviews of possible material savings in production can be found in Allwood and Cullen's (2012) appraisal of waste reductions of metals, particularly steel and aluminum. The authors trace these opportunities to six different sequential categories: using less metal by design, reducing yield losses, diverting manufacturing scrap, reusing metal components, striving for longer life products, and reducing final demand for metal services. Even material-sparing design, perhaps the most obvious pursuit that should be the aim of rational engineering, is a far from exhausted option. Allwood and Cullen (2012) estimate that lightweight design could reduce mass of steel beams by 20–50% and that the savings could be as high as 30% for reinforcing bars and pipes and 40% for car bodies, and that in aggregate they could add up to as much as 100 Mt of steel, or nearly as much as the annual pre-2008 U.S. consumption. Additive manufacturing (three-dimensional printing) offers the least wasteful way to produce complex shapes with minimal waste (Gibson et al. 2010).

Longer product life is another obvious material-sparing option but some of the greatest rewards could come from reducing yield losses and diverting manufacturing scrap. Most people are surprised to learn that substantial shares of newly produced metals (more than 25% of all steel, almost 50% of all aluminum) never make it into final products: 90% of the initially produced steel may end up in a simple I-beam, but the share is just over 50% for an aluminum can and mere 10% for aluminum aircraft wing skin. Better design could raise these shares and although these materials are recycled as internal scrap, their remelting consumes energy and generates atmospheric emissions.

Some of this internal waste is an inevitable by-product of manufacturing processes: waste comes from blanketing, cutting, and trimming of metal: turnings (swarf) are generated by drilling, milling, and grinding. Metals are also wasted due to surface and internal defects and exigencies of handling, and in some operation, a great deal of scarp comes from excessive orders as new material in excellent condition is treated as waste and returned for recycling. Allwood and Cullen (2012) estimated that up to 30 Mt of steel blanketing skeletons and up to 20 Mt of aluminum swarf would not have to be recycled by melting but could be reused in even less energy-intensive ways. Similarly, loss of materials on construction sites is commonly around 10% of initially delivered mass and it could be easily halved by appropriate management.

Opportunities for enhanced recycling remain large even in the case of paper and aluminum cans, the two materials whose recycling rates are the highest in all affluent countries (Japan's paper recycling may be the exception as it is already about as complete as is practical). Perhaps most notably, until 2008, paper was still the largest discarded material going into the U.S. landfills, surpassing even all discarded food and accounting for almost 21% of the total mass, compared to nearly 17% for plastics; subsequently, paper landfilling declined steadily and in 2018 (the last years for which USEPA numbers were available in 2022) it made up less than 12% of all mass, half of the mass of food waste with 24% (USEPA 2022d). Even so, the total mass of landfilled paper and paperboard was about 17 Mt, that is about 79% more than the annual production in such paper-making powers as Canada or Finland (FAO 2022).

Unfortunately, landfilling of plastics, despite their rising rate of recycling, has moved in the opposite direction, with 75% of all plastic waste (about 25 Mt) going into landfills in 2018 (making up 18% of all municipal waste) and with only 9% of generated waste recycled (and about 15% incinerated). And Europe, with its dedication to environmental management and with its far-reaching rules and regulations, has done better – but far worse than anticipated. EU's goal was for full diversion of plastic waste from landfills by 2020 (EPRO 2011) – but actual disposal of plastic waste in 2020 was about 23% to landfills, 35% recycled and the rest incinerated (Plastics Europe 2021).

Recycling should aim at maximum practicable rates. With readily recyclable items (paper, aluminum cans, and glass) the challenge is reduced primarily to voluntary (or deposit-induced) participation and well-organized collection. With complex products by far the most important universal step in the right direction would not require exceptional arrangements or ruinous investment: products should be designed with disassembly and recycling in mind, a task that has been made much easier by modern computer-assisted design procedures but one that is still not always seen as very important. Such rational, recycling-friendly design would be especially helpful in managing the rising mass of e-waste in general and of short-life mobile phones in particular.

An interesting innovation was the introduction of robot kiosks by ecoATM in the United States in 2008. The company has now more than 5,000 kiosks where customers can find the value of old mobiles and get instant cash for the devices that are sold to bidders: the firm acts as a broker for companies that can refurbish and resell the phones on the secondary market or recycle them for components or precious metals (ecoATM 2022). Unfortunately, it has not been a runaway success: by 2022, the company collected more than 30 million devices in the market where annual sales have been above 100 million units per year and the average lifespan of a device is no more than 2.5 years. Even so, this model should be widely adopted and extended to all other e-devices.

As I showed in the Chapter 2, history of material use is, to a large extent, history of material substitutions, and the process is now perhaps more important than ever and light-weighting has been its most common demonstration. Light-weighting could be used with the greatest rewards in all forms of metal-based manufacturing – and it may not require abandoning traditional materials with higher specific weight. For example, weight reductions can be achieved by using advanced high-strength steels processed by cold forming or hot stamping, or by forming lighter yet stiffer sandwich materials, such as car roofs with a less than 0.4 mm of polymers between two 0.2 and 0.3 mm steel sheets that weight nearly 40% less (Hoffmann 2012; Hara and Özgen 2016).

Substitutions of steel by aluminum have been progressing for decades, particularly in the car industry. In Chapter 3, I already noted how the aluminum content of America's average passenger vehicle rose by 35% between 2000 and 2020 and will likely increase by another 12% by 2026 (DuckerFrontier 2020) – but considerable light-weighting opportunities remain in the entire transportation sector: every form of passenger and freight transportation (cars, trucks, trains, vessel, and aircraft) remains a rewarding candidate. Light-weighting using carbon-fiber-reinforced plastic (CFRP) is particularly appealing because it can reduce the mass of a car chassis by 50% compared to a steel structure and by 30% compared to Al alloys (Ahmad et al. 2020). But light-weighting opportunities are everywhere, although some of them face nontrivial perception/preference barriers.

Putting wine into plastic pouches is a great example of such a resistance – but in aggregate, and considering the volume of intercontinental wine trade with millions of bottles carried by container ships, that substitution would entail large energy and material savings. Wine bottles can be made lighter (by as much as 25%) and quality of wine would not suffer if it were distributed in Tetra Pak, a paper container with a polymer lining that does (even when comparing 1 l pack with 0.75 l bottle) reduce mass by about 95% and whose life-cycle energy cost would be about 70% lower (Tetra Pak 2022). And speaking of wine, a major material replacement has been underway since the mid-1980s as traditional corks have been losing a growing part of the market to synthetics (some made from plant starches and sugars) and screw caps without degrading the quality of the final product and eliminating the loss of roughly 5% of cork-enclosed affected by cork taint (Goode 2010). Screw caps have been widely used in Australian wines but they have been outlawed in Spain.

But the packaging of liquids also shows that light-weighting may not always lead to the best functional solution. Weight ranking of common beverage containers (from the lightest to the heaviest) is aluminum cans, plastic bottles, steel cans, and glass bottles. When ranked by energy intensity of the material (MJ/kg of material), the ascending order is glass, steel, plastics, aluminum, and the energy/volume ranking is steel (2.4 MJ/l), plastics (3.2 MJ/l), glass (about 8.2 MJ/l), and aluminum (9 MJ/l). Glass and aluminum are the worst choices, and steel comes far ahead in energy spent to make a unit of fluid storage (ImpEE Project 2022).

But before I turn to making some closing observations regarding materials in the twenty-first century, I must make clear that circular economy (CE) – now often promoted "trending concept" whose achievement would offer the ultimate version of efforts to reduce material use – is an impossible objective. This conclusion is not a matter of personal opinion but it rests on fundamental natural realities and hence it is even more remarkable that the notion has received plenty of attention not just as a dubious academic concept but even on policy-making level with the EU promoting its Circular Economy Action Plan and promising "to lead the way to a circular economy at the global level" (European Commission 2015). Obviously, a truly CE would be an energetic, material, environmental, and social boon, but the best we can do is to move in its direction while knowing that such a state of affairs is unattainable.

6.4   Circular Economy

Here is a quote from a recent paper that is fairly typical of many available descriptions of the concept of CE:

> The key characteristic of a CE, which distinguishes a CE from other attempts to reduce energy and material consumption, is a holistic approach with the creation of circular loops of material, energy and waste flows encompassing all societal activities . . . The CE entered academic literature in 1966 when the ecological economist Kenneth E. Boulding criticised the linear "cowboy economy" of the past and described a future as a "spaceship economy" where all used resources were returned into the system . . . (Grafström and Aasma 2021, 1).

This, of course, is impossible unless we eliminate two natural processes that prevent any global human-directed circularity – dissipation of materials during their use and the generation of massive waste (hidden) material flows – and, above all, unless we abolish the second law of thermodynamics, the fundamental physical dictum that allows energies to flow only one way, toward disorder and dissipation.

Dissipation of materials is an extremely common phenomenon: most materials are not like steel in an old structural beam or in an old industrial machine or discarded vehicle that can be separated from other materials (wood, concrete, plastics, and aluminum) and recycled. These dissipative losses range from weathering and corrosion of exposed surfaces of stone, brick, cement, and metallic structures (in cities often accelerated by acid rain) to friction, abrasion, and pulverization during industrial processing, construction, transportation, and washing and to the use of toilet and tissue paper (NRDC 2019). Worldwide, abrasion of tires, shoe leather, and plastics (shoe soles, exterior paint, and road markings) and fibers detached from clothes during washing and drying add up to millions of tons of microparticles.

But by far the most fundamental example of dissipation is the use of water because this quintessential material of life is used by civilization in far larger quantities than any nonmetallic mineral or metal and because our means to govern the global water cycle are limited. As I explained, I have not included water in global or national totals of material flows because that inclusion would put all other flows into the category of rounding errors. Global withdrawals of fresh water now amount to about 4,000 km³ a year – that is 4 Tt, the mass nearly two orders of magnitude greater than the total annual use of all solid materials (Wada et al. 2016). But precisely because of its enormous circulating mass, water illustrates how utterly impossible is the idea of CE that will create "loops . . . encompassing all activities." Biosphere depends on incessant functioning of grand natural cycles and while their qualitative ranking is pointless (in order to thrive, plants need all three macronutrients, N and P and K), water cycle is obviously by far the largest relatively rapid quantitative transfer of an indispensable material.

Our interventions in the global water cycle have become more common as we withdraw more water for irrigation, industries, and households – but they have only marginal effect on the fundamental natural processes of evaporation, plant transpiration, precipitation, and runoff. We can cover a field with plastic and prevent local evaporation, but we have no control of the process – where the evaporated water will be eventually condensed and precipitated or what will happen to the runoff once it reaches the ocean. We recycle increasing shares of water directly used by people (municipal and rural water supplies) and by industries (nonconsumptive uses for cooling and processing, incorporation in products) but those two sectors claim, respectively, only about 12% and 19% of all water withdrawals, with agriculture accounting for nearly 70% of total water use. And while we can control the frequency and the level of irrigation, we have only very limited control of any subsequent runoff, leaching, and evaporation that put most of the world's water use beyond any possibility of circularization managed by humans rather than by the biosphere's grand water cycle.

As for the food production, there is absolutely no practical way how we could "circularize" the flows of key material inputs. As already noted, modern crops absorb often less than half of applied fertilizer nitrogen, with the losses due to the combination of erosion, leaching, volatilization, and denitrification. This means that every year we now lose on the order of 50–60 Mt of nitrogen to the environment and, moreover, these are not only economic losses of wasted investment in ammonia, urea, and nitrates but the escaped nitrates cause eutrophication of fresh and coastal waters (infamous seasonal dead zones deprived of oxygen due to excessive growth of algae), while nitrous oxide generated by denitrifying bacteria is the third most important greenhouse gas after CO_2 and CH_4. As for any closing of this extremely leaky loop, all we have done so far (and all we could do better) is to recover nitrates from municipal water waste, and an increasing number of cities has also adopted effective ways of reclaiming phosphates from urban sewage (Bunce et al. 2018).

Much as we cannot recycle all but a tiny share of water we withdraw for productive uses or only small fractions of nitrogen and phosphorus we use to fertilize our crops, we cannot recycle enormous flows of waste materials – soils, sediments, and rocks – that accompany production of energy, metals, and nonmetallic minerals, that are displaced during construction projects and that are eroded from fields. In aggregate, as already noted, these waste flows make up by far the largest share (>80%) of globally displaced raw materials. These massive (hidden because not accounted for) flows create a variety of undesirable environmental impacts (from acid drainage to extensive land use changes) but they have no acknowledged economic value and we try to dispose of them in the most inexpensive and the most convenient manner. Two kinds of massive material flows dominated this category.

The first flow consists of overburden materials, soils, sedimentary deposits, or solid rocks that have to be removed before we can begin profitable extractions of mineral deposits; the second one includes all processing wastes associated with the separation of targeted mineral (metals or nonmetallic compounds) from rocks. In the case of coal-mining overburden, seams exploited in the world's largest opencut coal mine in Wyoming (North Antelope Rochelle) are under 50–140 m of overburden, and as much as 500 m of overlay has to be removed in Hambach, Germany's largest lignite coal mine whose bottom is nearly 300 m below the sea level and whose overburden-to-seam ratio is as high as 6.2 : 1 (RWE 2022). Overburden is taken away by trucks or by conveyor belts for layered disposal in suitable areas adjacent to mines. The world's deepest metal mines have gone far deeper: Bingham Canyon in Utah (copper, gold, silver, and molybdenum) up to 1.2 km, Chuquicamata in Chile (copper) more than 850 m (MDO 2022). Mining in Bingham Canyon has been going out since the 1880s at it has produced nearly 6 Gt of overburden and ore (the ratio of the two masses is about 2 : 1).

After generations of mass-scale mining, extraction of metals from ores has to reckon with generally declining concentrations of elements in all commercially exploited ores of dominant metals (Fe, Al, and Cu) and with inherently low concentrations of many precious metals. I already noted that bauxite contains only 15–25% of aluminum, and that Chilean copper is now extracted from ore containing only 0.6% of the metal (ton of ore produces less than 6 kg of copper); taconite has just 30–40% of iron, zinc and lead about 5%, molybdenum, and silver ores contain just 0.1% of those elements, and typical economic grade gold ore has merely 0.006% of the metal.

The second flow is associated with all types of construction, from relatively small soil and rock displacement for building foundations and for final landscaping to massive transfers of waste material when building highways, railways, canals, and large dams, and also when using to reclaim coastal land.

Construction of the U.S. Interstate Highway System required the emplacement of about 1.5 Gt of natural aggregates (sand, gravel, and stone) as well as 48 Mt of cement, 35 Mt of asphalt, and 6 Mt of steel (Sullivan 2006). The world's most prominent land reclamation projects include Dubai's Palm Jumeirah (5.6 km²), Singapore's Marina Bay (3.6 km²), airports in Singapore (Changi, built with more than 40 Mm³ of sand taken from seabed), and Hong Kong (Chek Lap Kok), but China leads the world in aggregate reclamation (nearly 5,000 km² in thousands of projects along its coast, entailing the loss of half of the country's coastal wetlands) (An et al. 2007), followed by the Netherlands (2,700 km² with 970 km² Flevopolder reclaimed from the Ijsselmeer being the world's largest artificial island), South Korea, Bahrain, and Japan (Stauber et al. 2016).

Again, any idea that these masses of displaced material can be ever returned to their original sites is ludicrous. To these two massive "hidden" (or more appropriately simply dumped) and unrecyclable material flows, we should also add topsoil eroded from fields. Its global transfer now amounts to more than 40 Gt/year (Borrelli et al. 2020) and any idea of closing the loop by returning hundreds of millions of tons of silt disgorged annually by the Mississippi into the Gulf of Mexico onto the corn fields of Iowa of wheat field of Nebraska will always remain in the realm of science fiction – as will any recycling of topsoil lost by wind erosion in arid regions of Asia and Africa.

Finally, the inherently impossible task of circularizing energy uses. The second law of thermodynamics, formulated by Rudolf Clausius during the 1860s, recognizes that energy flow is not only a matter of quantity but also of quality, and that all energy conversion move only in one direction: energy is conserved, but its utility decreases. Combustion of a given mass of high-quality fuel generates heat that could be used to warm living spaces, process materials or (via prime movers, engines, or turbines) to move vehicles, planes, and ships. All of these conversions release heat and if its flow is hot enough, we can capture it before it reaches the atmosphere and use it profitably. Perhaps most notably, combined-cycle gas turbines work that way, increasing the overall efficiency of the process: as the gas that has rotated a gas turbine (to generate electricity) is still hot enough and can be used to heat steam to produce more electricity by a rotating steam turbine.

But all energy conversions eventually end up with low-temperature, low-quality heat that cannot be captured and reused: just think of capturing heat radiated by windows of a heated room, heat leaving the exhaust pipes of vehicles or heat generated by newly poured concrete as the material cures itself. This dissipated heat cannot be reused, the arrow of energy flow cannot be reversed, the low-temperature, low-quality heat cannot be turned into high-quality fuel or electricity (highly organized molecules of fuels, quanta of electricity) changes into low-temperature disorder. That is why any talk about "the creation of circular loops of material, **energy** and waste flows encompassing all societal activities" is plainly nonsense. And yet in 2019, the Royal Society organized a meeting on the topic of "Science to enable the circular economy" (Catlow et al. 2020): obviously, they could not mean this literally.

The second law is universally valid, and hence the low-quality heat generated by conversion of fuels or electricity (or by digestion of food) cannot be reused no matter if it comes from the stores of fossil carbon or from renewable (wind, solar, water, geothermal heat, and biofuels) conversions. But there is an important difference: combustion of coal, oil, and natural gas releases fossil carbon, while renewable conversions entail only a fraction of such releases, and they would generate no fossil carbon only once all wind turbines, PV cells, and hydro dams were built without using any metals or concrete produced with fossil fuels, or once all biofuels would be grown without synthetic fertilizers and harvest without using internal-combustion machines. Any truly sustainable economy would have to run solely on renewable energies – but, as explained earlier in this chapter, fossil-fuel-dependent material demands of these new conversions are actually higher than those of traditional techniques! But even if these new infrastructures could be built solely with renewable energies, they still could not reuse energies that would animate them.

The first attempt to quantify the actual rates of global or national levels of circularity was made by Haas et al. (2015) based on the best information for 2005. At that time 52% of EU's total material inputs (excluding water) were fossil fuels and food, that is energy inputs that are degraded in an irreversible way and cannot be recycled. Construction materials made up about 45% of all material inputs as they are embodied in long-lasting structures and infrastructures only a small share of those inputs (above all concrete, bricks, steel, and plastics) could be recycled in any given year. This left only relatively small (about 3%) material flows embedded in consumer products of which only some kinds (aluminum, steel, and glass) have high recycling rates.

And yet the authors concluded that overall recirculation in the EU economy was fairly high, at about 39% – only because they included all biomass flows. As Giampietro (2019) pointed out, such an inclusion is a misleading argument for two reasons. What gets recycled are individual nutrients (above all carbon and nitrogen) and not the biomass itself, and the recycling of biomass is not an anthropogenic but a purely biospheric (bacteria- and fungi-dependent) process whose rate and density cannot be regulated. More realistic assessments were launched in 2018 with the first Circularity Gap Report that traced the fate of all minerals, fossil fuels, and biomass and concluded only 9.1%, and the fifth annual report published in 2022 found the rate even lower at just 8.6% "which leaves a massive Circularity Gap of over 90%" (Circle Economy 2022).

Putting it this way is, as already explained, highly misleading because neither fuel or food energies can be ever recycled, and recycling of material wastes is inherently very limited: if we include all minerals, all fuels, and all biomass, we could never recycle most of these materials and hence it is misleading to talk about 90%+ circularity gap. On the other hand, the real situation is even worse than indicated by these reports: inclusion of water would lower the rate of annually recycled materials to a small fraction of 1%.

Whatever the actual rate might be, it is all too obvious that modern societies do not, indeed cannot, practice even remotely resembling a stable metabolic pattern governed by material recirculation and energized only by solar radiation and its natural transformations (wind, water, and biomass).

That conclusion does not change if we (as we must) exclude energy (whose flows are inherently impossible to be circular) and consider only all materials, or even if we were to exclude water, the most fundamental material enabling life on this planet. Our goals should be promoting the least wasteful material consumption alternatives, aiming at the highest practicable rates of reuse and recycling of all material inputs, not carrying on irrelevant discussion about global CE. Perhaps the only thing I should add is that even many instances where we have come close to circularizing some material flows thanks to relatively high rates of recycling we are, strictly speaking, engaged in down-cycling, with common examples being white printing paper turned packaging cardboard or PET drink containers becoming carpets.

6.5 Limits of Dematerialization

We live in a world where mere hints of a scientific or technical advance are interpreted as nearly accomplished facts and imminent prospects. This is evident when reading the news ranging from the progress of artificial intelligence to colonizing Mars and from the efficacy and affordability of personalized medicine to designing crops and humans via CRISPR. Realities are very different but that, now too often, seems irrelevant. Inevitably, assessments of material prospects have not been immune to these uncritical approaches, and there is no shortage of news about coming self-healing materials, 3D-printing of tall buildings, and miraculous nanomaterials endowed with new, remarkable properties and deployed on mass scales.

Of course, some of these claims are more sensible than others. Clearly, there was no first Mars mission in 2022 – but nanomaterials, engineered structures with at least one dimension of 100 nm or less, are already used as catalysts, in filters, semi-conductors, and cosmetics and as drug carriers, and will find many future uses in several major industries and in a growing range of everyday products. There are also many opportunities to endow classical materials with new desirable properties. Self-cleaning glass surfaces have a thin coating of TiO_2 and they help to clean the surface by a combination of photocatalysis and hydrophilicity (Pilkingtons's Activ™ glass), or they are made from super-hydrophobic materials. TiO_2 can be also incorporated into exposed materials, giving them self-cleaning properties as the UV-activated dioxide molecules serve as a catalyst to produce free radicals: unprotected surfaces will be eventually invaded by bacteria, mosses, algae, and fungi but the treated surface resists these degradations.

Experiments have shown that the fly-ash or metallurgical slag could replace as much as 65% of cement when combined with adequate dose of superplasticizer, and concrete's energy cost could be also greatly reduced by using a more durable variety of the material developed by Victor Li. This bendable material, more accurately know as ductile concrete or ECC (engineered cementitious composite), contains small polystyrene particles, and when a polymer fiber made from polyvinyl alcohol is added to concrete tiny plastic fibers bridging microcrack planes provide load-bearing capacity (Li 2012).

There is also a growing field of self-healing materials (polymers and composites) that can repair cracks or tears and recover their original functionality (Ghosh 2009; White et al. 2011; Wypych 2017). Another category of desirable accomplishments would be to make plastics not only completely biodegradable (oxo-degradables) but, even better, biocompostable, so they could be converted by microorganisms to CO_2 and H_2O in industrial composting plants in time similar to the decay of such common organic wastes. Of course, an even more radical step would be making biocompostable plastics, functionally identical to today's polymers, solely from plant feedstocks (for example PE and PP from ethanol derived from sugar cane). Next few decades should also see many mundane as well as unusual applications for new bio-inspired materials, structures designed by organisms but augmented by inserting synthetic materials (Meyers et al. 2013), and the use of synthetic printed materials made to act like living cells (Reiffel et al. 2013).

Lithium is perhaps the best example of an old material with an exceptionally promising future. This soft white metal has been used by ceramics and glass industries, in lubricants, polymers, and pharmaceuticals but its future use will be dominated by batteries, especially by rechargeable units in portable electronic gadgets (computers, tablets, and cellphones) and tools and in transportation (Goonan 2012). The metal's global production reached 82,000 t in 2020, with Australia producing nearly half of the total, and there is no shortage of exploitable resources.

Reserve/production ratio in 2020 was in excess of 250 years, and Bolivia, at Salar de Uyuni, has nearly a quarter of the world's resources of the metal (USGS 2022a).

As always, new solutions will also bring new problems. There are concerns about toxic potential of materials at the nanolevel because they can have undesirable interactions with living molecules or be outright toxic (Nel et al. 2006). Extracting lithium in order to make hundreds of millions of new batteries for future electric cars and then discarding the batteries will introduce a new environmental concern (Tahil 2006). Mass production of biopolymers will require considerable amounts of plant feedstock that will have to be grown in competition with food and feed crops and whose cultivation could have many unwelcome environmental impacts, well known from other intensive cropping practices.

6.5.1 Resources, Demand, Consumption

At the same time, many problems will be solved by substitutions. Copper is a perfect example of this possibility: it is not needed in such enormous quantities as steel but (as already explained) its presence in common consumer products is substantial. What would we do without new copper? Turning to wastes would be a limited option but, as Goeller and Weinberg (1976) pointed out decades ago, in the long run, electrical copper is almost entirely replaceable by aluminum and that element and steel, titanium, plastics, and composites can replace its structural uses.

And gold's use illustrates another adjustment shift, a huge opportunity for rationalized consumption. Ten-year average of demand for gold shows the global market dominated by jewelry (52%) and investment (27%% in bars, coins, and medals), with only 9% claimed by industry (World Gold Council 2022). During the coming decades, any conceivable increases in gold demand by the electronic industry could be met easily by a more rational division of the metal's final uses and by its near-complete recycling from gold-rich e-waste: surface gold mines yield 1–5 g Au/t of rock while computer circuit boards yield 250 g/t and mobile phones up to 350 g/t (Owens 2013).

In any case, matters look different from a longer perspective, particularly when assuming that the rest of the world will eventually want to consume as much as the affluent nations in general, or the United States in particular, have been claiming recently. For example, Gordon et al. (2006) estimate that the world's total copper resource is 1.6 Gt but to provide every person of a future global population of 10 billion people with 170 kg of copper stock (the per capita mean for North America around the year 2000) would require 1.7 Gt Cu, more than the estimated resource in the crust. Of course, even more of the metal would be needed to replace inevitable losses during the use and recycling. And the authors find that similar assumptions would also require the tapping of the lithosphere's entire stores of zinc and platinum.

Similar calculations demonstrating eventual resource limits can be done by assuming continuation of historic consumption growth rate. For copper (its pre-1900 global consumption added up to about 10 Mt) the 50 year consumption averages since 1900 are as follows: 70 Mt until 1950, 340 Mt until the year 2000, and (assuming continued post-1950 growth at 3.3%/year) 1,700 Mt until 2050. The cumulative total of more than 2 Gt would be greater than the best 2020 estimate of available reserves, and about as large as the total of estimates resources (USGS 2022a). Such calculations can be interpreted either as incontestable proofs of approaching global material deficits or, in darker versions, of the coming collapse of modern material-intensive civilization – or as interesting but simplistic exercises, because reasonable assumptions (based on known adjustments historically deployed by societies) could change those outcomes in fundamental ways.

The required stock of material can be greatly reduced by lower per capita consumption, particularly as the price of material will rise; better product design could maintain desired functionality with a fraction of deployed material; the need for new production could be lowered by assiduous recycling; these changes can greatly reduce the expected growth of demand. Moreover, declining fertilities indicate that future population totals might be significantly lower than the leading forecasts have indicated, and there is no reason why the North American or U.S. per capita stock should be used as the most desirable model to follow. Besides the availability of natural resources and the level of average per capita demand (annual use or circulating stock), there are three other fundamental variables (or constraints) that will determine the extent of future material production: our capability to extract them at affordable cost, degree of environmental impact accompanying material use, and the degree of material substitutability.

The last option has its inherent limited set by the evolution of the biosphere: hydrogen, oxygen, carbon, and nitrogen are the macro-constituents of every form of life that have no substitutes, but grand biogeochemical cycles assure their ready availability. Also required is a less massive presence of elements needed to build special organs or to enable biochemical processes: their list includes Ca, K, Na, Mg, Cl, S, and, in trace amounts, Co, Cu Cr, F, Fe, I, Mn, Mo, P, Se, Si, Sn, and V. But the total mass of these elements that need to be mobilized to support the Earth's life is minuscule compared to the mass of nonmetallic minerals and metallic elements extracted by modern civilization and so the concerns about substitutability apply mainly to those metallic elements that are needed in relatively large amounts but are present in the crust in relatively modest concentrations (unlike the enormous resources of Fe, Al, Ca, K, Mg, or Si) or in less accessible forms.

In reality, future material demands will be shaped by numerous dynamic linkages. Examples of these intertwined effects abound. Even a near-exhaustion of high-quality mineral deposits would be no cause for concern if a new and affordable extraction technique can tap resource considered utterly uneconomical in the past. Even a severely limited resource potential (basically un-extractable at any cost) ceases to be a problem where a ready substitution becomes available. Consumption of a still-abundant material may carry unacceptable environmental costs – but those, too, can be reduced (even eliminated) by a substitute or by a drastic shift in per capita demand.

This expected slowdown of per capita consumption of materials and energy (characteristic of all large, maturing systems) has been underway since the 1980s in all affluent, aging societies, often with shrinking populations. Even the U.S. primary energy demand (notwithstanding all SUVs, larger houses, more mobility, and much more electronics) has not shown any upward trend since the 1980s: first it had fluctuated between 320 and 340 GJ/year and since the year 2000, it has declined by nearly 20% to 280 GJ/capita, while the EU's average remained at around 150 GJ/capita for decades before declining slightly to about 140 GJ/capita in 2020. Similarly, American and European per capita demand for key materials (metals, cement, paper, and glass) has become largely saturated, and for a number of commodities, it has been declining but absolute demand has seen much smaller shifts.

Combination of relative dematerialization (per unit of GDP, per capita), minimal population growth (and often stagnation or actual decline), and rapid aging of affluent societies will be increasingly translated into weakening demand for materials. Average annual population growth rate of economically most developed regions has been below 0.6% since 2010, and by 2050–2055, it is forecast to be very close to zero; aging is already much in evidence in Japan and in the EU, and it is accelerating in China where a third of all people will be older than 65 years by 2050. As a result, future economic growth in affluent societies may not require substantial additional material inputs as any new requirements resulting from moderate GDP growth (1–2%/year) could be largely or completely met by a combination of continued lowering of average energy intensities, of relative dematerialization of products and services, and by intensified recycling and reuse.

And the greatest contributor to the rising material demand of the past four decades will not extend that record for another generation: Chinese economic growth rates have already come down from their peaks (10–14% a year in the early 2000s to 6–7% recently) and material consumption will follow. After all, in 2020, the country was already consuming 36% more steel per capita than EU-27 and 50% more than the United States, and in some of the country's regions, cement consumption has been as high, or even higher, than Spain's 2007 rate of 1,300 kg/capita that preceded the collapse of Spain's construction industry; all countries whose annual cement consumption surpassed for a while 1 t/capita had experienced sooner or later, the burst of their construction bubble (Bell 2012). These realities must be considered, yet again, in conjunction with the country's slowing rates of population growth and with the relatively rapid aging of its population: by 2040, China will have the same share of people above 60 years of age as Japan had in 2010.

Of course, even a near-stabilization of aggregate material requirements in affluent economies (accompanied by steadily falling per capita consumption) and considerable moderation of demand in China will still leave us with decades of potentially huge requirements for energy and for all kinds of commodities in India, other populous Asian countries, as well as in Africa. This means (barring unprecedented natural catastrophes or prolonged economic downturns) that we will not see significant declines in overall level of material extraction and production, and that reduced global rates of growth would count as a success during the coming generation. But within these confines of continuing (albeit moderated) growth there are many opportunities for change, for producing materials inputs much more efficiently, to design them for greater longevity and easier recyclability to last longer, to reuse all commodities much more effectively while creating fewer environmental impacts.

Continued light-weighting using new and traditional materials will combine with better design and smarter controls to offer better performance and superior functionality with reduced product mass, and although many descriptions of imminent infallible robotic diagnoses and personal assistance reflect wishful thinking more than they do tomorrow's quotidian realities, there is also no doubt that many services will entail mobilization of considerably smaller masses of materials. But we can be also fairly confident to say what these advances will not bring: their combined effect will not result in any absolute aggregate global dematerialization.

I have not seen a single forecast that would envisage any decline of global demand for any major material, and even low-case scenarios translate into considerable increases of consumption by 2050. For example, IEA sees 2050 steel growing by more than a third and cement production up to a quarter higher by the mid-century (IEA 2018, 2022), primary aluminum demand might go up by 50% (European Aluminium 2019), and the global demand for paper and paperboard is expected to reach nearly 940 Mt by 2050, 2.25 times the recent peak of 415 Mt in 2017. Similarly, even low- to moderate-growth scenarios foresee further substantial expansion of demand for materials ranging from fertilizers to polymers. The fastest gain will be in countries that combine still fairly low wealth per capita with relatively high population growth rates: India, Pakistan, Nigeria, and Ethiopia are the best examples in this expansive category.

Environment responds to absolute inputs and throughputs of pollutants resulting from these larger material flows: of course, without relative dematerialization global and national atmospheric emissions, water pollution, land degradation, inappropriate waste disposal, and toxic burden would be even worse, but the combination of continuing relative dematerialization, moderated population growth, and improved environmental protection would have to be seen as a great success if it could just slow down the overall rate of degradation or if it could, in a few specific cases, stop any further growth and even bring some noticeable declines (akin to the already achieved turnaround in global emissions of SO_2 or elimination of lead from common consumer products).

During the past generation, dematerialization, defined in terms of declining consumption per GDP of energy or of goods, has generally persisted in both affluent and modernizing countries (Ausubel and Waggoner 2008), and there are no insurmountable reasons why its pace should not be at least maintained, and even accelerated, in decades to come. But, as already stressed, this relative dematerialization will not result in any significant declines of aggregate global consumption, and just the opposite will be true as far many key fuels, raw materials, and manufactured products are concerned. And while further advances in electronics will be helpful, they will not be able to turn the aggregate material tide.

I will illustrate these limits by a few interesting comparisons. In Chapter 5 (in the section on relative dematerialization), I noted the list of devices whose ownership could be eliminated by owning an iPhone. Tupy (2012) drew this conclusion from the list:

While opinions regarding scarcity of resources in the future differ, dematerialization will better enable our species to go on enjoying material comforts and be good stewards of our planet at the same time. That is particularly important with regard to the people in developing countries, who ought to have a chance to experience material plenty in an age of rising environmental concerns (Tupy 2012/Cato Institute).

This is a perfect encapsulation of high hopes that people infatuated with e-gadgets have for the transformative powers of electronics in general and for its capacity as the agent of dematerialization in particular. How wrong that conclusion is can be best illustrated by looking at the material consequences of smartphones and other, concurrently diffusing, electronic devices. Nokia Communicator, released in 1996, was the first major smartphone and it took 16 years before the number of these devices surpassed one billion during the third quarter of 2013 and by 2018, the sales of smartphones peak at 1.4 billion units (Mongardini and Radzikowski 2020). Let us assume that every one of those 1.4 billion devices would have eliminated not only the need for a landline (true in many high-income countries, but in many low-income economies most people did not have any phone link until they bought their first mobile) but it would have also obviated the need to buy such small electronic devices as alarm clocks, transistor radios, and digital cameras.

With 170 g per mobile, annual sale of 1.4 billion smartphones would add up to about 240,000 of materials. When assuming 400 g per landline, the total mass of materials saved by not making 1.4 billion of landline sets would be about 560,000 t, and if half of the mobiles eliminated the need for other small devices, then, assuming their package would weigh 500 g, these savings would add another 350,000 t, for the net material saving of about 670,000 t a year. The total mass of saved materials (silicon, plastic, aluminum, steel, copper, and glass) would be only a tiny fraction of their specific aggregate demands. But that is merely a theoretical construct, a wishful assumption, as there have been no such eliminations, not even any substantial cuts in demand.

During the first two decades of the twenty-first century, worldwide sales of watches, and particularly of luxury models, continued to grow by about 2% a year, while by 2019, the sales of digital cameras declined by nearly 90% from their 2010 peak but were not eliminated. But whatever the net material savings attributable to mobile phones might be, they have been eliminated by higher sales of another electronic device. Global sales of more massive flat TVs sales remained within a narrow range of 210–219 million units a year since 2015 (Forrester 2022), and while not too long ago the best-selling display was 81-cm diagonal (weighing as little as 11 kg), now the 109-cm (43 in.) designs dominate (weighing about 16 kg) while the ownership of much heavier (27 kg) 70-in. diagonal units is rising. Taking just the difference of 5 kg between the smaller and the now average unit translates (with annual sales of 215 million units) to additional mass of 1 Mt: the larger mass of one electronic device (new TV sets) has completely eliminated weight savings resulting from assumed functional replacements by the other (smartphones).

And the chain of comparisons could be extended still further because soaring sales of large-diagonal LCD and OLED TVs have been accompanied by record sales of cars and luxury products many of which (private jetliners and yachts) represent some of the most concentrated embodiments of energy and materials in high-value electronics. Consequently, even if no alarm clocks, rolodexes, voice recorders, or digital cameras were ever bought because of smartphone's multifunctionality, material savings represented by such a loss of demand would have been obliterated by expanding claims for the same kind of resources ranging from aluminum and glass to wires and microprocessors.

6.5.2   Material Gaps and Attitudes

But there is a much more fundamental consideration when thinking about the global future of material consumption. Any realistic consideration must first acknowledge this: too many people still live in conditions of degrading and unacceptable material poverty that translates into poor nutrition, inadequate health care, frequent morbidity, fractured education, exhausting work, minimal leisure, absence of even small comforts, and shortened life expectancy. All of those people – and, depending on a specific category of that short list in the preceding sentence, they count in hundreds of million to a few billion – need to consume more materials per capita in order to enjoy decent life. But complications set right at this point: what is a decent life? To what extent we should define decent life by material consumption as opposed to measuring it by such universally valid indicators of physical quality of life as infant mortality, morbidity, and life expectancy? How do we include intangible indicators that measure quality of education or leisure? And what weight is to be given to subjective indicators of well-being, to satisfaction with life or to a relative state of one's happiness?

And there is more. What is the actual link between material consumption and objective and subjective quality of life once the basic needs for food, clothes, shelter, and mobility are well satisfied? Going from material misery to modest material comforts will make so many things in life better but, obviously, the link is not an endless escalator. But is so, where is a saturation point? Or can such a level be actually quantified in a meaningful way? These questions must be asked even if there are no easy answers mainly because of the situation that is the very opposite of the material poverty outlined at the outset of this section: too many people live in conditions of material excess that does not endow them with higher physical quality of life than that enjoyed by moderate consumers and that does not make them exceptionally happy.

On the most fundamental level, the question is about the very nature of modern economies. All but a tiny minority of economists (those of ecological persuasion) see a constant expansion of output as the fundamental goal. And not just any expansion: economies should preferably grow at annual rates in excess of 2%, better yet 3%. This is the only model, the only paradigm, the only precept as the economists in command of modern societies cannot envisage a system that would deliberately grow at a minimum rate, even less so one that would experience zero growth, and the idea of carefully managed decline appears to them outright unimaginable. The pursuit of endless growth is obviously a strategy that could work for many generations but that could not be sustained for centuries and millennia.

Alternatives are imaginable and unorthodox economists and ecologists have long argued for the decoupling of energy use and material consumption from standard perceptions of progress, for a transition toward a low-growth, then a zero-growth society and eventually even to a managed reduction of energy and material flows. The reason for this advocacy is obvious: the fundamental incompatibility of the growth imperative and of the second law of thermodynamics (Georgescu-Roegen 1971, 1975). That perspective dictates that minimizing entropy should be the foremost goal for a rational society. In practice, that would mean confronting excessive choice – 40,000+ items in an American food supermarket, or thousands of unique Android devices (smartphones and tablets) – that leads to throwaway culture: as much as 40% of food is wasted, millions of mobiles are discarded every day.

This calls for a new society where, once the basic material needs are taken care of, the sense of well-being and satisfaction would be derived from experiences that are not at all, or only marginally, correlated with higher energy flows and expanding material possessions. Modern affluent societies would seem to be perfect candidates for this shift because studies in the United States, Europe, and Australia have found that materialism is negatively related to life satisfaction, that the pursuit of wealth and possessions is associated with diminished subjective well-being: Ryan and Dziurawiec (2001) and Hudders and Pandelaere (2012) refer to earlier works, while the publications by Ifcher et al. (2019) and D'Ambrosio et al. (2020) are among the most recent findings. But checking the European data shows that even the continent where most countries now rank among the world's richest economies and where many populations are highly satisfied with their lives are not reducing their material consumption.

In 2021, EU consumed 15% more nonmetallic minerals and 40% more of metal ores than it did in 2012 while its population increase was less than 1.5% during those 10 years (Eurostat 2022a). None of this is surprising: as with all addictions, there is no easy way out of the habits of material acquisition, and the quest for possessions may have little to do with the ranking on the global satisfaction/happiness ladder. During his research into changing ownership and desires, Easterlin (2003, 11180–11181) found

> that as people acquire those goods for which aspirations were fairly high to start with (a home, a car, a TV), their aspirations increase for goods which were initially much less likely to be viewed as part of the good life . . . This finding suggests that new aspirations arise as previous ones are satisfied, and, to judge from the mean number of goods desired, to about the same extent.

This is why the calls for creating new economic arrangements where satisfaction and achievement would not be tied to material consumption will not be a winning election platform. So what practical steps – short of miraculously shifting the global economic system from a river with rising energy and material flows to a system designed to manage stability – can be taken, first in order to limit consumption of materials and then begin to move to absolute dematerialization? And does the best path toward this goal lead through collective enforcement or individual enlightenment?

Of course, history offers many examples of societies whose rulers kept populations in material misery, and even that was not the worst they could expect as famines took an incomparably greater toll than absence of material comforts. And there is no need to go for examples back into distant past: during the past three genera-tions, the most disturbing examples include the Maoist China (and Mao-made fam-ine that killed more than 30 million people between 1959 and 1961) and the still continuing madness of North Korea. But the evolution of modern societies has run in the opposite direction. Indeed, the states now exist to a large extent in order to maintain and to promote economic, technical, and legal foundations and infrastruc-tures of mass consumption. Even the former paragons of alternative options based on material restrictions or outright impoverishment – the Soviet Union and the Communist China – have embraced the consumption option with alacrity (the key difference between the two nations being that in the Chinese case the Party has not lost its control).

Consequently, it is exceedingly unlikely that we will soon see a major economy whose leaders will promise its citizen less and less, or at least no more than there is. Of course, there is always a possibility that enlightened self-interest of a large number of well-informed individuals will combine to subvert the old paradigm and that this will have a major long-term impact on material requirements. To this I would say: do not underestimate the appeal of possession, acquisition, ownership, and excess consumption. Without exaggeration, material acquisition in modern societies can be seen as a common form of addictive behavior that has been usually associated with alcohol, smoking, drugs, or gambling. A major dif-ference is that this addictive behavior has become even more pervasive as it has crossed all national and cultural barriers and evolved into a compulsive global phenomenon.

During that process, it got closely intertwined with the perceptions of class rank-ing and social status, it has offered opportunities for assuming vicarious identities, it has promoted individual and inter-generational aspirations, it has become a tool of desire, passion, and gratification, and it has brought the rewards of personal indulgence. Because of this, it is no exaggeration to see a fundamentally mundane act of shopping as an expression of self-identity pursued to negate rootlessness, role-lessness, and bureaucratic rationalities of modern society that traps its subjects and delimits their freedoms (Latimer 2001).

These realities have greatly blurred, if not erased, the boundaries between neces-sary and superfluous consumption as yesterday's unattainables become tomorrow's indispensables. They have also converted human desires into continuous process of collection and discarding as mass production and consumption succeeded in creat-ing a new ethos of ephemerality (Cooper 2001; Cultural Ecology 2014). Intentional obsolescence has become a key driver for many industries making consumer prod-ucts, a development going to GM's decision of the late 1920s to launch a new model every year (Slade 2006). As Lefebvre (1971) noted decades ago, besides the obso-lescence of products, there is also obsolescence of needs.

And, to make sure that the process does not stop, modern mass production is not only about pervasive quantity but also about mesmerizing abundance and variety, although that variety is mostly superficial, hiding essentially identical products in diversely packaged and promoted guises. In 2012, an estimate put the annual global output at about 10 billion different products (Marsh 2012). For comparison, Amazon offers more than 12 million products by mail, and when the third-party sellers are included, the total number that could be mail-ordered rise to about 350 million. Whatever the actual total of product diversity might be, it is likely to be greatly surpassed by the trend toward mass customization as the combination of CAD and 3-D printing allows many more companies to produce items perfectly tailored to individual needs. If the prices of this option fall enough, will this boost or lower the overall material demand as many people will choose the cachet of unique items?

6.5.3   Possible Changes

But there are many things we can do – while waiting for the combination of declining population growth, aging, and chronic unhappiness to put a sufficient downward pressure on material demand in affluent countries. There is no shortage of effective steps that could make a real difference, but the main problem with nearly all of them is the extent to which they will be broadly adopted and perpetuated by both producers and consumers in modern societies. Design for durability is perhaps the most obvious option. There is no reason why we could not design cars or refrigerators that last 20–25 years (their average lifespan is now 10–12 years, while a well-maintained jetliner can operate safely for 30 years). Similarly, lifespans of electronic devices could be doubled.

Technical challenges of doing so are minimal, but the products would have to cost more and the new way of doing things would have to break with the decades-long expectations of frequently changing models and designs and people. Is there a massive U.S. or EU constituency for proudly driving 17-year old cars? And in the case of e-devices, where new designs are introduced in intervals measured in months rather than in years and where having the latest design is, to some people, worth of queuing up for days before its release to be the first ones to own it, the idea of a durable cellphone or PC is an anathema. Not an impossible shift, but one that is not going to happen in a hurry.

An arrangement that could allow more frequent changes of models and that would assure complete recycling at the end of a product's life is not to sell any major manufactured items but merely rent them on long-term service contracts and then return them to their makers for disassembly and reuse: this approach can be applied to products ranging from computers to car tires and from refrigerators to air conditioners. Arrangements aimed at higher intensity of use are another kindred path to follow as leasing tries to address pervasively high idling capacity of many products. Leasing, rather than owning, has come to dominate such expensive industrial operations as offshore drilling rigs, transportation of massive objects by special ships and cargo plans, and by 2022, 49% of all large jetliners were also leased (Mackay 2022).

About a quarter of American cars were leased in 2021, and leasing could be used to lower pervasively high idling capacity of many consumer products ranging from lawn mowers to many specialized tools. These concepts of new rentalism – peer-to-peer consumerism, fractional ownership, or sharing economy – can be extended from machines and tools to housing, travel, and many services (McFedries 2012; Agarwal and Steinmetz 2019). New arrangements for leasing and take-back of products to be recycled, and in some cases even up-cycled rather than down-cycled, could also include household items ranging from carpets to mattresses (Braungart 2013). At the same time, rental services for infrequently used tools and machines have been around for decades, and so have been car leases (guaranteed take-back after an agreed period) and time-shares in vacation housing – but while these arrangements appeal to some, for many people the very idea of sharing their precious cars or toolboxes is unthinkable.

Clearly, the dominant model of endless-escalator economy has a deep foundation in individual desire of material possessions and given the global distribution of wealth, this consumption escalator has a long way to run. As Easterlin (2003) confirmed, consumers in affluent countries keep coming up with new objects of desire. SUVs – machines embodying commonly close to two and often even three tons of metals, plastics, glass, and rubber – have been the most massive must-haves of recent decades. Other common *desiderata* now include home gyms, ever larger screen TVs, and outdoor (patio or deck) furniture. The gyms are equipped with exercise machines weighing tens to hundreds of kilograms apiece (shipping weight of Bowflex Ultimate Home Gym is 272 kg). Sharp LC-90 (155-cm (diagonal) TV has the mass of 68 kg, outdoor furniture had progressed from simple plastic or rattan chairs to large dining tables and full-size sofas, to say nothing of adjacent large outdoor grills.

But the greatest progressing wave of material consumption is now engulfing hundreds of millions of relatively higher-income people in the cities of Asia and Latin America (and, to a much lesser extent, in Africa) whose often crassly ostentatious display of wealth surpasses many past European or North American excesses. Just check the size and the interiors of the world's largest private residence, Mukesh Ambani's Antilia, a 27-story skyscraper in Mumbai with 49 bedrooms, three helipads, and parking for 168 cars (Mishra 2022) or the size and the pseudo-European architecture (separated into "Paris," "Verona," and "Bruges") of Huawei's sprawling corporate headquarters in Shenzhen in China's Guangdong province (Taylor 2019).

Consumption pattern of China's *nouveaux riches* appears to be a particularly slavish copy of the worst American experiences as it consists of fake mansions within gated communities, obligatory large SUVs, and obsessive amassment of luxury goods ranging from Swiss watches to custom-made yachts and vineyards abroad. And right behind that wave we can already see a forming a lower but much wider crest of a much more massive flow that is beginning to elevate a few billion people in Asia and Latin America from bare subsistence in the countryside to incipient affluence in expanding cities: overwhelming majority of those material claims are helping to creating higher quality of life, but the share of frivolous uses is bound to rise.

And, eventually, another wave will also sweep Africa's masses, and by the time its mass consumption begins to rise, the continent may have more than two billion people (the total was just over 1.3 billion in 2020). The numbers are clear: during first two decades of the twenty-first century more than two billion people (most of them in China) saw a massive improvement in their quality of life – but (with more than one billion people in the economically most developed nations) that still leaves close to five billion waiting to get on the material escalator.

Comparison of average rates of car ownership is one of its best indicators of coming demand. In the rich countries, these rates have nearly saturated at more than 500 vehicles per 1,000 people (in 2020, the United States had about 850, Japan just over 600), but the rate was less than 400 in Brazil, just above 200 in China, and less than 150 in India. Even if the eventual ownership rates outside North America, Europe, Australia, and Japan would saturate at just the third of the Japanese level (about 200/1,000), the potential increase would amount to at least 1.4 billion new vehicles, doubling the total of vehicles that were on the roads in 2022.

A generation ago I could have done a good approximate calculation of material demands (steel, aluminum, magnesium, copper, plastics, and glass) implied by such an additions, in the early 2020s such a task is exceedingly uncertain because nobody can predict the actual speed of displacing IC machines by EVs. The two options have very different material claims and trajectories of aggressive vs. gradual electrification of the road transport would result in very different demands. Obviously, calculations resulting in substantial material gains are easier to do for refrigerators, TVs, or furniture but demands arising from the growing ownership of air conditioners will depend on the speed with which heat pumps, now seen as the preferred solution, will penetrate the market.

And so mass consumption of materials will continue, made of changing proportions of inputs necessary to provide decent quality of life and of materials turned into ephemeral junk; production of both will claim nonrenewable resources, will require considerable energy inputs (also for, now routinely, intercontinental transportation) and burden the environment with pollutants. To moderate and eventually to reduce this global wave of material consumption by appealing to reason – be it by stressing the untenability of existing economic arrangements, raising concerns about environmental burdens of rising consumption, preaching intergenerational responsibility for preserving natural resources or extolling monetary and security advantages restraint – remains a veritable Sisyphean task.

Deliberate individual choice of minimized material consumption – adoption of voluntary abnegation, limited wants, or utter simplicity – is much like the voluntary adherence to nutritional minima. Very few people (actually less than 1%) are willing to stick with veganism, the strictest form of vegetarian diet that does not permit even consumption of dairy products or eggs, and even less restrictive forms of what I call hyphenated vegetarianism (lacto- or lacto-ovo- or even lacto-ovo-pisci-vegetarianism) are consistently practiced by only a few percent of Western populations (Smil 2013). And so it will be with appeals to voluntary reductions of material demands.

As Kamdar (2011, 17) put it – writing about the pleasures of excess among India's newly rich – "voluntary simplicity can only appeal to those who have enough to choose to live with less." But that desire for simpler ways of life amidst material plenty is usually expressed in abandoning, largely in a symbolic way, just a few chosen claims and habits, and such actions will have only a marginal effect on the overall material demand. Perhaps we need powerful external drivers in order to significantly depress consumer demand and to change the course of the river economy, and two of them are already with us, already causing concerns but not yet sufficiently threatening to bring about relatively rapid and profound changes.

Obviously, the threat of excessive global warming has been the greatest planetary concern of the early twenty-first century, but it has not led to any drastic steps to lower the emissions of greenhouse gases: global meetings, from Kyoto to Copenhagen, from Paris to Edinburgh, have been ineffective and have done little to arrest the aggregate generation of greenhouse gases, with the CO_2 emissions rising from 25.9 Gt in 2000 to 37.9 Gt in 2019 before a small pandemic-induced dip (Olivier 2022). At the same time (as already noted), resolute acceleration of worldwide energy transition will result in a substantially higher demand for metals ranging from aluminum to zinc and for materials ranging from cement to sand.

In this book's first edition, I noted the possibility of two other unpredictable developments that could reduce global material flow: an unprecedented economic crisis witch consequence both much broader and much deeper than those following the downturn that began in 2008, and a major pandemic. The second event has come to pass. By September 2022, after two and half years of the latest pandemic, the best estimates put the real number of worldwide deaths at between 16.1 and 26.7 (median 22.3) million, up to 4.1 times the official claim (The Economist 2022). In many countries, Covid became temporarily the leading cause of mortality, but the toll could have been much higher: a pandemic whose mortality would merely replicate the population-adjusted toll of the 1918–1919 event would result in at least 200 million deaths.

The latest pandemic, most notably the temporary lockdowns put in place during its early stages and prolonged and severe reduction of travel and tourism, had seriously damaged some economic sectors, but its overall effect on energy and material uses has been relatively mild and temporary, with most indicators close to normal two years after the pandemic's official beginning in March 2020. The conclusion is inevitable: no concerns and no disruptions have translated into new impressive achievements of relatively rapid and profound dematerialization and hence we have not seen any absolute declines of material use on the global scale. And although both the rate of global population growth and wealth gap have been steadily declining, the global population is still growing and the gap between haves and have-nots remains.

Four necessities are imperative in order to perpetuate modern societies whose production meets more than just the basic existential human needs: food supply that supports normal development in childhood and adolescence and active life in adulthood, assurance of material provision compatible with decent quality of living, provision of health care that prevents premature mortality and helps to lengthen active years, and access to education. All of these requirements can be met only by well-functioning diversified economies whose advances, as we are repeatedly told, now depend above all on riding the endless escalator of scientific, technical, and managerial innovations in general and advances in electronics and artificial intelligence in particular. Importance of these innovations should not be underestimated but none of these advances will eliminate the need for mass-scale inputs of energies (including food) and raw materials.

Even if major steps are taken to reduce today's food waste by a third, by 2050, the global population of nearly 10 billion people would require at least 20% more food just to maintain the current level of supply, and at least a third more in order to eliminate, and to prevent, existing undernourishment. Even if improvements in energy conversion efficiency run considerably above the historic rate, the world in 2050 would need at least 40% higher supply of primary energy than in 2020, and an even higher increase in electricity generation. Even if relative dematerialization (reduced use of materials per unit of output) will progress, the total need for basic metals and nonmetallic minerals will expand significantly.

Two key examples illustrate such needs: if Africa were to reach the average per capita level of Asia's 2020 steel consumption, it would have to boost its use of finished steel products more than 10-fold (WSA 2022a). And if Africa were merely to match Asia's current level of nitrogen use in cropping its demand for synthetic fertilizer would have to go up 10–fold. Obviously, equally stunning comparisons could be cited for aluminum, glass, plastics, or microprocessors. Any forecasts of near-term absolute declines in material consumption could be thus realized only when the affluent countries would substantially reduce their material demand – and when low-income countries could be left to sink into even greater poverty without developing their infrastructures and improving their quality of life.

Without further increases in energy and material consumption, it would be impossible to raise the poorest third of the humanity (mostly in Asia and Africa) to at least a modicum of prosperity. This means that the global demand for more energy and materials should not be seen as very close to its highest level. How many people would be affected by this hardship forced on them by absolute global dematerialization? World Bank's latest calculations, published in 2022 and referring to 2018, had about 660 million people living at less than $1.90/day, 1.76 billion at less than $3.20/day, and 3.3 billion at less than $5.50/day (World Bank 2022c). The latter total was about two-fifths of humanity, living mostly in Asia and Africa. Lifting more than three billion people just to the mean of the global affluence would require increases of energy and material consumption at least twice as large as we have seen during China's post-1980 economic rise.

And replicating just the EU's average of primary energy and material use (the United States mean is more than twice as high) would mean that the poorest fifth of humanity would have to increase its consumption roughly 10–fold. But many *nouveaux riches* in Asia now aspire to go much beyond those moderately affluent EU averages as they aim to replicate material riches enjoyed by America's upper middle class, the richest fifth of the country's population. That richest fifth consumes annually more than 500 GJ of primary energy per capita a year, while the humanity's poorest fifth has to do with less than 20 GJ/capita. This is more than a 25-fold gap that could not be closed during the twenty-first century but that will, in the world with globally shared information, motivate the aspiration of millions in Asia, Africa, and Latin America. Obviously, these realities are not conducive to any declines of energy and material use in low-income countries. We will thus see further bifurcations of demand for energy and materials driven by different rates of economic growth composed of different sectoral needs.

I refuse to forecast the course of these complex affairs and prefer to be agnostic about all matters whose outcome will be determined by complex interactions of population dynamics, human desires and attitudes, economic growth and product costs, technical innovations, international relations, and global environmental change. They will determine how long the humanity can pursue what has worked for the past few centuries but what cannot work across millennia: its intensive pursuit of river economy, of ever-larger one-way use of energy and material. As I have shown, true CE is a delusion, but we can, and should, do better than to keep on intensifying our wasteful linear arrangements.

Nothing has been irretrievably foreclosed and at this point it is possible to imagine rational futures of minimized energy and material use aimed at high global quality of life for a stationary, even slowly declining population – as well as a further indiscriminate quest for energy and materials that ends, to a large extent, wasted in ephemeral consumption, perpetuates the great global gap in the average standard of living, and weakens the fundamental biospheric functions, the only irreplaceable foundations of any civilization. We must hope that human ingenuity (so admirably deployed particularly during the past two centuries) and adaptability (displayed, unfortunately, not well ahead of anticipated crises but only when they are upon us) will, sooner rather than later, guide us along the first path – but even in that case the transformation of humanity's material uses will be a gradual and difficult process with an uncertain outcome.

References

AA (Aluminum Association). 2021. The Aluminum Can Advantage: Sustainability Key Performance Indicators. https://www.aluminum.org/sites/default/files/2021-11/KPI_Report_2021.pdf

Aas, W. et al. 2019. Global and regional trends of atmospheric sulfur. *Scientific Reports* 9:953. doi: 10.1038/s41598-018-37304-0.

Abbott, R.T. and S.P. Dance. 2000. *Compendium of Seashells: A Color Guide to More Than 4,200 of the World's Marine Shells*. El Cason, CA: Odyssey Publications.

Abeles, P.W. 1949. *The Principles and Practice of Prestressed Concrete*. London: Charles Lockwood.

Abramson, A. 2003. *The History of Television, 1942 to 2000*. Jefferson, NC: McFarland & Company.

Adam, J.-P. 1994. *Roman Building: Materials and Techniques*. London: Routledge.

Adriaanse, A. et al. 1997. *Resource Flows: The Material Basis of Industrial Economies*. Washington, DC: World Resources Institute.

Adshead, S.A.M. 1997. *Material Culture in Europe and China, 1400–1800*. London: Macmillan.

Agarwal, N. and R. Steinmetz. 2019. Sharing economy: A systematic literature review. *International Journal of Innovation and Technology Management* 16:1–17.

Agrawal, G. 2010. *Fiber-Optic Communication Systems*. New York: John Wiley.

Ahmad, H. et al. 2020. A review of carbon fiber materials in automotive industry. *ICMTMTE 2020 IOP Conference Series: Materials Science and Engineering* 971:032011. doi: 10.1088/1757-899X/971/3/032011.

Air Products. 2022. On-Site Generated Gas. https://www.airproducts.com/supply-modes/gen-gas-on-site#:~:text=Air%20Products%20pioneered%20the%20on,of%20on%2Dsite%20gas%20generation

Aktas, C.B. and M.M. Bilec. 2012. Impact of lifetime on US residential building results. *International Journal of Life Cycle Assessment* 17:37–349.

Albanese, T. 2012. Rio Tinto CEO: China steel production to reach 1 billion metric tons by 2030. *BrightWire News*, September 14, 2012, 5:37.

Allwood, J.M. and J.M. Cullen. 2012. *Sustainable Materials with Both Eyes Open.* Cambridge: UIT.

Almond, J.K. 1981. A century of basic steel: Cleveland's place in successful removal of phosphorus from liquid iron in 1879, and development of basic converting in ensuing 100 years. *Ironmaking and Steelmaking* 8:1–10.

Almqvist, E. 2003. *History of Industrial Gases.* Berlin: Springer.

Amato, I. 2013. Concrete solutions. *Nature* 494:300–301.

An, S. et al. 2007. China's natural wetlands: Past problems, current status, and future challenges. *Ambio* 36:335–342.

Anderson, M.S. 1988. *War and Society in Europe of the Old Regime, 1618–1789.* New York: St. Martin's Press.

Anderson, R. and R.C. Anderson. 1926. *A Short History of the Sailing Ship.* New York: Robert M. McBride.

Andrae, A.S.G. and O. Andersen. 2010. Life cycle assessments of consumer electronics – are they consistent? *International Journal of Life Cycle Assessment* 15:827–836.

Andrew, R.M. 2019. Global CO_2 emissions from cement production, 1928–2018. *Earth System Science Data.* doi: 10.5194/essd-2019-152.

Ansari, A.A. and S.S. Gill. 2013. *Eutrophication: Causes, Consequences and Control.* Berlin: Springer.

APA (The Engineered Wood Association). 2012. Concrete Forming: Design/Construction Guide. https://www.ufpconcrete.com/-/media/UFP-Concrete/Literature/APA-CF-Design-Consrution-Guide.pdf?la=en

APA (The Engineered Wood Association). 2022. Glulam. https://architectuul.com/architecture/home-insurance-building; https://www.apawood.org/glulam

Architectuul. 2022. Home Insurance Building. https://architectuul.com/architecture/home-insurance-building

ASCE (American Society of Civil Engineers). 2021. 2013 Report Card for America's Infrastructure. https://infrastructurereportcard.org/

Atsushi, U. 1995. The riddle of Japan's quakeproof pagodas. *Japan Echo* 22(1):70–77.

Atterbury, P., ed. 1982. *The History of Porcelain*. New York: Morrow.

Austin, P. 2010. Reducing Energy Consumption in Paper Making Using Advanced Process Control and Optimisation. https://www.tappi.org/content/events/11papercon/documents/318.567%20ppt.pdf

Ausubel, J.H. 2003. *Decarbonization: The Next 100 Years*. New York: Rockefeller University.

Ausubel, J.H. and P.E. Waggoner. 2008. Dematerialization: Variety, caution, and persistence. *Proceedings of the National Academy of Sciences* 105:12774–12779.

AWWS (American Water Works Service). 2004. *Deteriorating Buried Infrastructure Management Challenges and Strategies*. Washington, DC: US EPA.

Ayres, R.U. and A.V. Kneese. 1969. Production, consumption and externalities. *American Economic Review* 59:282–297.

Baekeland, L.H. 1909. *Condensation Product and Method of Making Same*. Specification of Letters Patent 942,809, December 7, 1909. Washington DC: USPTO.

Bajpai, P. 2010. *Environmentally Friendly Production of Pulp and Paper*. Hoboken, NJ: John Wiley.

Bandaranayake, S. 1974. *Sinhalese Monastic Architecture: The Vihāras of Anurādhapura*. Leiden: E.J. Brill.

Barcelo, L. et al. 2014. Cement and carbon emissions. *Materials and Structures* 47(6):1055–1065. doi: 10.1617/s11527-013-0114-5.

Barney, G.O., ed. 1980. *The Global 2000 Report to the President*. Washington, DC: Council on Environmental Quality.

Barrett, W. 1990. World bullion flows, 1450–1800. In: J.D. Tracy, ed., *The Rise of Merchant Empires: Long Distance Trade in the Early Modern World, 1350–1750*, Cambridge: Cambridge University Press, pp. 224–254.

Barreveld, W.H. 1989. *Rural Use of Lignocellulosic Residues*. Rome: FAO.

Barros, E. et al. 2002. Effects of land-use system on the soil macrofauna in western Brazilian Amazonia. *Biology and Fertility of Soils* 35:338–347.

Bartelmus, P. 2003. Dematerialization and capital maintenance: Two sides of the sustainability coin. *Ecological Economics* 46:61–81.

Beardsley, T.M. 2011. Peak phosphorus. *BioScience* 61:91.

Beaumont, W.W. 1902. *Motor Vehicles and Motors: Their Design Construction and Working by Steam Oil and Electricity*, vol. 2. Westminster: Archibald Constable & Company.

Beck, T.R. 2001. *Electrolytic Production of Aluminum*. Seattle, WA: Electrochemical Technology Corporation.

Beer, D. et al. 1998. Future technologies for energy-efficient iron and steel making. *Annual Review of Energy and the Environment* 23:123–205.

Behrens, A. et al. 2007. The material basis of the global economy. *Ecological Economics* 64:444–453.

Bell, I.L. 1884. *Principles of the Manufacture of Iron and Steel*. London: George Routledge & Sons.

Bell, P. 2012. China: Ready for a new era? *International Cement Review*, March 2012. http://www.cemnet.com/Articles/story/149175/china-ready-for-a-new-era-.html

Bentley-Condit, V.K. and E.O. Smith. 2009. Animal tool use: Current definitions and an updated comprehensive catalog. *Behaviour* 147:185–221.

Berard, A., ed. 1991. *History, Science, Theology, and the Shroud*. Amarillo, TX: Man in the Shroud Committee of Amarillo.

Berge, B. 2009. *The Ecology of Building Materials*. Amsterdam: Elsevier.

Bernardini, O. and R. Galli. 1993. Dematerialization: Long-term trends in the intensity of use of materials and energy. *Futures* 25:431–448.

Berners-Lee, M. 2022. *The Carbon Footprint of Everything*. Vancouver: Greystone Books.

Berry, R.S. et al. 1975. An international comparison of polymers and their alternatives. *Energy Policy* 3:144–155.

Berry, B. et al. 1999. A retrospective of twentieth-century steel. *Iron Age New Steel* 15:1–14.

Bessa, J. et al. 2021. First evidence of chimpanzee extractive tool use in Cantanhez, Guinea-Bissau: Cross-community variation in honey dipping. *Frontiers in Ecology and Evolution* 9:625303. doi: 10.3389/fevo.2021.625303.

Bessemer, H. 1905. *Autobiography*. London: Offices of Engineering.

Bill, J. 2011. Revisiting Gokstad: Interdisciplinary Investigations of a Find Complex Excavated in the 19th Century. https://www.researchgate.net/publication/281118137

Billington, D.P. 1989. *Robert Maillart's Bridges: The Art of Engineering*. Princeton, NJ: Princeton University Press.

Binczewski, G.J. 1995. The point of a monument: A history of the aluminum cap of the Washington Monument. *JOM* 47(11):20–25.

BIR (Bureau of International Recycling). 2016. Environmental Benefits of Recycling – 2016 Edition. https://www.bir.org/publications/facts-figures/download/172/174/36?method=view

BIR (Bureau of International Recycling). 2020. Paper and Board Recycling in 2019. https://www.bir.org/the-industry/paper

BIR (Bureau of International Recycling). 2021. *World Steel Recycling in Figures 2016–2020*. Brussels: BIR. https://www.bir.org/publications/facts-figures/download/821/175/36?method=view.

Birch, A. 1967. *The Economic History of the British Iron and Steel Industry 1784–1879*. London: Cass.

Birdsey, R.A. 1996. Carbon storage in United States forests. In: N. Sampson and D. Hair, eds., *Forests and Global Change*, vol. II, Washington, DC: American Forests, pp. 1–25.

Bisanzio, D. et al. 2019. Mapping changes in housing in sub-Saharan Africa from 2000 to 2015. *Nature* 568:391–395.

Blanc, A. et al. 1998. *Lutetian Limestones in the Paris Region: Petrographic and Compositional Examination*. Brookhaven, NY: Brookhaven National Laboratory.

Boeckel, B. and K.-H. Baumann. 2008. Vertical and lateral variations in coccolithophore community structure across the subtropical frontal zone in the South Atlantic Ocean. *Marine Mircopaleontology* 67:255–273.

Boesch, C. and M. Tomasello. 1998. Chimpanzee and human cultures. *Current Anthropology* 39:591–614.

Bolin, C.A. and S. Smith. 2011. Life cycle assessment of ACQ-treated lumber with comparison to wood plastic composite decking. *Journal of Cleaner Production* 19:620–629.

Borchers, W. 1904. *Electric Smelting and Refining*. London: Charles Griffin and Company.

Börjesson, P. and L. Gustavsson. 2000. Greenhouse gas balances in building construction: Wood versus concrete from lifecycle and forest land-use perspectives. *Energy Policy* 28:575–588.

Borrelli, P. et al. 2020. Land use and climate change impacts on global soil erosion by water (2015–2070). *Proceedings of the National Academy of Sciences* 117:21994–22001.

Bosc, J.-L. 2001. *Joseph Monier et la naissance du ciment armé*. Paris: Editions du Linteau.

Boskey, A.L. 2003. Biomineralization: An overview. *Connective Tissue Research* 44(Supplement 1):5–9.

Boulding, K. 1970. *Economics as Science*. New York: McGraw Hill.

BP (British Petroleum). 2022. Statistical Review of World Energy. https://www.bp.com/content/dam/bp/business-sites/en/global/corporate/pdfs/energy-economics/statistical-review/bp-stats-review-2022-full-report.pdf

Brauch, M.D. et al. 2020. Shared-Use Infrastructure Along the World's Largest Iron Ore Operation: Lessons Learned from the Carajás Corridor. https://scholarship.law.columbia.edu/cgi/viewcontent.cgi?article=1179&context=sustainable_investment_staffpubs

Braungart, M. 2013. Upcycle to eliminate waste. *Nature* 494:174–175.

Brice, G. 1752. *Description de la ville de Paris, et de tout ce qu'elle contient de plus remarquable*. Paris: Librairies Associés.

Bringezu, S. and H. Schütz. 2001. *Total Material Requirement of the European Union*. Copenhagen: European Environment Agency.

Brodribb, G. 1987. *Roman Brick and Tile.* Gloucester: Ian Sutton.

Broecker, W.S. 1970. A boundary condition on the evolution of atmospheric oxygen. *Journal of Geophysical Research* 75:3553–3557.

Broudy, E. 2000. *The Book of Looms: A History of the Handloom from Ancient Times to the Present.* Lebanon, NH: University Press of New England.

Brown, K.S. et al. 2009. Fire as an engineering tool of early modern humans. *Science* 325:859–862.

Brunetta, L. and C.L. Craig. 2010. *Spider Silk: Evolution and 400 Million Years of Spinning, Waiting, Snagging, and Mating.* New Haven, CT: Yale University Press.

Brydson, J.A. 1975. *Plastic Materials.* London: Newnes-Butterworths.

BTS (Bureau of Transportation Statistics). 2022. U.S. Passenger Miles. https://www.bts.gov/content/us-passenger-miles

Buckingham, D.A. 2006. *Steel Stocks in Use in Automobiles in the United States.* Denver, CO: USGS.

Buehler, E. and G.K. Teal. 1956. *Process for Producing Semiconductive Crystals of Uniform Resistivity.* US Patent 2,768,914, October 30, 1956. Washington, DC: USPTO.

Bunce, J.T. et al. 2018. A review of phosphorus removal technologies and their applicability to small-scale domestic wastewater treatment systems. *Frontiers in Environmental Science* 6:8. doi: 10.3389/fenvs.2018.00008.

Burke, J. 2012. *Nest: The Art of Birds.* Crows Nest, NSW: Allen & Unwin.

Caletti, C.C. 2020. Göbekli Tepe and the sites around the Urfa Plain (SE Turkey): Recent discoveries and new interpretations. *Asia Anteriore Antica. Journal of Ancient Near Eastern Cultures* 2:95–123. doi: 10.13128/asiana-679.

Campbell, C.J. 1997. *The Coming Oil Crisis.* Brentwood, UK: Multi-Science Publishing and Petroconsultants.

Canadian Wood Council. 2004. *Wood-Frame Housing – A North American Marvel.* Ottawa: Canadian Wood Council.

CanTech International. 2012. Ball Reduces Weight of Beverage Cans. https://www.cantechonline.com/news/3522/ball-reduces-weight-of-beverage-cans/

Carmona, M. and P. Camiller. 2002. *Haussmann: His Life and Times and the Making of Modern Paris.* Lanham, MD: Ivan R. Dee.

Carothers, W.H. 1937. *Linear Condensation Polymers.* US Patent 2,071,250, February 16, 1937. Washington, DC: USPTO.

Carrara, S. et al. 2020. *Raw Materials Demand for Wind and Solar PV Technologies in the Transition Towards a Decarbonised Energy System.* Brussels: European Union https://op.europa.eu/en/publication-detail/-/publication/19aae047-7f88-11ea-aea8-01aa75ed71a1/language-en.

Cartwright, J. et al. 2011. *Assessing the Environmental Impacts of Industrial Laundering: Life Cycle Assessment of Polyester/Cotton Shirts.* Santa Barbara, CA: Donald Bren School of Environmental Science & Management.

Cathcart, R.B. 2011. Anthropic rock: A brief history. *History of Geo- and Space Sciences* 2:57–74.

Catlow, C.R. et al. 2020. Science to enable the circular economy. *Philosophical Transactions of the Royal Society A* 378:20200060. doi: 10.1098/rsta.2020.0060.

CEC (Commission for Environmental Cooperation). 2016. *Quantitative Characterization of Domestic and Transboundary Flows of Used Electronic Products. Case Study: Used Computers and Monitors in North America.* Montreal: CEC. http://www.cec.org/files/documents/publications/11673-quantitative-characterization-domestic-and-transboundary-flows-used-electronic-en.pdf.

CEIC. 2021. Number of Motor Vehicle: Passenger: Beijing. https://www.ceicdata.com/en/china/no-of-motor-vehicle-passenger-car/cn-no-of-motor-vehicle-passenger-beijing

Cement Association of Canada. 2006. A Life Cycle Perspective on Concrete and Asphalt Roadways: Embodied Primary Energy and Global Warming Potential. http://www.cement.ca/images/stories/athena%20report%20Feb.%202%202007.pdf.

Chapman, P.F. et al. 1974. The energy cost of fuels. *Energy Policy* 2:231–243.

Charette, R.N. 2009. This car runs on code. *IEEE Spectrum*, February 2009. https://spectrum.ieee.org/this-car-runs-on-code

Chatterjee, K. 2020. *Cement Production Technology.* London: Routledge.

Chen, C. et al. 2022. China economy-wide material flow account database from 1990 to 2020. *Scientific Data* 9:502. doi: 10.1038/s41597-022-01611-z.

Choy, C.A. et al. 2019. The vertical distribution and biological transport of marine microplastics across the epipelagic and mesopelagic water column. *Scientific Reports* 9:7843. doi: 10.1038/s41598-019-44117-2.

Circle Economy. 2022. *The Circularity Gap Report 2022.* Amsterdam: Circle Economy. https://www.circularity-gap.world/2022.

Cleveland, C.J. and M. Ruth. 1999. Indicators of dematerialization and the materials intensity of use. *Journal of Industrial Ecology* 2:15–50.

CMI (Can Manufacturers Institute). 2017. Cans are the Most Recycled Drinks Package in the World. https://www.cancentral.com/media/news/cans-are-most-recycled-drinks-package-world#:~:text=The%20study%20identified%20aluminum%20recycling,percent%20in%20the%20United%20States.

CMI (Can Manufacturers Institute). 2020. 1960–2020 CMI Beverage Can Shipments. https://www.cancentral.com/media/publications/1960-2020-cmi-beverage-can-shipments

Cobb, H.M. 2010. *The History of Stainless Steel*. Materials Park, OH: ASM International.

Cole, M. et al. 2011. Microplastics as contaminants in the marine environment: A review. *Marine Pollution Bulletin* 62:2588–2597.

Composites World. 2006. Advanced Materials for Aircraft Interiors. http://www.compositesworld.com/articles/advanced-materials-for-aircraft-interiors

Cooper, E. 2000. *Ten Thousand Years of Pottery*. London: British Museum.

Cooper, R. 2001. Interpreting mass: Collection/dispersion. In: N. Lee and R. Munro, eds., *The Consumption of Mass*, Oxford: Blackwell Publishers, pp. 16–43.

Corning. 2019. Understanding Fiber Optics & Local Area Networks. https://www.corning.com/catalog/coc/documents/brochures/LAN-737-EN.pdf

Costello, S. 2021. Where is the iPhone Made? https://www.lifewire.com/where-is-the-iphone-made-1999503

Courland, R. 2011. *Concrete Planet*. New York: Prometheus Books.

CRI (Container Recycling Institute). 2004. Trillions of Cans. https://www.container-recycling.org/index.php/trillions-of-cans

Crown. 2021. Beverage Can Weight Reduction: Celebrating a Milestone Achievement. https://www.crowncork.com/news/all-about-cans/beverage-can-weight-reduction-celebrating-milestone-achievement

Cultural Ecology. 2014. Consumerism. http://www.culturalecology.info/version2/Consumerism.html

Cwalina, B. 2008. Biodeterioration of concrete. *Architecture, Civil Engineering, Environment* 1(4):133–140.

Czochralski, J. 1918. Ein neues Verfahren zur Messung des Kristallisationsgeschwindigkeit der Metalle. *Zeitschrift der physikalischen Chemie* 92:219–221.

D'Ambrosio, C. et al. 2020. Money and happiness: Income, wealth and subjective well-being. *Social Indicators Research* 148:47–66. doi: 10.1007/s11205-019-02186-w.

D'Amico, B. et al. 2021. Global potential for material substitution in building construction: The case of cross laminated timber. *Journal of Cleaner Production* 279:123487.

Dahiya, S. and L. Myllyvirta. 2019. *Global SO$_2$ Emission Hotspot Database*. Amsterdam: Greenpeace.

Daigo, I. et al. 2009. Material stocks and flows accounting of copper and copper-based alloys in Japan. *Resources Conservation and Recycling* 53:208–217.

Daigo, I. et al. 2010. Substance flow analysis of chromium and nickel in material flow of stainless steel in Japan. *Resources Conservation and Recycling* 54:851–863.

Daniel, G. 1980. Megalithic monuments. *Scientific American* 263(1):78–90.

Daniels, C.J. et al. 2018. A global compilation of coccolithophore calcification rates. *Earth Systems Scientific Data* 10:1859–1876. doi: 10.5194/essd-10-1859-2018.

Darwin, C.R. 1881. *The Formation of Vegetable Mould, Through the Action of Worms, With Observations on Their Habits.* London: John Murray.

Das, S. 2011. Life cycle assessment of carbon fiber-reinforced polymer composites. *International Journal of Life Cycle Assessment* 16:268–282.

Data USA. 2022. Pulp, Paper, & Paperboard Mills. https://datausa.io/profile/naics/pulp-paper-paperboard-mills

Davidovits, J. 2002. *Ils ont bâti les pyramides.* Paris: Jean-Cyrille Godefroy.

Davidson, I. and W.C. McGrew. 2005. Stone tools and the uniqueness of human culture. *Journal of the Royal Anthropological Institute* 11:793–817.

Deffeyes, K.S. 2001. *Hubbert's Peak: The Impending World Oil Shortage.* Princeton, NJ: Princeton University Press.

Derwent, R. et al. 2006. Global environmental impacts of the hydrogen economy. *International Journal of Nuclear Hydrogen Production and Applications* 1(1):57–67.

Des Cars, J. 1988. *Haussmann, la gloire du Second Empire.* Paris: Perrin.

Di Blasi, C. et al. 2010. Biomass screening for the production of furfural via thermal decomposition. *Industrial & Engineering Chemistry Research* 49:2658–2671.

Di Mauro, B. et al. 2018. Saharan dust events in the European Alps: Role on snowmelt and geochemical characterization. *The Cryosphere Discussions* 13(4):1147–1165. doi: 10.5194/tc-2018-241.

Diderot, D. and J.L.R. D'Alembert. 1751–1777. *L'Encyclopedie ou Dictionnaire Raisonne des Sciences des Arts et des Metiers.* Paris: Ave Approbation and Privilege du Roy.

Diesel, R. 1913. *Die Entstehung des Dieselmotors.* Berlin: Verlag von Julius Springer.

Djukanovic, G. 2018. Why are Anode Production Costs Rising? https://aluminiuminsider.com/why-are-anode-production-costs-rising/

Dominish, E. et al. 2019. *Responsible Minerals Sourcing for Renewable Energy.* Sydney: University of Technology Sydney. https://www.uts.edu.au/sites/default/files/2019-04/ISFEarthworks_Responsible%20minerals%20sourcing%20for%20renewable%20energy_Report.pdf.

Drehobl, A. et al. 2020. How High are Household Energy Burdens? https://www.aceee.org/sites/default/files/pdfs/u2006.pdf

Duan, H. et al. 2013. *Quantitative Characterization of Domestic and Transboundary Flows of Used Electronics Analysis of Generation, Collection, and Export in the United States*. Cambridge, MA: MIT. https://www.step-initiative.org/files/_documents/other_publications/MIT-NCER%20US%20Used%20Electronics%20Flows%20Report%20-%20December%202013.pdf.

DuckerFrontier. 2020. North American Light Vehicle Aluminum Content and Outlook. https://drivealuminum.org/wp-content/uploads/2022/05/DuckerFrontier-Aluminum-Association-2020-Content-Study-Summary-Report-FINAL.pdf

Duffar, T., ed. 2010. *Crystal Growth Processes Based on Capillarity: Czochralski, Floating Zone, Shaping and Crucible Techniques*. Oxford: Wiley-Blackwell.

Duffy, C. 1985. *The Fortress in the Age of Vauban and Frederick the Great, 1660–1789*. London: Routledge & Kegan Paul.

Durrer, R. 1948. Sauerstoff-Frischen in Gerlafingen. *Von Roll Werkzeitung* 19(5):73–74.

Easterlin, R.A. 2003. Explaining happiness. *Proceedings of the National Academy of Sciences* 100:1176–1183.

EC (European Commission). 2001. Economy-Wide Material Flow Accounts and Derived Indicators a Methodological Guide. https://ec.europa.eu/eurostat/documents/3859598/5855193/KS-34-00-536-EN.PDF.pdf/411cd453-6d11-40a0-b65a-a33805327616?t=1414780409000

EC (European Commission). 2022. European Platform on Life Cycle Assessment. https://ec.europa.eu/environment/ipp/lca.htm

ecoATM. 2022. The ecoATM Story. https://www.ecoatm.com/pages/our-story

Ecoinvent. 2022. Ecoinvent Databse. https://ecoinvent.org/the-ecoinvent-database/

Edinburgh Center for Carbon Management. 2006. Forestry Commission Scotland Greenhouse Gas Emissions Comparison Carbon Benefits of Timber in Construction. https://forestry.gov.scot/images/corporate/pdf/carbon-benefits-of-timber-in-construction-2006.pdf

Edmondson, J. and A. Holland. 2020. Materials for Electric Vehicles: Electric Motors, Battery Cells & Packs, HV Cabling 2020–2030. https://www.idtechex.com/en/research-report/materials-for-electric-vehicles-electric-motors-battery-cells-and-packs-hv-cabling-2020-2030/770#:~:text=Motor%20materials%20forecasted%20include%20aluminium,%2Dsteel%2C%20terbium%20and%20praseodymium.&text=Materials%20demand%20from%20the%20following,Battery%20Cells

Ehrlich, P.R. and J. Holdren, eds. 1988. *The Cassandra Conference: Resources and the Human Predicament*. College Station, TX: Texas A&M University Press.

Eil, A. et al. 2020. *Dirty Stacks, High Stakes: An Overview of Brick Sector in South Asia*. Washington, DC: World Bank. https://documents1.worldbank.org/curated/en/685751588227715919/pdf/Dirty-Stacks-High-Stakes-An-Overview-of-Brick-Sector-in-South-Asia.pdf.

El Khalloufi, M. et al. 2021. Titanium: An overview of resources and production methods. *Minerals* 11(12):1425. doi: 10.3390/min11121425.

Elvidge, C.D. et al. 2007. Global distribution and density of constructed impervious surfaces. *Sensors* 7:1962–1979.

Endler, J.A. et al. 2010. Great bowerbirds create theaters with forced perspective when seen by their audience. *Current Biology* 20:1679–1684.

Energetics. 2000. *Energy and Environmental Profile of the U.S. Chemical Industry.* Columbia, MD: Energetics. http://www1.eere.energy.gov/manufacturing/resources/chemicals/pdfs/profile_chap1.pdf.

Engel, E. 1857. Die Productions- und Consumtionsverhaltnisse des Königreichs Sachsen. *Zeitschrift des Statistischen Bureaus des Königlich-Sächsischen, Ministerium des Innern* 8(9):1–54.

Engineering ToolBox. 2008. Fossil and Alternative Fuels – Energy Content. https://www.engineeringtoolbox.com/fossil-fuels-energy-content-d_1298.htm

Engström, S. et al. 2010. Tall Towers for Large Wind Turbines. https://www.osti.gov/etdeweb/servlets/purl/992745

EPRO (European Association of Plastics Recycling and Recovery Organisations). 2011. Plastic Packaging Statistics 2010. http://www.plasticseurope.org/documents/document/20111107101127-final_pe_factsfigures_uk2011_lr_041111.pdf

Equinor. 2022. The Statfjord Area. https://www.equinor.com/energy/statfjord

Eriksson, P.-E. 2004. Comparative LCAs for Wood and Other Construction Methods. http://www.ewpa.coem/Archive/2004/jun/Paper_032.pdf

European Aluminium. 2019. Vision 2050. https://www.european-aluminium.eu/media/2552/sample_executive-summary-vision-2050_web_pages_20190408.pdf

European Commission. 2015. Circular Economy Action Plan. https://ec.europa.eu/environment/circular-economy/pdf/new_circular_economy_action_plan.pdf

European Commission. 2022. European Platform on Life Cycle Assessment (LCA). https://ec.europa.eu/environment/ipp/lca.htm#:~:text=Life%20Cycle%20Assessment%20(LCA)%20is,life%2Dcycle%20of%20the%20product

Eurostat. 2018. Economy-Wide Material Flow Accounts. 2018 Edition. https://ec.europa.eu/eurostat/documents/3859598/9117556/KS-GQ-18-006-EN-N.pdf/b621b8ce-2792-47ff-9d10-067d2b8aac4b?t=1537260841000

Eurostat. 2022a. Material Flow Accounts. https://ec.europa.eu/eurostat/databrowser/view/env_ac_mfa/default/table?lang=en

Eurostat. 2022b. Material Flow Accounts and Resource Productivity. https://ec.europa.eu/eurostat/statistics-explained/index.php?title=Material_flow_accounts_and_resource_productivity

Eurostat. 2022c. Resource Productivity Statistics. https://ec.europa.eu/eurostat/statistics-explained/index.php?title=Resource_productivity_statistics#Resource_productivity_of_the_EU_and_across_Member_States_over_time

Fact MR. 2022. Electronic Grade Silicon Market Size, Share & Growth Report 2032. https://www.factmr.com/report/electronic-grade-silicon-market

Fan, Z. and J. Friedmann. 2021. Low-carbon production of iron and steel: Technology options, economic assessment, and policy. *Joule* 5:829–862.

FAO (Food and Agriculture Organization). 1990. *Energy Conservation in the Mechanical Forest Industries*. Rome: FAO. http://www.fao.org/docrep/T0269E/T0269E00.htm.

FAO (Food and Agriculture Organization). 2022. FAOSTAT. https://www.fao.org/faostat/en/#data

FCH (Fuel Cells and Hydrogen). 2019. *Hydrogen Roadmap Europe*. Luxembourg: FCH.

Felton, N. 2008. Consumption Spreads Faster Today. *New York Times*, February 10, 2008. http://www.nytimes.com/2008/02/10/opinion/10cox.html?ex=1360299600&en=9ef4be7de32e4b53&ei=5090&partner=rssuserland&emc=rss&pagewanted=all

Fernandes, F.M. et al. 2010. Ancient clay bricks: Manufacture and properties. In: M. Bostenaru and R. Přikryl, eds., *Materials, Technologies and Practice in Historic Heritage Structures*, Berlin: Springer, pp. 29–48.

Fernández, E. et al. 1993. Production of organic and inorganic carbon within a large-scale coccolithophore bloom in the northeast Atlantic Ocean. *Marine Ecology-Progress Series* 97:271–285.

Fernández-González, F. 2006. *Ship Structures Under Sail and Under Gunfire*. Madrid: Universidad Politécnica de Madrid. http://oa.upm.es/1520/1/PONEN_FRANCISCO_FERNANDEZ_GONZALEZ_01.pdf.

Fischer-Kowalski, M. et al. 2011. Methodology and indicators of economy-wide material flow accounting. *Journal of Industrial Ecology* 15:855–876.

Fitchen, J. 1961. *The Construction of Gothic Cathedrals: A Study of Medieval Vault Erection*. Chicago: The University of Chicago Press.

Fleming, J.A. 1934. *Memories of a Scientific Life: An Autobiography*. London: Marshall, Morgan & Scott.

Fleming, S. 2021. Your Smartphone has Become Your Home, Anthropologists Say. https://www.weforum.org/agenda/2021/05/how-we-interact-with-smartphones-report/

Fleming, P.A. and J.P. Loveridge. 2003. Miombo woodland termite mounds: Resource islands for small vertebrates? *Journal of Zoology, London* 259:161–168.

Flemings, M.C. and D.V. Ragone. 2009. Puddling: A new look at an old process. *ISIJ International* 49:1960–1966.

FLSmidth. 2011. Rotary Kilns for Cement Plants. https://fdocuments.net/document/rotary-kilns-for-cement-plants.html

Flynn, D.O. and A. Giráldez. 1995. Born with a "Silver Spoon": The origin of world trade in 1571. *Journal of World History* 6:202–221.

Föll, H. 2000. *Electronic Materials*. Kiel: University of Kiel.

Forbes, R. 1972. Copper. In: *Studies in Ancient Technology*, vol. 6, Leiden: E. J. Brill, pp. 1–133.

Ford. 2010. Model T Facts. https://web.archive.org/web/20101023232110/http://media.ford.com/article_display.cfm?article_id=858

Forest Products Laboratory. 2010. *Wood Handbook*. Madison, WI: USDA. https://www.fs.usda.gov/treesearch/pubs/37440.

Forrester, C. 2022. Samsung, LG Dominate Global TV Market. Advanced Television. https://advanced-television.com/2022/02/22/samsung-lg-dominate-global-tv-market/

Forth Bridges. 2022. Three Bridges Spanning Three Centuries. https://www.theforthbridges.org/

Franits, W. 2004. *Dutch Seventeenth-Century Genre Painting*. New Haven, CT: Yale University Press.

Frank, A.G. 1998. *ReOrient: Global Economy in the Asian Age*. Berkeley, CA: University of California Press.

Franz, J.E. 1974. *N-phosphonomethyl-glycine Phytotoxicant Compositions*. US Patent 3,799,758, March 26, 1974. Washington, DC: USPTO.

Frison, G.C. 1987. Prehistoric, plains-mountain, large-mammal, communal hunting strategies. In: M.H. Nitecki and D.V. Nitecki, eds., *The Evolution of Human Hunting*, New York: Plenum Press, pp. 177–223.

Fruehan, R., ed. 1998. *The Making, Shaping and Treating of Steel Steelmaking and Refining Volume*. Pittsburgh, PA: The AISE Steel Foundation.

Fujitsuka, N. et al. 2013. Dynamic Modeling of World Steel Cycle Toward 2050. http://conferences.chalmers.se/index.php/LCM/LCM2013/paper/viewFile/734/333

Furuhashi, T. et al. 2009. Molluscan shell evolution with review of shell calcification hypothesis. *Comparative Biochemistry and Physiology Part B* 154:351–371.

Garrett, R.W. et al. 2016. Leaf processing behaviour in Atta leafcutter ants: 90% of leaf cutting takes place inside the nest, and ants select pieces that require less cutting. *Royal Society Open Science* 3:150111. doi: 10.1098/rsos.150111.

Garrison, V.H. et al. 2006. Saharan dust – a carrier of persistent organic pollutants, metals and microbes to the Caribbean? *Revista de Biología Tropical* 54:9–21.

Gartner. 2022. Gartner Says Worldwide PC Shipments Declined 5% in Fourth Quarter of 2021 but Grew Nearly 10% for the Year. https://www.gartner.com/en/newsroom/press-releases/2022-01-12-gartner-says-worldwide-pc-shipments-declined-5-percent-in-fourth-quarter-of-2021-but-grew-nearly-10-percent-for-the-year#:~:text=Worldwide%20PC%20shipments%20totaled%20339.8,2020%20(see%20Table%203)

GE Aerospace 2022. GE 90 Engine Family. https://www.geaerospace.com/propulsion/commercial/ge90.

Gediga, J. et al. 2019. Life cycle assessment of zircon sand. *International Journal of Life Cycle Analysis* 24:1976–1984.

Gelencsér, A. et al. 2011. The red mud accident in Ajka (Hungary): Characterization and potential health effects of fugitive dust. *Environmental Science and Technology* 45:1608–1615.

Georgescu-Roegen, N. 1971. *The Entropy Law and the Economic Process.* Cambridge, MA: Harvard University Press.

Georgescu-Roegen, N. 1975. Energy and economic myths. *Ecologist* 5(164–174):242–252.

Geschichte-Club VÖEST. 1991. *Geschichte der VÖEST: Rückblick auf die wechselhaften Jahre des grössten österreichischen Industrieunternehmens.* Linz: Geschichte-Club VÖEST.

GESP (The Global E-Waste Statistics Partnership). 2022. What is E-Waste? https://globalewaste.org/

GHK. 2006. A Study to Examine the Benefits of the End of Life Vehicles Directive and the Costs and Benefits of a Revision of the 2015 Targets for Recycling, Re-use and Recovery Under the ELV Directive. http://ec.europa.eu/environment/waste/pdf/study/final_report.pdf

Ghosh, S.K., ed. 2009. *Self-Healing Materials Fundamentals, Design Strategies, and Applications.* Weinheim: Wiley-VCH.

Giampietro, M. 2019. On the circular bioeconomy and decoupling: Implications for sustainable growth. *Ecological Economics* 162:153–156.

Gibson, I. et al. 2010. *Additive Manufacturing Technologies: Rapid Prototyping to Direct Digital Manufacturing.* New York: Springer.

Gielen, D. 2021. *Critical Minerals for the Energy Transition.* Abu Dhabi: International Renewable Energy Agency. https://irena.org/-/media/Files/IRENA/Agency/Technical-Papers/IRENA_Critical_Materials_2021.pdf.

Gilbert, P.U.P.A. et al. 2022. Biomineralization: Integrating mechanism and evolutionary history. *Science Advances* 8:eabl9653.

Giljum, S. et al. 2008. *Global Dimensions of European Natural Resource Use.* Vienna: Sustainable Europe Research Institute. http://seri.at/wp-content/uploads/2009/08/SERI-Working-Paper-7.pdf.

Ginsburg, M., ed. 1991. *The Illustrated History of Textiles.* London: Studio Editions.

Glassglobal. 2022. Plants. https://www.glassglobal.com/

Global Environmental Facility. 2012. Energy Efficiency Improvements in the Indian Brick Industry. http://www.resourceefficientbricks.org/background.php

Goeller, H.E. and A.M. Weinberg. 1976. The age of substitutability. *Science* 191:683–689.

Gold, B. et al. 1984. *Technological Progress and Industrial Leadership: The Growth of the U.S. Steel Industry, 1900–1970*. Lexington, MA: Lexington Books.

Gong, X. et al. 2012. Life cycle energy consumption and carbon dioxide emission of residential building designs in Beijing. *Journal of Industrial Ecology* 16:576–587.

González, L.F.V. et al. 2021. *Operations and Basic Processes in Steelmaking*. Cham: Springer.

Goode, J. 2010. Alternatives to cork in wine bottle closures. In: *Managing Wine Quality, Woodhead Publishing Series in Food Science*, Woodhead Publishing, pp. 255–270. doi: 10.1533/9781845699987.1.255.

Goodman, S.H., ed. 1998. *Handbook of Thermoset Plastics*. Westwood, NJ: Noyes Publications.

Goonan, T.G. 2012. Lithium Use in Batteries. http://pubs.usgs.gov/circ/1371/pdf/circ1371_508.pdf

Gordon, R.B. et al. 2006. Metals stocks and sustainability. *Proceedings of the National Academy of Sciences* 103:1209–1214.

Gould, J.L. and C.G. Gould. 2007. *Animal Architects: Building and the Evolution of Intelligence*. New York: Basic Books.

Graedel, T.E. 2011. On the future availability of the energy metals. *Annual Review of Material Research* 41:323–335.

Graedel, T.E. et al. 2011. Recycling Rates of Metals. http://www.unep.org/resourcepanel/Portals/24102/PDFs/Metals_Recycling_Rates_110412-1.pdf

Graedel, T.E. et al. 2015. Criticality of metals and metalloids. *Proceedings of the National Academy of Sciences* 112:4257–4262.

Grafström, J. and S. Aasma. 2021. Breaking circular economy barriers. *Journal of Cleaner Production* 292:126002.

Grand View Research. 2021. Industrial Gases Market Size, Share & Trends. https://www.grandviewresearch.com/industry-analysis/industrial-gases-market

Grantham, J. 2012. Be persuasive. Be brave. Be arrested (if necessary). *Nature* 491:303.

Gray, H. 1903. Solid Rolled Metal i-beam of Nine Inches and Under Twelve Inches in Height. https://patents.google.com/patent/US1013649A/en

Green Hydrogen Systems. 2022. Accelerating Global Energy Transition with Green Hydrogen. https://greenhydrogensystems.com/

Greenpeace. 2003. PVC: The Poison Plastic. https://www.greenpeace.org/usa/wp-content/uploads/legacy/Global/usa/report/2009/4/pvc-the-poison-plastic.html

GRID. 2017. An Average Day in a Large-Sized Copper Mine. https://www.grida.no/resources/11419

Grotte, J. and B. Marrey. 2000. *Freyssinet: la précontrainte et l'Europe, 1930–1945*. Paris: Éditions du Linteau.

Gruber, T. et al. 2011. Community-specific evaluation of tool affordances in wild chimpanzees. *Scientific Reports* 1:128. doi: 10.1038/srep00128.

GSM Arena. 2022. Mobile Phone Evolution: Story of Shapes and Sizes. https://www.gsmarena.com/mobile_phone_evolution-review-493p6.php

Gunston, B. 1986. *World Encyclopedia of Aero Engines*. Wellingborough: Patrick Stephens.

Gutowski, T.G. et al. 2013. The energy required to produce materials: Constraints on energy-intensity improvements, parameters of demand. *Philosophical Transactions of the Royal Society A* 371:20120003. doi: 10.1098/rsta.2012.000.

Gypsum Association. 2022. Your Technical Center for Gypsum Panel Products. https://gypsum.org

Haas, W. et al. 2015. How circular is the global economy? An assessment of material flows, waste production and recycling in the European Union and the world in 2005. *Journal of Industrial Ecology* 19:765–777.

Haberl, H. et al. 2007. Quantifying and mapping the human appropriation of net primary production in earth's terrestrial ecosystems. *Proceedings of the National Academy of Sciences* 104:12942–12947.

Habu, J. 2004. *Ancient Jomon of Japan*. Cambridge: Cambridge University Press.

Hackney, C.R. et al. 2021. Sand mining far outpaces natural supply in a large alluvial river. *Earth Surface Dynamics* 9:1323–1334.

Hahn, P.O. 2001. The 300 mm silicon wafer – a cost an technology challenge. *Microelectronic Engineering* 56:3–13.

Hale, J. 2006. Boeing 787 from the Ground Up. https://www.boeing.com/commercial/aeromagazine/articles/qtr_4_06/AERO_Q406_article4.pdf

Hall, E.C. 1996. *Journey to the Moon: The History of the Apollo Guidance Computer*. Washington, DC: American Institute of Aeronautics & Astronautics.

Hammond, G.P. and C.I. Jones. 2008. Embodied energy and carbon in construction materials. *Proceedings of the Institution of Civil Engineers – Energy* 161:87–98.

Hansell, M.H. 2007. *Built by Animals: The Natural History of Animal Architecture*. Oxford: Oxford University Press.

Hansell, M. 2011. Houses made by protists. *Current Biology* 21(13):R485–R487. doi: 10.1016/j.cub.2011.05.050.

Hara, D. and G.O. Özgen. 2016. Investigation of weight reduction of automotive body structures with the use of sandwich materials. *Transport Research Arena* 14:18–21.

Harmand, S. et al. 2015. 3.3-million-year-old stone tools from Lomekwi 3, West Turkana, Kenya. *Nature* 521:310–315.

Harper, D. 1992. *Eutrophication of Freshwaters: Principles, Problems and Restoration*. New York: Springer.

Hatayama, H. et al. 2009. Assessment of the recycling of aluminum in Japan, the United States, Europe and China. *Materials Transactions* 50:650–656.

Hatayama, H. et al. 2010. Outlook of the world steel cycle based on the stock and flow dynamics. *Environmental Science & Technology* 44:6457–6463.

Hattori, T. 2015. From 20 mm to 450 mm. *Telescope Magazine*. https://www.tel.com/museum/magazine/material/150430_report04_03/index.html

Hauschild, M.Z. et al. 2018. *Life Cycle Assessment: Theory and Practice*. Berlin: Springer.

He, K. et al. 2020. Review of the energy consumption and production structure of China's steel industry: Current situation and future development. *Metals* 10:302. doi: 10.3390/met10030302.

Heal, D.W. 1975. Modern perspectives on the history of fuel economy in the iron and steel industry. *Ironmaking and Steelmaking* 4:222–227.

Hebbert, F.J. 1990. *Soldier of France: Sébastien le Prestre de Vauban, 1633–1707*. New York: P. Lang.

Heinberg, R. 2010. *Peak Everything: Waking Up to the Century of Declines*. Gabriola Island, BC: New Society Publishers.

Herman, A. 2012. *Freedom's Forge: How American Business Produced Victory in World War II*. New York: Random House.

Herman, R. et al. 1990. Dematerialization. *Technological Forecasting and Social Change* 37:333–347.

Hermes, M.E. 1996. *Enough for One Lifetime: Wallace Carothers, Inventor of Nylon*. Washington, DC: American Chemical Society and the Chemical Heritage Foundation.

Hernández, A.S. et al. 2020. Coccolithophore biodiversity controls carbonate export in the Southern Ocean. *Biogeosciences* 17(1):245–263. doi: 10.5194/bg-17-245-2020-supplement.

Hess, A. 2008. *Frank Lloyd Wright the Buildings*. New York: Rizzoli.

Hills, R.L. 1993. *Power from Steam: A History of the Stationary Steam Engine*. Cambridge: Cambridge University Press.

Hischier, R. 2014. Life cycle assessment of manufactured nanomaterials: Inventory modelling rules and application example. *International Journal of Life Cycle Assessment* 19:941–943.

Hoffmann, O. 2012. Steel Lightweight Materials and Design for Environmental Friendly Mobility. ThyssenKrupp Steel Europe. http://www.industrialtechnologies2012.eu/sites/default/files/presentations_session/03_Oliver_Hoffmann.pdf

Hogan, W.T. 1971. *Economic History of the Iron and Steel Industry in the United States*. Lexington, MA: Lexington Books.

Holden, R.N. 2014. The origins of the power loom revisited. *International Journal for the History of Engineering & Technology* 84(2):135–159.

Hölldobler, B. and E.O. Wilson. 1990. *The Ants*. Cambridge, MA: Harvard University Press.

Holliday, R. et al. 2012. *Simply No Substitute?* Birmingham: Cientifica.

Holt, J.A. and J.F. Easy. 1993. Numbers and biomass of mound-building termites (Isoptera) in a semiarid tropical woodland near Charters Towers, North Queensland, Australia. *Sociobiology* 21:281–286.

Hong, S. et al. 1994. Greenland ice evidence of hemisphere lead pollution two millennia ago by Greek and Roman civilizations. *Science* 265:1841–1843.

Hopkins, J. et al. 2015. Phenological characteristics of global coccolitho-phore blooms. *Global Biogeochemical Cycles* 29:239–253.

Horn, R.E. et al. 2009. *Life Cycle Assessment: Principles, Practice and Prospects*. Canberra: CSIRO Publishing.

Hosford, W.F. and J.L. Duncan. 1994. The aluminum beverage can. *Scientific American* 271(3):48–53.

Hsu, F. et al. 2015. Exploring and estimating in-use steel stocks in civil engineering and buildings from night-time lights. *International Journal of Remote Sensing* 34:490–504.

Huang, W. et al. 2021. Human body burden of heavy metals and health consequences of Pb exposure in Guiyu, an e-waste recycling town in China. *International Journal of Environmental Research and Public Health* 18(23):12428. doi: 10.3390/ijerph182312428.

Hudders, L. and M. Pandelaere. 2012. The silver lining of materialism: The impact of luxury consumption on subjective well-being. *Journal of Happiness Studies* 13:411–437.

Huffa, C.D. et al. 2010. Mobile elements reveal small population size in the ancient ancestors of Homo sapiens. *Proceedings of the National Academy of Sciences* 107:2147–2152.

Humar, M., ed. 2020. *Wood Properties and Processing*. Basel: MDPI.

Hund, K. et al. 2020. *Minerals for Climate Action*. https://pubdocs.worldbank.org/en/961711588875536384/Minerals-for-Climate-Action-The-Mineral-Intensity-of-the-Clean-Energy-Transition.pdf

Hyde, C.K. 1977. *Technological Change and the British Iron Industry 1700–1870*. Princeton, NJ: Princeton University Press.

IAI (International Aluminium Institute). 2018. Lightweight. https://packaging.world-aluminium.org/benefits/lightweight/

IAI (International Aluminium Institute). 2021. Aluminum Recycling. https://international-aluminium.org/wp-content/uploads/2021/01/wa_factsheet_final.pdf

IAI (International Aluminium Institute). 2022a. Current Statistics. https://international-aluminium.org/

IAI (International Aluminium Institute). 2022b. Primary Aluminum Smelting Energy Intensity. https://international-aluminium.org/statistics/primary-aluminium-smelting-energy-intensity/

IBM. 2022. IBM Archives: The Birth of the IBM PC. https://www.ibm.com/ibm/history/exhibits/pc25/pc25_birth.html

ICCT (International Council on Clean Transportation). 2021. European Vehicle Market Statistics 2020/2021. https://theicct.org/wp-content/uploads/2021/12/ICCT-EU-Pocketbook-2021-Web-Dec21.pdf

IDTechEx. 2017. The Electric Vehicle Market and Copper Demand. https://copperalliance.org/wp-content/uploads/2017/06/2017.06-E-Mobility-Factsheet-1.pdf

IEA. 2022. Cement. https://www.iea.org/reports/cement

IEA (International Energy Agency). 2007. *Tracking Industrial Energy Efficiency and CO₂ Emissions*. Paris: IEA. https://www.iea.org/reports/tracking-industrial-energy-efficiency-and-co2-emissions.

IEA (International Energy Agency). 2018. Technology Roadmap. https://www.iea.org/reports/technology-roadmap-low-carbon-transition-in-the-cement-industry

IEA (International Energy Agency). 2021a. Cement. https://www.iea.org/reports/cement

IEA (International Energy Agency). 2021b. Global Hydrogen Review 2021b. https://www.iea.org/reports/global-hydrogen-review-2021

IEA (International Energy Agency). 2021c. Trends in Photovoltaic Applications 2021. https://tecsol.blogs.com/files/iea-pvps-trends-report-2021-1.pdf

IEA (International Energy Agency). 2021d. Iron and Steel. https://www.iea.org/reports/iron-and-steel

IEA (International Energy Agency). 2022a. Global Hydrogen Review 2022. https://www.iea.org/events/global-hydrogen-review-2022

IEA (International Energy Agency). 2022b. The Role of Critical World Energy Outlook Special Report Minerals in Clean Energy Transitions. https://www.iea.org/reports/the-role-of-critical-minerals-in-clean-energy-transition

IEA (International Energy Agency). 2022c. World Balance 2020. https://www.iea.org/sankey/

IEA (International Energy Agency). 2022d. Global EV Outlook 2022. https://www.iea.org/reports/global-ev-outlook-2022

Ifcher, J. et al. 2019. Income inequality and well-being in the U.S.: Evidence of geographic-scale- and measure-dependence. *Journal of Economic Inequality* 17:415–434. doi: 10.1007/s10888-018-9404-z.

IFIAS (International Federation of Institutes for Advanced Study). 1974. *Energy Analysis Workshop on Methodology and Conventions.* Stockholm: IFIAS.

ILZSG (International Lead and Zinc Study Group). 2022. Lead and Zinc Statistics. https://www.ilzsg.org/static/statistics.aspx

IMF (International Monetary Fund). 2022. People's Republic of China. https://www.imf.org/en/Countries/CHN

ImpEE Project. 2013. Plastic Waste in the UK. University of Cambridge: The Cambridge-MIT Institute. http://www-g.eng.cam.ac.uk/impee/topics/RecyclePlastics/files/Recycling%20Plastic%20v3%20PDF.pdf

ImpEE Project. 2022. *Plastic waste in the UK*. University of Cambridge: The Cambridge-MIT Institute. http://www-g.eng.cam.ac.uk/impee/topics/RecyclePlastics/files/Recycling%20Plastic%20v3%20PDF.pdf.

Infante-Amate, J. et al. 2015. The Spanish transition to industrial metabolism long-term material flow analysis (1860–2010). *Journal of Industrial Ecology* 19(5):866–876. doi: 10.1111/jiec.12261.

Inoue, T. et al. 2001. The abundance and biomass of subterranean termites (Isoptera) in a dry evergreen forest of northeast Thailand. *Sociobiology* 37:41–52.

Intel. 2022. The Intel 4004. https://www.intel.com/content/www/us/en/history/virtual-vault/articles/the-intel-4004.html

Intrakamhaeng, V. et al. 2019. Initiatives to reduce lead from electronic devices: Evidence of success from the toxicity characteristic leaching procedure. *Journal of the Air & Waste Management Association* 69:1116–1121.

ISAAA (International Service for the Acquisition of Agri-biotech Applications). 2021. Breaking Barriers with Breeding. https://www.isaaa.org/resources/publications/briefs/56/default.asp

ISO (International Organization for Standardization). 2006a. *ISO 14044:2006 Environmental Management – Life Cycle Assessment – Requirements and Guidelines*. Geneva: ISO. http://www.iso.org/iso/catalogue_detail?csnumber=38498.

ISO (International Organization for Standardization). 2006b. *ISO 14040:2006 Environmental Management – Life Cycle Assessment – Principles and Framework*. Geneva: ISO. http://www.iso.org/iso/catalogue_detail?csnumber=37456.

Ixer, R. and R. Bevins. 2017. The bluestones of Stonehenge. *Geology Today* 33:183–187.

Jacobs and IPST (Institute of Paper Science and Technology). 2006. *Pulp and Paper Industry: Energy Bandwidth Study*. Atlanta, GA: Georgia Institute of Technology.

James, H. 2001. Use and production of solid sawn timber in the United States. *Forest Products Journal* 51(7/8):23–28.

Jamieson, A.J. et al. 2019. Microplastics and synthetic particles ingested by deep-sea amphipods in six of the deepest marine ecosystems on Earth. *Royal Society Open Science* 6:180667. doi: 10.1098/rsos.180667.

Jenkins, D., ed. 2003. *The Cambridge History of Western Textiles*. Cambridge: Cambridge University Press.

Jevons, W.S. 1865. *The Coal Question: An Inquiry Concerning the Progress of the Nation, and the Probable Exhaustion of Our Coal Mines*. London: Macmillan.

Jolliet, O. et al. 2013. *Environmental Life Cycle Assessment*. London: Waterstones.

Jönsson, Å. et al. 1997. Life cycle assessment of flooring materials: Case study. *Building and Environment* 32:245–255.

JPA (Japan Paper Association). 2021. Environmental Efforts Made by Japanese Paper Industry. https://www.jpa.gr.jp/en/env/recycle/segregation/#:~:text=Japan%20has%20a%20long%20history,system%20of%20the%20Japanese%20society

Kahhat, R. and E. Williams. 2012. Materials flow analysis of e-waste: Domestic flows and exports of used computers from the United States. *Resources, Conservation and Recycling* 67:67–74.

Kalliala, E.M. and P. Nousiainen. 1999. Environmental profile of cotton and polyester-cotton fabrics. *AUTEX Research Journal* 1:8–20.

Kallis, G. 2017. Radical dematerialization and degrowth. *Philosophical Transactions of the Royal Society A* 375:20160383. doi: 10.1098/rsta.2016.0383.

Kamdar, M. 2011. The pleasures of excess. *World Policy Journal* 28(2):15–19.

Kaneko, T. et al. 2022. Fujisan. https://www.data.jma.go.jp/svd/vois/data/tokyo/STOCK/souran_eng/volcanoes/055_fujisan.pdf

Karg, S. 2011. New research on the cultural history of the useful plant *Linum usitatissimum* L. (flax), a resource for food and textiles for 8,000 years. *Vegetation History and Archaeobotany* 20:507–508.

Kato, K. et al. 1997. Energy payback time and life-cycle CO2 emission of residential PV power system with silicon PV module. *Appendix B-8: Environmental Aspects of PV Power Systems*, IEA PVPS Task 1 Workshop. Utrecht University, Utrecht.

Katsikantami, I. et al. 2016. A global assessment of phthalates burden and related links to health effects. *Environment International* 97:212–236.

Kauppi, P.E. et al. 2010. Changing stock of biomass carbon in a boreal forest over 93 years. *Forest Ecology and Management* 259:1239–1244.

Kawaoka, K. et al. 2006. Latest blast furnace relining technology at Nippon Steel. *Nippon Steel Technical Report* 94:127–132.

Kawashima, C. 1986. *Minka: Traditional Houses of Rural Japan*. Tōkyō: Kōdansha.

Kaye, G.D. 2002. Using GIS to Estimate the Total Volume of Mauna Loa Volcano, Hawai'i. https://gsa.confex.com/gsa/2002CD/finalprogram/abstract_34712.htm

Kazan, H. et al. 2020. Life cycle assessment of cotton woven shirts and alternative manufacturing techniques. *Clean Technologies and Environmental Policy* 22:849–864.

Kelly, T.D., and Matos, G.R. 2016. Historical Statistics for Mineral and Material Commodities in the United States (2016 Version): U.S. Geological Survey Data Series 140. https://www.usgs.gov/centers/national-minerals-information-center/historical-statistics-mineral-and-material-commodities

Kenny, C. 2012. Paving paradise. *Foreign Policy* 2012:31–32.

Kermeli, K. et al. 2011. *Energy Efficiency Improvement and Cost Saving Opportunities for the Concrete Industry*. Berkeley, CA: Ernest Orlando Lawrence Berkeley National Laboratory.

Kerr, R.A. 2012. Is the world tottering on the precipice of peak gold? *Science* 335:1038–1039.

King, C.D. 1948. *Seventy-five Years of Progress in Iron and Steel*. New York: American Institute of Mining and Metallurgical Engineers.

King, B. 2006. *Design of Straw Bale Buildings: The State of the Art*. San Rafael, CA: Green Building Press.

Kinnaman, T.C. et al. 2014. The socially optimal recycling rate: Evidence from Japan. *Journal of Environmental Economics and Management* 68:54–70.

Klare, M.T. 2012. The end of easy everything. *Current History* 111(741):24–28.

Klein Goldewijk, K. et al. 2010. Long-term dynamic modelling of global population and built-up area in a spatially explicit way: HYDE 3.1. *The Holocene* 20:565–573.

Klimont, Z. et al. 2013. The last decade of global anthropogenic sulfur dioxide: 2000–2011 emissions. *Environmental Research Letters* 8:1–6.

Klöpffer, W. and B. Grahl. 2014. *Life Cycle Assessment (LCA): A Guide to Best Practice*. Chichester: John Wiley.

Kovanda, J. and T. Hak. 2011. Historical perspectives of material use in Czechoslovakia in 1855–2007. *Ecological Indicators* 11:1375–1384.

Krausmann, F. et al. 2008. Global patterns of socioeconomic biomass flows in the year 2000: A comprehensive assessment of supply, consumption and constraints. *Ecological Economics* 65:471–487.

Krausmann, F. et al. 2009. Growth in global materials use, GDP and population during the 20th century. *Ecological Economics* 68:2696–2705.

Krausmann, F. et al. 2011. The metabolic transition in Japan. *Journal of Industrial Ecology* 15:877–892.

Krausmann, F. et al. 2017. Global socioeconomic material stocks rise 23-fold over the 20th century and require half of annual resource use. *Proceedings of the National Academy of Sciences* 114:1880–1885. doi: 10.1073/pnas.1613773114.

Krumhardt, K.M. et al. 2019. Coccolithophore growth and calcification in an acidified ocean: Insights from community earth system model simulations. *Journal of Advances in Modeling Earth Systems* 11:1418–1437.

Kuriki, S. et al. 2010. Recycling potential of platinum group metals in Japan. *Journal of Japan Institute of Metals* 74:801–805.

Kvavadze, E. et al. 2009. 30,000-year-old wild flax fibers. *Science* 325:1359.

Lacitis, E. 2019. 50 Years Ago, the First 747 Took Off and Changed Aviation. https://www.seattletimes.com/ seattle-news/50-years-ago-today-the-first-747-took-off-and-changed-aviation/

Lamon, N. et al. 2018. Wild chimpanzees select tool material based on efficiency and knowledge. *Proceedings of the Royal Society B* 285:20181715. doi: 10.1098/rspb.2018.1715.

Lampert, L. et al. 2002. Spatial variability of phytoplankton composition and biomass on the eastern continental shelf of the Bay of Biscay (north-east Atlantic Ocean). Evidence for a bloom of *Emiliania huxleyi* (Prymnesiophyceae) in spring 1998. *Continental Shelf Research* 22:1225–1247.

Landsberg, H.H. 1964. *Natural Resources for U.S. Growth: A Look Ahead to the Year 2000*. Baltimore, MD: Johns Hopkins Press.

Landsberg, H.H. et al. 1963. *Resources in America's Future: Patterns of Requirements and Availabilities, 1960-2000*. Baltimore, MD: Johns Hopkins Press.

Langlois, S. 2020. 9 things that Americans couldn't possibly live without, according to the rest of the world. *Market Watch*. https://www.marketwatch. com/story/9-things-americans-couldnt-possibly-live-without-according-to-the-rest-of-the-world-11606708045

Latimer, J. 2001. All-consuming passions: Materials and subjectivity in the age of enhancement. In: N. Lee and R. Munro, eds., *The Consumption of Mass*, Oxford: Blackwell Publishers, pp. 158–173.

Lau, W.W.Y. et al. 2020. Evaluating scenarios toward zero plastic pollution. *Science* 369:1455–1461.

Lauk, C. et al. 2012. Global socioeconomic carbon stocks in long-lived products 1900–2008. *Environmental Research Letters* 7:034023. http://iopscience.iop. org/1748-9326/7/3/034023.

Leach, E.R. 1959. Hydraulic society in Ceylon. *Past and Present* 15:2–26.

Lefebvre, H. 1971. *Everyday Life in the Modern World*. London: Allen Lane.

Lehner, M. and Z. Hawass. 2017. *Giza and the Pyramids: The Definitive History*. Chicago: University of Chicago Press.

Lei, Q. 2011. The Development of China's Cement Industry. http://www.tcma.org. tr/images/file/Cin%20Cimento%20Birligi%20Baskani%20Lei%20 QIANZHI%20.pdf

Lenzen, M. and C.J. Dey. 2000. Truncation error in embodied energy analysis of basic iron and steel products. *Energy* 25:577–585.

Lenzen, M. and G. Treloar. 2002. Differential convergence of life-cycle inventories toward upstream production layers, implications for life-cycle assessment. *Journal of Industrial Ecology* 6:3–4.

Li, J. 2012. World Bank experience on clean brick production in South Asia Region. In: *INE Proceedings of the Workshop on Public Policies to Mitigate Environmental Impact of Artisanal Brick Production.* Guanajuato, Mexico, September 4-6, 2012.

Li, J. et al. 2006. Energy Demand in Wood Processing Plants. https://citeseerx.ist.psu.edu/viewdoc/download?doi=10.1.1.553.3440&rep=rep1&type=pdf

Li, Y. et al. 2006. TCLP heavy metal leaching of personal computer components. *Journal of Environmental Engineering* 132:497–504.

Li, C. et al. 2010. Recent large reduction in sulfur dioxide emissions from Chinese power plants observed by the Ozone Monitoring Instrument. *Geophysical Research Letters* 37. doi: 10.1029/2010GL042594.

Li, B. et al. 2019. Peak phosphorus, demand trends and implications for the sustainable management of phosphorus in China. *Resources, Conservation and Recycling* 146:316–328.

Linde, C.P. 1916. *Aus meinem Leben und von meiner Arbeit.* München: R. Oldenbourg.

Linde. 2019. Air Separation Plants. https://www.linde-engineering.com/en/images/Air-separation-plants-history-and-technological-progress-2019_tcm19-457349.pdf

Linde. 2022. Making Our World More Productive. https://www.linde.com/

Lindner, M. and H. Verkerk. 2022. How has Climate Change Affected EU Forests and What Might Happen Next? https://efi.int/forestquestions/q4

Linebaugh, P. 1993. *The London Hanged: Crime and Civil Society in the Eighteenth Century.* Cambridge: Cambridge University Press.

Liu, S. et al. 2018. Exploring the driving forces of energy consumption and environmental pollution in China's cement industry at the provincial level. *Journal of Cleaner Production* 184:274–285. doi: 10.1016/j.jclepro.2018.02.277.

Liu, X. et al. 2019. Impact of anthropogenic activities on global land oxygen flux. *Earth System Science Data.* doi: 10.5194/essd-2019-36.

Longhurst, A.H. 1979. *The Story of the Stūpa.* New Delhi: Asian Educational services.

Lowenstam, H.A. 1981. Minerals formed by organisms. *Science* 211:1126–1130.

Lu, Z. et al. 2010. Sulfur dioxide emission in Chin and sulfur trends in East Asia since 2000. *Atmospheric Chemistry and Physics* 10:6311–6331.

Luo, Z. and A. Soria. 2008. *Prospective Study of the World Aluminium Industry.* Seville: European Commission. ftp://ftp.jrc.es/pub/EURdoc/JRC40221.pdf.

Macfarlane, A. and G. Martin. 2002. *Glass: A World History*. Chicago: University of Chicago Press.

Mackay, A. 2022. Why are Airlines Leasing more Aircraft? *ICF*. https://www.icf.com/insights/transportation/why-are-airlines-leasing-more-aircraft

Madlool, N.A. et al. 2011. A critical review on energy use and savings in the cement industries. *Renewable and Sustainable Energy Reviews* 15:2042–2060.

Magnaghi, G. 2010. Recovered Paper Market in 2010. http://www.bir.org/assets/Documents/industry/MagnaghiReport2010.pdf

Malpass, D.B. 2010. *Introduction to Industrial Polyethylene: Properties, Catalysts, and Processes*. New York: Wiley-Scrivener.

Man, Y. et al. 2019. Life cycle energy consumption analysis and green manufacture evolution for the papermaking industry in China. *Green Chemistry* 21:1011–1020. doi: 10.1039/C8GC03604G.

Marsh, P. 2012. *The New Industrial Revolution: Consumers, Globalisation and the End of Mass Production*. New Haven, CT: Yale University Press.

Martin, S.J. et al. 2018. A vast 4,000-yearold spatial pattern of termite mounds. *Current Biology* 28:R1283–R1295.

Material Flows Net. 2018. The Material Flows Analysis Portal. http://www.materialflows.net/

Matos, G.R. 2009. *Use of Minerals and Materials in the United States from 1900 through 2006*. Washington, DC: USGS. http://pubs.usgs.gov/fs/2009/3008/.

Matos, G. and L. Wagner. 1998. Consumption of materials in the United States, 1900–1995. *Annual Review of Energy* 23:107–122. http://pubs.usgs.gov/annrev/ar-23-107/aerdocnew.pdf.

Matthews, E. et al. 2000. *The Weight of Nations: Material Outflows from Industrial Economies*. Washington, DC: World Resources Institute.

MB Stone International. 2022. Sain Maximin. https://mbstone-international.com/portfolio/saint-maximin/

McFedries, P. 2012. Consumption 2.0. *IEEE Spectrum* June 2012, 26.

McKelvey, V.E. 1973. Mineral resource estimates and public policy. In: D.A. Brobst and W.P. Pratt, eds., *United States Mineral Resources*, Washington, DC: USGS, pp. 9–19.

McLaren, D.J. and B.J. Skinner, eds. 1987. *Resources and World Development*. Chichester: John Wiley.

McMahan, S. 2019. New materials may reduce or replace rare-earth elements in strong permanent magnets. *IEEE Power*. https://eepower.com/news/new-materials-may-reduce-or-replace-rare-earth-elements-in-strongest-permanent-magnets/

McWhan, D. 2012. *Sand and Silicon: Science that Changed the World*. New York: Oxford University Press.

MDO (Mining Data Online). 2022. Chuquicamata Mine. https://miningdataonline. com/property/405/Chuquicamata-Mine.aspx

Meadows, D.H. et al. 1972. *The Limits to Growth.* New York: Universe Books.

Meil, J. et al. 2009. *Status of Energy Use in the Canadian Wood Products Sector.* Vancouver, BC: Forintek.

Meinrenken, C.J. et al. 2022. The Carbon Catalogue, carbon footprints of 866 commercial products from 8 industry sectors and 5 continents. *Scientific Data* 9:87. doi: 10.1038/s41597-022-01178-9.

Mendels, F.F. 1972. Proto-industrialization: The first phase of the industrialization process. *Journal of Economic History* 32:241–261.

Mendoza, J.M.F. et al. 2012. Life cycle assessment of granite application in sidewalks. *International Journal of Life Cycle Assessment* 17:580–592.

Menzel, P. 1995. *Material World: A Global Family Portrait.* San Francisco: Sierra Club.

Metalshub. 2020. New Copper Scrap Import Regulations. https://www.metals-hub. com/blog/New-copper-scrap-import-regulations-in-China-will-create-commercial-opportunities/

Meyers, M.A. et al. 2013. Structural biological materials: critical mechanics-materials connections. *Science* 339:773–779.

MHI RJ Aviation. 2022. CRG Series. https://mhirj.com/en/products-and-services/ crj-series

Miettinen, A. et al. 2008. *The Palaeoenvironment of the Antrea Net Find.* Helsinki: Finnish Antiquarian Society.

Mills, M.P. 2020. *Mines, Minerals and "Green" Energy; A Reality Check.* New York: Manhattan Institute.

Mishra, S. 2022. Antilia: All You Want to Know about Mukesh Ambani's Home. https://housing.com/news/mukesh-ambani-house-antilia/

Mitscherlich, G. 1963. *Zustand, Wachstum und Nutzung des Waldes im Wandel der Zeit.* Freiburg: Freiburger Universitätsreden.

Modern Airliners. 2022. Boeing 737 Specifications. https://modernairliners.com/ boeing-737/boeing-737-specifications/

Mongardini, J. and A. Radzikowski. 2020. *Global Smartphone Sales May Have Peaked: What Next?* Washington, DC: IMF Working Paper Research Department.

Monteiro, F.M. et al. 2016. Why marine phytoplankton calcify. *Science Advances.* doi: 10.1126/sciadv.1501822.

Montgomery Ward & Company. 1895. *Catalogue and Buyers' Guide 1895.* New York: Skyhorse Publishing 2008.

Moore, G. 1965. Cramming more components onto integrated circuits. *Electronics* 38(8):114–117.

Moore, G.E. 1975. Progress in digital integrated electronics. *Technical Digest, IEEE International Electron Devices Meeting*, 11–13.

Moore, C. 2022. Germany leads Europe with target to reach 100% clean power by 2035. *Euractiv*. https://www.euractiv.com/section/energy/opinion/ germany-leads-europe-with-target-to-reach-100-clean-power-by-2035/

Moore, C.G. and C. Phillips. 2011. *Plastic Ocean*. London: Penguin.

Morita, Z. and T. Emi, eds. 2003. *An Introduction to Iron and Steel Processing*. Tokyo: Kawasaki Steel 21st Century Foundation.

Mukerji, C. 1983. *From Graven Images: Patterns of Modern Materialism*. New York: Columbia University Press.

Müller, P. H. 1948. Dichloro-Diphenyl-Trichloroethane and Newer Insecticides. *Nobel Lecture*, December 11, 1948. https://www.nobelprize.org/uploads/2018/ 06/muller-lecture.pdf

Müller, G. and J. Watling. 2016. The engineering in beaver dams. *River Flow 2016: Eighth International Conference on Fluvial Hydraulics*. Saint Louis, July 12-15, 2016. https://eprints.soton.ac.uk/400282/

Müller, D.B. et al. 2006. Exploring the engine of anthropogenic iron cycles. *Proceedings of the National Academy of Sciences* 103:16111–16116.

Murakami, S. et al. 2004. Material flow accounting for metals in Japan. *Materials Transactions* 45:3184–3193.

Myhre, G. et al. 2013. Anthropogenic and natural radiative forcing. In: T.F. Stocker et al., eds., *Climate Change 2013: The Physical Science Basis. Contribution of Working Group I to the Fifth Assessment Report of the Intergovernmental Panel on Climate Change*, Cambridge: Cambridge University Press.

NAPCOR (National Association for PET Container Resources). 2022. https:// napcor.com/

NAS (National Academy of Sciences). 1969. *Resources and Man*. San Francisco: W.H. Freeman.

National Archives. 2012. British Empire: Living in the British Empire – India. http://www.nationalarchives.gov.uk/education/empire/pdf/g2cs4s2.pdf

Natta, G. 1963. Biographical. https://www.nobelprize.org/prizes/chemistry/1963/ natta/biographical/

NBSC (National Bureau of Statistics of China). 2022. Statistical Data. http://www. stats.gov.cn/english/

Ndoro, W. 1997. Great Zimbabwe. *Scientific American* 277(5):94–99.

Nel, A. et al. 2006. Toxic potential of materials at the nanolevel. *Science* 311:622–627.

Nemoto, T. et al. 2011. Resource recycling for sustainable industrial development. *Hitachi Review* 60:335–340.

Newby, F., ed. 2001. *Early Reinforced Concrete*. Burlington, VY: Ashgate.

Nicolai, F.N.P. and S.L.B. Lana. 2019. Urban mining: A process designed to recover Au from e-waste. *International Journal of Sustainable Development Research* 5(2):30–40.

Nielsen, S. 2003. Boeing turns to Russian programming talent in massive database project. *Serverworld* 2003(1):1–4.

Norgate, T. and S. Jahanshahi. 2011. Reducing the greenhouse gas footprint of primary metal production: Where should the focus be? *Minerals Engineering* 24:1563–1570.

NRC. 2009. *Energy Benchmarking: Canadian Potash Production Facilities.* Ottawa: NRC. https://publications.gc.ca/site/eng/9.667649/publication.html.

NRDC (National Resources Defence Council). 2019. *The Issue with Tissue: How Americans are Flushing Forests Down the Toilet.* New York: NRDC.

NREL (National Renewable Energy Laboratory). 2022a. Best Research-Cell Efficiencies. https://www.nrel.gov/pv/assets/pdfs/best-research-cell-efficiencies-rev220630.pdf

NREL (National Renewable Energy Laboratory). 2022b. U.S. Life Cycle Inventory Database. https://www.nrel.gov/lci/

Nuttall, G.H. 1967. *Theory and Operation of the Fourdrinier Paper Machine.* London: Phillips.

O'Brien, P. 2019. *The Precocious Mechanization of a Global Industry: English Cotton Textile Production from the Flying Shuttle (1733) to the Self-Acting Mule (1825): A Bibliographical Survey and Critique.* London: London School of Economics.

O'Brien, C.J. et al. 2012. Global marine phytoplankton functional type biomass distributions: Coccolithophores. *Earth Systems Scientific Data Discussions* 5:491–520.

O'Connor, S. 2019. How Chinese Companies Facilitate Tech Transfer from the US. https://www.uscc.gov/sites/default/files/Research/How%20Chinese%20Companies%20Facilitate%20Tech%20Transfer%20from%20the%20US.pdf

Oakes, J. 2005. *Inuit Annuraangit. Our Clothes: A Travelling Exhibition of Inuit Clothing.* Winnipeg. MB: Aboriginal Issues Press.

OEC (The Observatory of Economic Complexity). 2022 The Observatory of Economic Complexity. https://oec.world/#:~:text=The%20Economic%20complexity%20Index%20(ECI,and%20provinces%20using%20the%20OEC

OECD (Organization for Economic Cooperation and Development). 1995. *The Life Cycle Approach: An Overview of Product/Process Analysis.* Paris: OECD.

OECD (Organization for Economic Cooperation and Development). 2022. Material Resources. https://www.oecd-ilibrary.org/environment/data/oecd-environment-statistics/material-resources_data-00695-en

Olivier, J.G.J. 2022. *Trends in Global CO_2 and Total Greenhouse Gas Emissions 2021 Summary Report.* The Hague: PBL. https://www.pbl.nl/sites/default/files/

downloads/pbl-2022-trends-in-global-co2-and-total-greenhouse-gas-emissions-2021-summary-report_4758.pdf

Omodara, L. et al. 2019. Recycling and substitution of light rare earth elements, cerium, lanthanum, neodymium, and praseodymium from end-of-life applications: A review. *Journal of Cleaner Production* 236:117573.

OpenLCA Nexus. 2022. openLCA Nexus

Oswalt, S.N. 2021. Annual Harvest of Wood Products by Volume and as a Percentage of Net Growth or Sustained Yield. https://www.fs.usda.gov/research/sites/default/files/2022-03/Indicator2.13.pdf

Otis. 2022. History. https://www.otis.com/en/us/our-company/history

Owens, B. 2013. Mining: Extreme prospects. *Nature* 495:S4–2S6.

Pabortsava, K. and R.S. Lampitt. 2020. High concentrations of plastic hidden beneath the surface of the Atlantic Ocean. *Nature Communications* 11:4073. doi: 10.1038/s41467-020-17932-9.

Pant, S. 1976. *The Origin and Development of Stūpa Architecture in India*. New Delhi: Bharata Manisha.

Pascale, A. et al. 2016. E-waste informal recycling: An emerging source of lead exposure in South America. *Annals of Global Health* 82:197–201.

Path. 2022. Covid-19 Oxygen Tracker. https://www.path.org/programs/market-dynamics/covid-19-oxygen-needs-tracker/

Patterson, C.C. 1972. Silver stocks and losses in ancient and medieval times. *The Economic History Review* 25:205–233.

Paulsson, B.S. 2019. Radiocarbon dates and Bayesian modeling support maritime diffusion model for megaliths in Europe. *Proceedings of the National Academy of Sciences* 116:3460–3465.

PCA (Portland Cement Association). 2019. How Concrete is Made. http://www.cement.org/tech/cct_concrete_prod.asp

PCA (Portland Cement Association). 2022. Recycled Aggregates. https://www.cement.org/learn/concrete-technology/concrete-design-production/recycled-aggregates

Penner, M. et al. 1997. *Canada's Forest Biomass Resources: Deriving Estimates from Canada's Forestry Inventory*. Victoria, BC: Pacific Forestry Centre. http://warehouse.pfc.forestry.ca/pfc/4775.pdf.

Peray, K.E. 1986. *The Rotary Cement Kiln*. New York: Chemical Publishing.

Petersen, A.K. and B. Solberg. 2002. Greenhouse gas emissions, life-cycle inventory and cost-efficiency of using laminated wood instead of steel construction. Case: Beams at Gardermoen airport. *Environmental Science and Policy* 5:169–182.

Petersen, A.K. and B. Solberg. 2004. Greenhouse gas emissions and costs over the life cycle of wood and alternative flooring materials. *Climatic Change* 64:143–167.

Petrasch, J. 2020. Settlements, migration and the break of tradition: The settlement patterns of the Earliest Bandkeramik and the LBK and the formation of a neolithic lifestyle in Western Central Europe. *Quaternary International* 560–561:248–258.

Petrides, D. et al. 2018. Dematerialization and environmental sustainability: Challenges and rebound effects. *Procedia CIRP* 72:845–849.

Pilkington, L.A.B. 1969. Review lecture: The float glass process. *Proceedings of the Royal Society of London. Series A, Mathematical and Physical Sciences* 314:1–25.

Plastics Europe. 2021. *Plastics - The Facts 2021*. Brussels: Plastics Europe. https://plasticseurope.org/wp-content/uploads/2021/12/Plastics-the-Facts-2021-web-final.pdf.

Ponseti, M. and J. López-Pujol. 2006. The Three Gorges Dam Project in China: History and Consequences. https://www.researchgate.net/publication/26447373_The_Three_Gorges_Dam_Project_in_China_History_and_Consequences

Porada, E. 1965. *The Art of Ancient Iran: Pre-Islamic Cultures*. New York: Crown Publishers.

Prak, M. 2011. Mega-structures of the Middle Ages: The construction of religious buildings in Europe and Asia, c. 1000-1500. *Journal of Global History* 6:381–406.

Praseeda, K.I. et al. 2017. Life-cycle energy assessment in buildings: Framework, approaches, and case studies. In: *Encyclopedia of Sustainable Technologies*, pp. 113–136. doi: 10.1016/B978-0-12-409548-9.10188-5.

Pratt & Whitney. 2022. GP 7200 Engine. https://prattwhitney.com/products-and-services/products/commercial-engines/gp7200

President's Materials Policy Commission. 1952. *Resources for Freedom, A Report to the President by the President's Materials Policy Commission*. Washington, DC: US Government Printing Office.

PSP (Plastic Soup Foundation). 2022. Production of Plastic. https://www.plasticsoupfoundation.org/en/plastic-facts-and-figures/

Rafiqul, I. et al. 2005. Energy efficiency improvements in ammonia production – perspectives and uncertainties. *Energy* 30:2487–2504.

Reeves, M. et al. 2006. *Clay Materials used in Construction*. London: Geological Society.

Reiffel, A.J. et al. 2013. High-fidelity tissue engineering of patient-specific auricles for reconstruction of pediatric microtia and other auricular deformities. *PLoS One* 8(2):e56506. doi: 10.1371/journal.pone.0056506. Epub 2013 Feb 20.

Revitasari, R. and B.H. Susanto. 2018. Analysis of energy consumption of Indonesian flat glass industry PT. X based on green industry standards. *MATEC*

Web of Conferences 187:03005. doi: 10.1051/matecconf/201818703005 ICCMP 201.

Ricciardi, M. et al. 2021. Microplastics in the aquatic environment: Occurrence, persistence, analysis, and human exposure. *Water* 13:973. doi: 10.3390/w13070973.

Richardson, B.A. 1978. *Wood Preservation*. New York: Construction Press.

Roche, D. 2000. *A History of Everyday Things: The Birth of Consumption in France, 1600-1800*. Cambridge: Cambridge University Press.

Rogers, J.G. 2018. Paper making in a low carbon economy. *Energy* 6:187–202.

Römer, L. and T. Scheibel. 2008. The elaborate structure of spider silk. *Prion* 2:154–161.

Ruthven, D.M. et al. 1993. *Pressure Swing Adsorption*. New York: John Wiley.

RWE. 2022. Hambach Mine. https://www.rwe.com/en/the-group/countries-and-locations/hambach-mine-site

Ryan, L. and S. Dziurawiec. 2001. Materialism and its relationship to life satisfaction. *Social Indicators Research* 55:185–197.

Ryu, Y. et al. 2019. What is global photosynthesis? History, uncertainties and opportunities. *Remote Sensing of Environment* 223:95–114.

Sahan, M. et al. 2019. Determination of metal content of waste mobile phones and estimation of their recovery potential in Turkey. *International Journal of Environmental Research and Public Health* 16:887. doi: 10.3390/ijerph16050887.

Salmon, E.T. 1999. *Roman Coins and Public Life under the Empire*. Ann Arbor, MI: The University of Michigan Press.

Sampson, B. 2019. GE9X Recognized as Most Powerful Jet Engine in the World. https://www.aerospacetestinginternational.com/videos/ge9x-recognised-as-most-powerful-jet-engine-in-the-world.html#:~:text=The%20GE9X%20engine%20for%20the

Sanz, C.M., ed. 2013. *Tool Use in Animals*. Cambridge: Cambridge University Press.

Schandl, H. and N. Eisenmenger. 2006. Regional patterns in global resource extraction. *Journal of Industrial Ecology* 10:133–147.

Schandl, H. and N. Schulz. 2002. Changes in the United Kingdom's natural relations in terms of society's metabolism and land-use from 1850 to the present day. *Ecological Economics* 41:203–221.

Schandl, H. et al. 2008. Australia's resource use categories. *Journal of Industrial Ecology* 12:669–685.

Schandl, H. et al. 2016. *Global Material Flows and Resource Productivity*. Nairobi: UNEP. https://www.unep.org/resources/report/global-material-flows-and-resource-productivity-assessment-report-unep.

Schandl, H. et al. 2017. Global material flows and resource productivity: Forty years of evidence. *Journal of Industrial Ecology*. doi: 10.1111/jiec.12626.

Scheel, H.J. and T. Fukuda, eds. 2004. *Crystal Growth Technology*. New York: Wiley.

Schlesinger, M.F. et al. 2022. Overview. In: *Extractive Metallurgy of Copper* (6th edition). https://www.sciencedirect.com/topics/earth-and-planetary-sciences/copper-ore.

Schmidt, M. et al. 2012. Life cycle assessment of silicon wafer processing for microelectronic chips and solar cells. *International Journal of Life Cycle Assessment* 17:126–144.

Schutz, B. 2002. *Great Cathedrals*. New York: Abrams.

SEMI. 2022. Silicon Shipment Statistics. https://www.semi.org/en/products-services/market-data/materials/si-shipment-statistics

Semon, W.L. 1933. *Synthetic Rubber-Like Composition and Method of Making Same*. US Patent 1,929,453. Washington, DC: USPTO.

Sethuraj, M.R. and N.M. Mathew, eds. 1992. *Natural Rubber: Biology, Cultivation, and Technology*. New York: Elsevier.

Shadman, F. and T.J. McManus. 2004. Comment on "The 1.7 kilogram microchip: Energy and material use in the production of semiconductor devices". *Environmental Science & Technology* 38:1915.

Shaeffer, R.E. 1992. *Reinforced Concrete: Preliminary Design for Architects and Builders*. New York: McGraw-Hill.

Shafizadeh, F. 1981. Basic principles of direct combustion. In: S.S. Sofer and O.R. Zabrosky, eds., *Biomass Conversion Process for Energy and Fuels*, New York: Plenum Press, pp. 103–112.

Shao, Q. et al. 2017. The high 'price' of dematerialization: A dynamic panel data analysis of material use and economic recession. *Journal of Cleaner Production* 167:1201–1132.

Shawe-Taylor, D. and Q. Buvelot. 2015. *Masters of the Everyday: Dutch Artists in the Age of Vermeer*. London: Royal Collection Trust.

Shephard, S. 2000. *Pickled, Potted, and Canned: How the Art and Science of Food Preserving Changed the World*. New York: Simon & Schuster.

Shumaker, R.W. et al. 2011. *Animal Tool Behavior: The Use and Manufacture of Tools by Animals*. Baltimore, MD: The Johns Hopkins University Press.

SIA (Semiconductor Industry Association). 2021. Semiconductor Market Data. https://www.semiconductors.org/data-resources/market-data/

Siemens. 2019. We Power the World with Innovative Gas Turbines. https://assets.new.siemens.com/siemens/assets/api/uuid:ab8578bf-d86f-45d9-a26b-7ac7a274fadd/siemens-gas-turbine-portfolio.pdf

Sika. 2020. *Concrete Handbook*. Zurich: Sika https://www.sika.com/content/dam/dms/corporate/t/glo-sika-concrete-handbook.pdf.

da Silva, V.P. et al. 2010. Variability in environmental impacts of Brazilian soybean according to crop production and transport scenarios. *Journal of Environmental Management* 91:1831–1839.

Sim, G. and Y. Rusman. 2011. Tin's Bear Market Bottoming as Shortages Boost Timah Profit. https://www.bloomberg.com/news/articles/2011-07-04/bear-market-in-tin-ending-as-shortages-mean-pt-timah-s-profit-advances-55-

Simon, J. 1981. *The Ultimate Resource.* Princeton, NJ: Princeton University Press.

Simon, J., ed. 1995. *The State of Humanity.* New York: Blackwell-Wiley.

Singh, R. 2006. *Lac Culture.* Varanasi: Department of Zoology, Udai Pratap Autonomous College.

Singh, S.J. et al. 2012. India's biophysical economy, 1961–2008. Sustainability in a national and global context. *Ecological Economics* 76:60–69.

Sivak, M. and O. Tsimhoni. 2009. Fuel efficiency of vehicles on US roads: 1936-2006. *Energy Policy* 37:3168–3170.

Slade, G. 2006. *Made to Break: Technology and Obsolescence in America.* Cambridge, MA: Harvard University Press.

Smil, V. 1999. Crop residues: Agriculture's largest harvest. *BioScience* 49:299–308.

Smil, V. 2000. *Feeding the World.* Cambridge, MA: The MIT Press.

Smil, V. 2001. *Enriching the Earth.* Cambridge, MA: The MIT Press.

Smil, V. 2003. *Energy at the Crossroads.* Cambridge, MA: The MIT Press.

Smil, V. 2005. *Creating the Twentieth Century.* New York: Oxford University Press.

Smil, V. 2006a. *Transforming the Twentieth Century.* New York: Oxford University Press.

Smil, V. 2006b. Peak oil: A catastrophist cult and complex realities. *World Watch* 19:22–24.

Smil, V. 2008. *Energy in Nature and Society.* Cambridge, MA: The MIT Press.

Smil, V. 2010. *Prime Movers of Globalization.* Cambridge, MA: The MIT Press.

Smil, V. 2012. Jeremy Grantham, Starving for Facts. *The American* December 5, 2012.

Smil, V. 2013. *Harvesting the Biosphere.* Cambridge, MA: The MIT Press.

Smil, V. 2016. *Still the Iron Age: Iron and Steel in the Modern World.* Oxford: Elsevier.

Smil, V. 2017. *Energy and Civilization: A History.* Cambridge, MA: MIT Press.

Smil, V. 2019. *Growth: From Microorganisms to Megacities.* Cambridge, MA: MIT Press.

Smil, V. 2022. *How the World Really Works.* London: Viking.

Smil, V. 2023. *Invention and Innovation.* Cambridge, MA: MIT Press.

Smil, V. et al. 1983. *Energy Analysis in Agriculture.* Boulder, CO: Westview.

Smith, D.C. 1970. *History of Papermaking in the United States*. New York: Lockwood Publishing.

Smith, S.J. et al. 2010. Anthropogenic sulfur dioxide emissions: 1850–2005. *Atmospheric Chemistry and Physics Discussions* 10:16111–16151.

Social Compare. 2022. Apple iPhone Product Line Comparison. https://socialcompare.com/en/comparison/apple-iphone-product-line-comparison

Sommath, S.S. et al. 1995. A review of wood railroad ties performance. *Forest Products Journal* 45(9):55.

Spalding, M.A. and A.M. Chatterjee. 2017. *Handbook of Industrial Polyethylene and Technology: Definitive Guide to Manufacturing, Properties, Processing, Applications and Markets*. New York: Scrivener Publishing.

Stanley, S.M. et al. 2005. Seawater chemistry, coccolithophore population growth, and the origin of Cretaceous chalk. *Geology* 33:593–596.

Stauber, J.L. et al. 2016. Global change. In: *Marine Ecotoxicology*, Amsterdam: Elsevier, pp. 273–313.

Steel, W. 2017. Constructing the construction state: Cement and postwar Japan. *The Asia Pacific Journal Japan Focus* 15(11). https://apjjf.org/2017/11/Steele.html.

Steinberger, J.K. et al. 2010. Global patterns of materials use: A socioeconomic and geophysical analysis. *Ecological Economics* 69: 1148–1158.

Strauven, H. 2016. Energy consumption during float glass annealing. *Glass International*: 40–41.

Strawbale. 2022. StrawBale. http://www.strawbale.com/

Strom, E.T. and S.C. Rasmussen, eds. 2011. *100+ Years of Plastics: Leo Baekeland and Beyond*. New York: Oxford University Press.

Stuart, M. 2012. *Concrete Deterioration*. Fairfx, VA: PDH Center. http://www.pdhonline.org/courses/s155/s155content.pdf.

Sullivan, D.E. 2005. Metal Stocks in Use in the United States. https://pubs.usgs.gov/fs/2005/3090/2005-3090.pdf

Sullivan, D.E. 2006. Materials in Use in U.S. Interstate Highways. https://pubs.usgs.gov/fs/2006/3127/2006-3127.pdf

Sullivan, D.E. et al. 2000. *20th Century U.S. Mineral Prices Decline in Constant Dollars*. Washington, DC: USGS. http://pubs.er.usgs.gov/publication/ofr00389.

Sun, G. et al. 2022. High-Resolution and Multitemporal Impervious Surface Mapping in the Lancang-Mekong Basin with Google Earth Engine. https://essd.copernicus.org/preprints/essd-2022-251/

Suwanmanee, U. et al. 2013. Life cycle assessment of single use thermoform boxes made form polystyrene (PS), polylactic acid (PLA), and PLA starch: Cradle to consumer gate. *International Journal of Life Cycle Assessment* 18:401–417.

Svarc, J. 2022. Most Efficient Solar Panels. https://www.cleanenergyreviews.info/blog/most-efficient-solar-panels

Tahil, W. (2006). The Trouble with Lithium: Implications of Future PHEV Production for Lithium Demand, http://tyler.blogware.com/lithium_shortage.pdf

Takamatsu, N. et al. 2014. Steel recycling circuit in the world. *Tetsu to hagane* 100:740–749.

Takiguchi, H. and K. Morita. 2011. Global flow analysis of crystalline silicon. In: S. Basu, ed., *Crystalline Silicon - Properties and Uses*, pp. 329–355.

Tallavaaraa, M. et al. 2015. Human population dynamics in Europe over the Last Glacial Maximum. *Proceedings of the National Academy of Sciences* 112:8232–8237.

Tanikawa, H. et al. 2015. The weight of society over time and space a comprehensive account of the construction material stock of Japan, 1945–2010. *Journal of Industrial Ecology* 19:778–791. doi: 10.1111/jiec.12284.

Tanner, A.H. 1998. *Continuous Casting: A Revolution in Steel*. Fort Lauderdale: Write Stuff Enterprises.

Taylor, A. 2019. Photos of Huawei's European-Themed Campus in China. https://www.theatlantic.com/photo/2019/05/photos-of-huaweis-european-themed-campus-in-china/589342/

Taylor, P. et al. 2006. *Luxury or Necessity? Things We Can't Live Without: The List has Grown in the Past Decade*. Washington, DC: Pew Research Center. http://www.pewsocialtrends.org/files/2010/10/Luxury.pdf.

Terakado, R. et al. 2009. In-use stock of copper in Japan estimated by bottom-up approach. *Journal of the Japan Institute of Metals* 73:713–719.

Tetra Pak. 2022. Wine and Spirits – Carton Packaging on the Rise. https://www.tetrapak.com/en-ca/insights/food-categories/wine-and-spirits

The Constructor. 2022. Classification of Aggregates. https://theconstructor.org/building/classification-of-aggregates-size-shape/12339/

The Economist. 2022. The Pandemic's True Death Toll. https://www.economist.com/graphic-detail/coronavirus-excess-deaths-estimates

Thieme, H. 1997. Lower Paleolithic hunting spears from Germany. *Nature* 385:807–810.

Thomas, J., ed. 1979. *Energy Analysis*. Boulder, CO: Westview Press.

Thomsen, C.J. 1836. *Ledetraad til nordisk oldkyndighed, udg. af det Kongelige nordiske oldskrift-selskab*. Copenhagen: S. L. Møllers.

Thorpe, R.S. and O. Williams-Thorpe. 1991. The myth of long-distance megalith transport. *Antiquity* 65:64–73.

Tickner, J.A. et al. 2001. Health risks posed by use of di-2-ethylhexyl phthalate (DEHP) in PVC medical devices: A critical review. *American Journal of Industrial Medicine* 39:100–111.

Tilahun, A. et al. 2012. Quantifying the masses of *Macrotermes subhyalinus* mounds and evaluating their use as a soil amendment. *Agriculture, Ecosystems and Environment* 157:54–59.

Torres, A. et al. 2017. A looming tragedy of the sand commons. *Science* 357:970–971.

Tréguer, P. et al. 1995. The silica balance in the world ocean: A reestimate. *Science* 268:375–379.

Trinkhaus, E. 2005. Early modern humans. *Annual Review of Anthropology* 34:207–230.

Tsetlin, Y. 2018. The origin of ancient pottery production. *Journal of Historical Archeology & Anthropological Sciences* 3:193–198.

Tupy, M.L. 2012. The Miracle that Is the iPhone (or How Capitalism Can Be Good for the Environment). http://www.cato.org/blog/miracle-iphone-or-how-capitalism-can-be-good-environment

Turner, J.S. 2000. *The Extended Organism.* Cambridge, MA: Harvard University Press.

Twede, D. 2002. The packaging technology and science of ancient transport amphoras. *Packaging Technology and Science* 15:181–195.

Twede, D. 2005. The cask age: The technology and history of wooden barrels. *Packaging Technology and Science* 18:253–264.

Ulm, F.-J. 2012. Innovationspotenzial Beton:Von Atomen zur Grünen Infrastruktur. *Beton- und Stahlbetonbau* 107:504–509.

UNCTAD. 2018. Gum Arabic: Growing Demand Means New Opportunities for African Producers. https://unctad.org/news/gum-arabic-growing-demand-means-new-opportunities-african-producers

Underwood, J.R. 2001. Anthropic rocks as a fourth basic class. *Environmental and Engineering Geoscience* 7:104–110.

UNECE (United Nations Economic Commission for Europe). 2010. *Forest Product Conversion Factors for the UNECE Region.* Geneva: UNECE. http://www.unece.org/fileadmin/DAM/timber/publications/DP-49.pdf.

UNEP (United Nations Environment Programme). 2018. *Single-Use Plastics: A Roadmap for Sustainability.* Nairobi: UNEP.

UNEP (United Nations Environment Programme). 2022a. Global Material Flows Database. https://www.resourcepanel.org/global-material-flows-database

UNEP (United Nations Environment Programme). 2022b. Resolution adopted by the United Nations Environment Assembly on 2 March 2022. https://wedocs.unep.org/xmlui/bitstream/handle/20.500.11822/39764/END%20PLASTIC%20POLLUTION%20-%20TOWARDS%20AN%20INTERNATIONAL%20LEGALLY%20BINDING%20INSTRUMENT%20-%20English.pdf?sequence=1&isAllowed=y

UNEP (United Nations Environment Programme). 2022c. Sand and Sustainability: 10 Strategic Recommendations to Avert a Crisis. https://unepgrid.ch/en/resource/2022SAND

Unger, P.W., ed. 1994. *Managing Agricultural Residues*. Boca Raton, FL: Lewis Publishers.

University of Stuttgart. 2007. *The Sustainability of Packaging Systems for Fruit and Vegetable Transport in Europe Based on Life-Cycle Analysis*. Stuttgart: University of Stuttgart.

UNOOSA (United Nations Office for Outer Space Affairs). (2022). UNOOSA. https://www.unoosa.org/.

US DOE. 2022. Achieving American Leadership in the Solar Photovoltaics Supply Chain. https://www.energy.gov/sites/default/files/2022-02/Solar%20Energy%20Supply%20Chain%20Fact%20Sheet.pdf

USDA (United States Department of Agriculture). 2022. Food Prices and Spending. https://www.ers.usda.gov/data-products/ag-and-food-statistics-charting-the-essentials/food-prices-and-spending/#:~:text=Total%20food%20budget%20share%20increased,from%20home%20(5.1%20percent)

USEPA. 2021b. *The 2021 EPA Automotive Trends Report Greenhouse Gas Emissions, Fuel Economy, and Technology since 1975*.

USEPA (US Environmental Protection Agency). 2003. *Background Document for Life-Cycle Greenhouse Gas Emission Factors for Clay Brick Reuse and Concrete Recycling*. Washington, DC: USEPA.

USEPA (US Environmental Protection Agency). 2021a. Plastics: Material-Specific Data. https://www.epa.gov/facts-and-figures-about-materials-waste-and-recycling/plastics-material-specific-data

USEPA (US Environmental Protection Agency). 2021b. The 2021 EPA Automotive Trends Report Greenhouse Gas Emissions, Fuel Economy, and Technology Since 1975. https://nepis.epa.gov/Exe/ZyPDF.cgi?Dockey=P1013L1O.pdf

USEPA (US Environmental Protection Agency). 2022a. Aluminum: Material-Specific Data. https://www.epa.gov/facts-and-figures-about-materials-waste-and-recycling/aluminum-material-specific-data

USEPA (US Environmental Protection Agency). 2022b. Paper and Paperboard: Material-Specific Data. https://www.epa.gov/facts-and-figures-about-materials-waste-and-recycling/paper-and-paperboard-material-specific-data#:~:text=Approximately%2046%20million%20tons%20of,recycling%20rate%20of%2064.8%20percent

USEPA (US Environmental Protection Agency). 2022c. Air Pollutant Emissions Trends Data. https://view.officeapps.live.com/op/view.aspx?src=https%3A%2F%2Fwww.epa.gov%2Fsites%2Fdefault%2Ffiles%2F2021-03%2Fnational_tier1_caps.xlsx&wdOrigin=BROWSELINK

USEPA (US Environmental Protection Agency). 2022d. National Overview: Facts and Figures on Materials, Wastes and Recycling. https://www.epa.gov/

facts-and-figures-about-materials-waste-and-recycling/national-overview-facts-and-figures-materials#R&Ctrends

USGS (US Geological Survey). 2001. Obsolete Computers, "Gold Mine," or High-Tech Trash? http://pubs.usgs.gov/fs/fs060-01/

USGS (US Geological Survey). 2006. Recycled Cell Phones – A Treasure Trove of Valuable Materials. http://pubs.usgs.gov/fs/2006/3097/

USGS (US Geological Survey). 2008. Demand for Minerals in the United States. In: *The Encyclopedia of Earth.* http://www.eoearth.org/article/Demand_for_minerals_in_the_United_States?topic=49532

USGS (US Geological Survey). 2018. Water Use in the United States in 2015. https://www.usgs.gov/mission-areas/water-resources/science/water-use-united-states

USGS (US Geological Survey). 2022a. Mineral Commodity Summaries 2022. https://pubs.er.usgs.gov/publication/mcs2022

USGS (US Geological Survey). 2022b. Recycling-Metals. https://pubs.usgs.gov/myb/vol1/2018/myb1-2018-recycling.pdf

Vale. 2022. Valemax. http://www.vale.com/en/initiatives/innovation/valemax/pages/default.aspx

Van Ewijk, S. and J.A. Stegemann. 2021. Limited climate benefits of global recycling of pulp and paper. *Nature Sustainability* 4(2):1–8. doi: 10.1038/s41893-020-00624-z.

Van Kauwenbergh, S.J. 2010. World Phosphate Rock Reserves and Resources. http://pdf.usaid.gov/pdf_docs/PNADW835.pdf

Vandiver, P.B. et al. 1989. The origins of ceramic technology at Dolni-Vestonice, Czechoslovakia. *Science* 246:1002–1008.

Vasconcellos, A. 2010. Biomass and abundance of termites in three remnant areas of Atlantic forest in northeastern Brazil. *Revista Brasileira de Entomologia* 54:455–461.

van der Velden, N.M. et al. 2014. LCA benchmarking study on textiles made of cotton, polyester, nylon, acryl, or elastane. *International Journal of Life Cycle Assessment* 19:331–356.

Verbraeck, A., ed. 1976. *The Energy Accounting of Materials, Products, Processes and Services.* Rotterdam: TNO (Netherlands Institute for Applied Scientific Research).

Vlachopoulos, J. 2009. *An Assessment of Energy Savings from Mechanical Recycling of Polyethylene Versus New Feedstock.* Washington, DC: World Bank.

van der Voet, E. et al. 2005. Dematerialization: Not just a matter of weight. *Journal of Industrial Ecology* 8:121–137.

Wada, Y. et al. 2016. Modeling global water use for the 21st century: The Water Futures and Solutions (WFaS) initiative and its approaches. *Geoscientific Model Development* 9:175–222.

Wang, H. et al. 2021. Effect of lead exposure from electronic waste on haemoglobin synthesis in children. *International Archives of Occupational and Environmental Health* 94:911–918.

Wärtsilä. 2006. The World's Most Powerful Engine Enters Service. https://www.wartsila.com/media/news/12-09-2006-the-world%27s-most-powerful-engine-enters-service

Warwick, N. et al. 2022. *Atmospheric Implications of Increased Hydrogen Use.* London: UK Government. https://assets.publishing.service.gov.uk/government/uploads/system/uploads/attachment_data/file/1067144/atmospheric-implications-of-increased-hydrogen-use.pdf.

Watari, T. and R. Yokoi. 2021. International inequality in in-use metal stocks: What it portends for the future. *Resources Policy* 70:101968.

Watts, P. 1905. *The Ships of the Royal Navy as They Existed at the Time of Trafalgar.* London: Institution of Naval Architects.

WEEE Forum. 2019. International E-Waste Day: 57.4 M Tonnes Expected in 2021. https://weee-forum.org/ws_news/international-e-waste-day-2021/

Weisberg, D.E. 2008. Engineering Design Revolution. http://images.designworldonline.com.s3.amazonaws.com/CADhistory/85739614-The-Engineering-Design-Revolution-CAD-History.pdf

Weisz, H. et al. 2006. The physical economy of the European Union: Cross-country comparison and determinants of material consumption. *Ecological Economics* 58:676–698.

Wendel, J.F. et al. 1999. Genes, jeans, and genomes: Reconstructing the history of cotton. In: L.W.D. van Raamsdonk and J.C.M. den Nijs, eds., *Plant Evolution in Man-Made Habitats*, Amsterdam: University of Amsterdam, pp. 133–161.

Wernick, I.K. et al. 1996. Materialization and dematerialization: Measures and trends. *Daedalus* 125(3):171–198.

Wesseler, J.H.H., ed. 2005. *Environmental Costs and Benefits of Transgenic Crops.* Berlin: Springer.

White, S.R. et al. 2011. Self-healing polymers and composites. *American Scientists* 99:392–399.

Whiten, A. et al. 1999. Cultures in chimpanzees. *Nature* 399:682–685.

WHO (World Health Organization). 2021. Children and Digital Dumpsites. https://www.who.int/publications/i/item/9789240023901

Wilburn, D.R. and T.G. Goonan. 2013. *Aggregates from Natural and Recycled Sources.* Washington, DC: USGS. http://pubs.usgs.gov/circ/1998/c1176/c1176.html.

Wilkins, J. et al. 2012. Evidence for early hafted hunting technology. *Science* 338:942–946.

Wilkinson, P.H. 1936. *Diesel Aircraft Engines*. Brooklyn, NY: P.H. Wilkinson.

Wilkinson, C.F. et al. 1999. The potential health effects of phthalate esters in children's toys: A review and risk assessment. *Regulatory Toxicology and Pharmacology* 30:140–155.

Williams, E. 2004. Energy intensity of computer manufacturing: Hybrid assessment combining process and economic input–output methods. *Environmental Science & Technology* 38:6166–6174.

Williams, E. 2011. Environmental effects of information and communication technologies. *Nature* 479:354–358.

Williams, E. et al. 2002. The 1.7 kilogram microchip: Energy and material use in the production of semiconductor devices. *Environmental Science & Technology* 36:5504–5510.

Williams, E. et al. 2003. Forecasting material and economic flows in the global production chain for silicon. *Technological Forecasting & Social Change* 70:341–357.

Wilson, A. and J. Boehland. 2005. Small is beautiful: U.S. house size, resource use, and the environment. *Journal of Industrial Ecology* 9:277–287.

van Winkle, T.L. et al. 1978. Cottom versus polyester. *American Scientist* 66:280–290.

Wirsenius, S. 2000. *Human Use of Land and Organic Materials*. Göteborg: Chalmers University of Technology.

Wood, R. et al. 2009. A material history of Australia: Evolution of material intensity and drivers of change. *Journal of Industrial Ecology* 13:847–862.

World Bank. 2022a. Poverty. https://data.worldbank.org/topic/11

World Bank. 2022b. Foreign Direct Investment, Net Inflows. https://data.worldbank.org/indicator/BX.KLT.DINV.CD.WD

World Bank. 2022c. April 2022 Global Poverty Update from the World Bank. https://blogs.worldbank.org/opendata/april-2022-global-poverty-update-world-bank

World Gold Council. (2022). Data. https://www.gold.org/goldhub/data

Worrell, E. and C. Galitsky. 2008. *Energy Efficiency Improvement and Cost Saving Opportunities for Cement Making*. Berkeley, CA: Ernst Orlando Lawrence Berkeley National Laboratory. http://www.energystar.gov/ia/business/industry/LBNL-54036.pdf.

Worrell, E. and M.A. Reuter, eds. 2014. *Handbook of Recycling*. Amsterdam: Elsevier.

Worrell, E. et al. 2008. *World Best Practice Energy Intensity Values for Selected Industrial Sectors*. Berkeley, CA: Ernst Orlando Lawrence Berkeley National Laboratory. http://ies.lbl.gov/iespubs/62806.pdf.

Wrangham, R.W. et al., eds. 1996. *Chimpanzee Cultures*. Cambridge, MA: Harvard University Press.

WSA (World Steel Association). 2020. Methodology Report: Life Cycle Inventory Study for Steel Products. https://worldsteel.org/wp-content/uploads/Life-cycle-inventory-LCI-study-2020-data-release.pdf

WSA (World Steel Association). 2021. Life Cycle Inventory (LCI) Study 2020 Data Release. https://worldsteel.org/wp-content/uploads/Life-cycle-inventory-LCI-study-2020-data-release.pdf

WSA (World Steel Association). 2022a. World Steel in Figures. https://worldsteel.org/wp-content/uploads/World-Steel-in-Figures-2022.pdf

WSA (World Steel Association). 2022b. *Raw Materials*. Brussels: WSA. https://worldsteel.org/steel-topics/raw-materials/.

WSA (World Steel Association). 2022c. Steel – A Permanent Material in the Circular Economy. https://worldsteel.org/circulareconomy/

Wu, X. et al. 2012. Early pottery at 20,000 years ago in Xianrendong Cave, China. *Science* 336:1696–1700.

Wypych, G. 2017. *Self-Healing Materials*. Amsterdam: Elsevier.

Xu, M. and T. Zhang. 2007. Material flows and economic growth in developing China. *Journal of Industrial Ecology* 11:121–140. doi: 10.1162/jiec.2007.1105.

Yamato, S. 2006. Buddhist Buildings in the Horyu-ji Area: Monuments on the World Heritage List. https://www.nara.accu.or.jp/el/textpdf/Buddhist_Buildings_in_the_Horyu-ji_Area_(2006).pdf

Yellishetty, M. et al. 2010. Iron ore and steel production trends and material flows in the world: Is this really sustainable? *Resources, Conservation and Recycling* 54:1084–1094.

Yenne, B. 2010. *The American Aircraft Factory in WWII*. Minneapolis, MN: Zenith Press.

Yogananda, P. 1946. *Autobiography of a Yogi*. New York: The Philosophical Library. http://www.crystalclarity.com/yogananda/chap30.php.

Yokoi, R. et al. 2018. Dynamic analysis of in-use copper stocks by the final product and end-use sector in Japan with implication for future demand forecasts. *Resources, Conservation and Recycling* 180:106153.

Yoshida, A. 2022. China's ban of imported recyclable waste and its impact on the waste plastic recycling industry in China and Taiwan. *Journal of Material Cycles and Waste Management* 24:73–82.

Zafar, S. 2021. Straw Utilization in Denmark. *Journal of Cleaner Production* 274:771–779.

Zhang, B. et al. 2021. Embodied energy in export flows along global value chain: A case study of China's export trade. *Frontiers in Energy Research* 9:649163. doi: 10.3389/fenrg.2021.64916.

Zhang, X. et al. 2022. GISD30: global 30 m impervious-surface dynamic dataset from 1985 to 2020 using time-series Landsat imagery on the Google Earth Engine platform. *Earth Systems Scientific Data* 14:1831–1856. doi: 10.5194/essd-14-1831-2022.

Ziegler, K. 1963. Consequences and development of an invention. Nobel Lecture, December 12, 1963. In: *The Nobel Prize in Chemistry 1963*, Stockholm: Nobel e-Museum.

Zier, M. et al. 2021. A review of decarbonization options for the glass industry. *Energy Conversion and Management X* 10:100083.

Zulehner, W. 2003. Historical overview of silicon crystal pulling development. *Materials Science and Engineering* 73:1–12.

Index

Note-Locators followed by *t* refer to tables
